ENVIRONMENTAL SCIENCE, ENGINEERING AND TECHNOLOGY

IRRIGATION MANAGEMENT, TECHNOLOGIES AND ENVIRONMENTAL IMPACT

Environmental Science, Engineering and Technology

Additional books in this series can be found on Nova's website under the Series tab.

Additional e-books in this series can be found on Nova's website under the E-book tab.

Water Resource Planning, Development and Management

Additional books in this series can be found on Nova's website under the Series tab.

Additional e-books in this series can be found on Nova's website under the E-book tab.

ENVIRONMENTAL SCIENCE, ENGINEERING AND TECHNOLOGY

IRRIGATION MANAGEMENT, TECHNOLOGIES AND ENVIRONMENTAL IMPACT

M. HOSSAIN ALI
EDITOR

nova publishers
New York

Library of Congress Cataloging-in-Publication Data

ISBN: 978-1-62417-862-7

Published by Nova Science Publishers, Inc. † New York

CONTENTS

Preface		**vii**
Chapter 1	Irrigation Management of Field Crops: Rice *M. Maniruzzaman and Debjit Roy*	**1**
Chapter 2	The Sap Flow Technique: A Precise Means to Estimate Water Consumption of Young Olive Trees (Olea Europaea L.) *Chiraz Charfi Masmoudi, Moncef Masmoudi, Jihène Abid Karray and Nétij Ben Mechlia*	**41**
Chapter 3	Responses of Apple to Partial Root-zone Drying: A Review *Riccardo Lo Bianco*	**71**
Chapter 4	Field Studies on the Use of Saline Water for Deficit Irrigation in Arid Tunisia *Kamel Nagaz, Mohamed Moncef Masmoudi and Netij Ben Mechlia*	**87**
Chapter 5	Water Conservation Interventions in Pakistan Agriculture for Achieving Food Security: Synthesis and New Evidence *Rai Niaz Ahmad, M. Arshad and Munir A. Hanjra*	**129**
Chapter 6	Assessing Suitability of Groundwater for Irrigation Purpose in Coastal Jorf Aquifer, Southeastern Tunisia: Geostatistical Approach *Belgacem Agoubi, Adel Kharroubi and Habib Abida*	**159**
Chapter 7	Assessing the Factors Influencing Groundwater Irrigation Technology Adoption in Ghana *Victor Owusu, Alberta Vincentia Asante and Paul Pavelic*	**181**
Chapter 8	Irrigation Management: Effects of Soybean Diseases on Seed Composition in Genotypes Differing in Disease Resistance under Irrigated and Nonirrigated Conditions *Nacer Bellaloui, Alemu Mengistu, Daniel K. Fisher, L. H. S. Zobiole and Hamed K. Abbas*	**193**

Chapter 9	Micro-irrigation Scheduling Using Clean Energy *Ahmed Al-Busaidi and Tahei Yamamoto*	**229**
Chapter 10	Multiple Use of Irrigation Reservoir Water Resource in the Upper East Region of Ghana *Makarius V. Mdemu*	**255**
Chapter 11	Capital Budgeting Methods for Irrigation Technology Adoption Decisions: A Review *Chi Huu Truong*	**273**
Chapter 12	Irrigation with Wastewater: Impacts, Scope and Challenges *M. G. Mostofa Amin, Mohammed Mizanur Rahman and Moudud Hasan*	**285**
Index		**301**

PREFACE

The importance of irrigation for agricultural production (food, fiber, fodder, and for bio-fuel) needs no elaboration. Today's irrigation is under challenging condition having water demand from multiple sectors (household, industry, and environment), quality degradation, and uncertain climatic condition. This situation demands for precise estimation/determination of irrigation need for crop, advance technique for water application, consideration for associated factors of irrigation technology adoption, and feasibility study for using marginal quality water in irrigation (such as saline water, waste-water). This new book presents irrigation management strategies for rice crop, sap flux technique for estimating irrigation need (application in young olive trees), partial root-zone wetting technique for water-saving irrigation (in Apple), synthesis of water conservation and land preservation (with particular reference to Pakistan), deficit irrigation with saline water in saline soil (with reference to Tunisia), assessing the suitability of groundwater for irrigation purposes through geo-statistical approach (with reference to Jorf aquifer, Tunisia), factors affecting groundwater irrigation technology adoption (with reference to Ghana), irrigation management in soybean for disease control (with reference to USA), simulation techniques for micro- irrigation scheduling (with applications in Japan, Iran, and China), multiple uses of medium and small reservoir water resources (with reference to Upper East Region of Ghana), capital budgeting methods for irrigation technology adoption decisions, and scope and challenges of irrigation with wastewater.

Chapter 1 - Rice feeds more than half of the world population, and it is the second largest crop in planting acreage after wheat. Rice is the major water consumer crop in the world. Rice plant can be grown in wide range of soil and climatic variability. This chapter describes the history of rice, its morphology, rooting behavior and characteristics of different genotypes/variety with different water regimes, growth phases and stages of rice plants, rice cultivation methods, water requirements, water losses, field water balance, irrigation management strategies, irrigation application and drainage systems, factors affecting irrigation practices and technology adoption, water conservation and effects of irrigation management on yield and water productivity.

Chapter 2 - This chapter evaluates the sap flux technique for its applicability with young olive trees characterized by low soil coverage. Dissipation sap flow technique of 'Granier' was used to study water consumption of young olive trees. Results show satisfactory behavior of the system when working under field conditions. The technique proved as robust and requires little maintenance. It gave valuable results which are coherent and in agreement with others made under different environments. However, many constraints and problems were found, amongst, sensor calibration, large variability of responses and determination of the water conductive stem section. This technique can be used with confidence in young olives. However, when sap flow technique is adopted to estimate transpiration for stands of vegetation in isolated trees over a long period, appropriate methods of scaling from plant to unit area of land must be used to take into account errors made in scaling transpiration from plant to stand.

Chapter 3 - Irrigation water productivity tends to increase with deficit compared to conventional irrigation. Partial root-zone drying (PRD) is a relatively recent strategy of deficit irrigation. Apple cultivars with different vigor, period of fruit growth, and crop load perform differently in response to PRD and generalizations on the potential use of PRD as a valuable irrigation strategy for apple orchards result difficult. Fruit size and number per tree seem to be sensitive to some extent to the reduction of irrigation volumes operated with PRD. On the contrary, vegetative growth is mostly unaffected by PRD, and fruit quality may be somewhat improved. These general responses are associated with a relatively good water status of trees under PRD. This chapter reviews all of these aspects.

Chapter 4 - In areas with limited water resources, better management strategies of irrigation water are required for farmers seeking a viable means to maximize crop production. Options to improve water productivity and reduce salt input to soils could be sought through the practice of drip irrigation, precise scheduling, and deficit irrigation. Building on available knowledge, participatory investigations have been conducted on deficit irrigation practices with saline waters in the arid areas of Medenine, located in south-eastern Tunisia. Research concerned the combined impacts of cultural practices and irrigation methods on selected crops: potato, carrot, barley and millet. Scheduling consisted in the total and/or partial replacement of accumulated crop evapotranspiration, as derived from meteorological data and dual crop coefficient. Moreover, yield improvements associated with conversion to subsurface drip irrigation system has been substantial. Interactive effects of salinity and fertilizers are found to be significant, insinuating that the effect of salinity could be mitigated by increasing nitrogen fertilizer. Future work should focus on the integration of data on water salinity and rainfall for better selection of the cultivar and management of fertilizers.

Chapter 5 - Conserving water and preserving land resources can make a sterling contribution towards achieving food security in many regions of the world. Water competition among agriculture and other sectors is intensifying the need to conserve irrigation water through productivity enhancing interventions for boosting food production and supporting rural livelihoods. This chapter presents a synthesis of water conservation interventions with emphasis on the Indus Basin where bulk of Pakistan agriculture is concentrated. The interventions include: watercourse improvement program; laser land leveling; bed and furrow irrigation/bed planting; raised bed technology; permanent raised beds; reduced or zero tillage; and drip/irrigation system. Synthesis and new evidence is drawn from a number of resource conservation projects and interventions implemented at farmer fields in a number of settings across the Basin. New evidence shows that bed furrow planting system improves crop yield for wheat, rice, sugarcane and cotton and offers practical opportunities for producing "more crop per drop" and contributing towards achieving food security in Pakistan.

Chapter 6 - Groundwater represents the main water resource to satisfy the needs of various sectors in many regions of the world. The Jorf coastal aquifer (Tunesia) is subjected to seawater intrusion and groundwater quality degradation. The object of this study is to assess the suitability of groundwater of Jorf aquifer for irrigation purposes through geostatistical approach. With this aim, water samples were collected from hand-dug wells, and chemical parameters such as electrical conductivity, pH, total dissolved solids (TDS), Na, K, Ca, Mg, Cl, HCO_3 and SO_4 were analyzed. Sodium adsorption ratio (SAR), magnesium adsorption ratio (MAR), sodium percentage (%Na) and permeability index (PI) were computed. Results showed that groundwater salinity was controlled by sodium and chloride. Groundwater was dominated by Na–Cl–Ca–SO_4 hydrochemical facieses, mainly as a result of mineral dissolution (halite and gypsum) and seawater intrusion. The water of coastal zone is characterized by fair to poor quality, and is therefore just suitable for chemical amendments in open soils and with excessive leaching. A refined geostatistical assessment map of the groundwater quality indicated that only the water of central and northwest of the study area is suitable for irrigation purposes.

Chapter 7 - Groundwater irrigation is believed to have the highest level of water control necessary for smallholder irrigation farmers to overcome household-level poverty and food insecurity. This study analyses the effects of the determinants of groundwater irrigation technology adoption with a data from 240 smallholder irrigation farmers in Ghana. The empirical results show that household size and membership of farmer-based organization tend to have positive significant impacts on the probability of adopting groundwater manual irrigation whilst extension contacts and farm size have negative impacts. In addition the age, gender and education of the farmer, household size, access to credit, farm size and owner-cultivation tend to have significant positive effects on the probability of adopting groundwater motor pump irrigation technology whilst experience in irrigation farming and extension contacts have negative effects. To promote the groundwater irrigation technology adoption, the update of skills and knowledge of the farmers on the existing irrigation technologies must be facilitated through frequent participation in workshops and training programs on irrigation.

Chapter 8 - Soybean seed is a major source of protein and oil in the world. Nutritional qualities of soybean seed are determined by the quantity and quality of seed composition components (protein, oil, fatty acids, isoflavones, and minerals). Charcoal rot is a disease caused by the fungus *Macrophomina phaseolina* (Tassi) Goid and results in yield loss and poor seed quality, especially under drought conditions. Phomopsis, caused by *Phomopsis longicolla* Hobbs, is also one of the major soybean seed decay pathogens, resulting in poor seed quality, especially under humid, warm conditions. Disease infection is influenced by irrigation management. The research demonstrated that seed composition components were significantly altered by irrigation treatment and disease (charcoal rot or *Phomopsis*), and the degree of alteration of these components was depended on the level of resistance of the genotype to the disease. The research highlights the significance of irrigation management and soybean varietal selection in maintaining high seed nutritional qualities. Special section devotes to soybean irrigation and current irrigation techniques.

Chapter 9 - Micro-irrigation system is classified as the small scale irrigation which can be applied for thousands square meters of agricultural fields. With limited water resources for crop production, there is a need to reduce micro-irrigation facilities while maximizing the efficiency of local environmental condition such as harvested rain water that can be used for field irrigation. This chapter aims to develop a series of program packages of simulation models on irrigation scheduling for two cases of small scale irrigation systems, which are supported by rain harvesting catchment apron, one or two water storage tanks and a groundwater pump operated by wind or solar energy. The computer programming used in this study was issued by Microsoft 2008 visual basic language. The selected fields were distributed in three locations; Mu Us Shamo desert in the Inner Mongolia Autonomous, China (Mu Us), Shavour Soil Institute Station in the Khuzestan Province, Iran (Shavour) and Arid Land Research Center of Tottori University, Japan. It is expected that the simulation results would be used as a guide to design a sustainable micro-irrigation projects.

Chapter 10 - Multiple uses of artificial water reservoirs are increasingly becoming an important aspect in planning and development of water resources management in semi-arid areas of sub-Saharan Africa. This paper documents and analyzes the dry season multiple uses of medium and small reservoir water resources that are widely distributed throughout the Upper East Region of Ghana. Temporal reservoir water uses and existing reservoir water resources management arrangements are elaborated. In the description and analysis, attention is given to both large and small reservoirs. Although the later are many in number and spatially distributed, the former are few and site specific. Finally, the constraints and potentials for improving reservoir water resources management are discussed.

Chapter 11 - Policies to encourage the uptake of efficient irrigation technology have played an important role in achieving water conservation goals in arid area and in reducing polluted return flows from irrigation farms. A good policy depends significantly on the method used to determine farmers' irrigation technology adoption decision. This chapter provides a critical review of capital budgeting methods applicable to modeling the irrigation technology adoption decision. Elements affecting the value of investment, including the roles of risk preferences, risk transformation, the value of information, are analyzed and real option frameworks are discussed. The chapter is concluded with future research suggestions.

Chapter 12 - World's water resource is under threatened state as numerous water consuming sectors are competing for limited useable water of the earth. To combat with the growing agricultural water demand, wastewater from different sources is considered as a potential replacement of fresh irrigation water. Use of wastewater in irrigation in a sustainable manner can increase water security in agricultural sector. On the other hand, it can spread harmful contaminants to soil and water environment if not managed properly. Impacts of wastewater irrigation on soil and water environment, strategic measures for using wastewater in irrigation, and challenges of wastewater use in irrigation have been discussed in this chapter.

In: Irrigation Management, Technologies ...
Editor: M. Hossain Ali
ISBN: 978-1-62417-862-7

Chapter 1

IRRIGATION MANAGEMENT OF FIELD CROPS: RICE

M. Maniruzzaman **[*]** ***and Debjit Roy***
Irrigation and Water Management Division,
Bangladesh Rice Research Institute (BRRI), Gazipur, Bangladesh

ABSTRACT

Rice is the major water consumer crop in the world. The rice plant can be grown in a wide range of soil and climatic variability. This chapter describes the history of rice, its morphology, rooting behavior and characteristics of different genotypes/variety with different water regimes, growth phases and stages of rice plants, rice cultivation methods, water requirements, water losses, field water balance, irrigation management strategies, irrigation application and drainage systems, factors affecting irrigation practices and technology adoption, water conservation and effects of irrigation management on yield and water productivity.

1. ORIGIN, HISTORY, AND CLASSIFICATION OF RICE

1.1. General Overview

Rice has fed more people than any other crop has for thousands of years. The ancient Indian name for rice, *Dhanya*, means "sustenance for the human race." Especially in much of Asia, life without rice has been unthinkable. Rice feeds more than half of the world's population.

Rice is the second largest crop in planting acreage after wheat. The major rice growing regions are found in more than a hundred countries in Asia, Latin America and Africa, but major rice exporting countries only include Thailand, the United States, Vietnam, Pakistan and India. About 85 percent of total rice production is for human consumption. Rice provides 23 percent of the global human per capita energy and 16 percent of the per capita protein

[*] E-mail: mmziwmbrri@gmail.com.

(IRRI, 1997). In Asia, where people typically eat rice two or three times a day, 250 million rice farms (the average rice land per farm is less than 1 ha) produce more than 90 percent of the world's rice.

1.2. Origin and Diffusion

Origin and diffusion of rice are still unsettled issues (Huke and Huke, 1990). Asian cultivated rice originated in the region from South China to the Ganges in South and Southeast Asia, including the river valleys and deltas of the Brahmaputra of northern India, the Irrawaddy of Burma, the Mekong of Vietnam and the Yangtze of China. Based on the number of wild rice species and the evidence of rice glumes in the burnt clay from the late Neolithic period, Ting (1949) concluded that rice might have originated in South China and then spread northwards. Copeland (1924), Chang (1975) and some Japanese rice scientists asserted that rice might have originated from South or Southeast Asia, including India, China, Thailand and Indonesia. Thermo-luminescence and carbon-14 tests of the pottery shards with the imprints of rice grains in Thailand indicate that rice could be dated back to at least 4000 B.C.E. (IRRI, 1997). The three oldest pieces of archaeological evidence for rice origin are from Maharaga of India (6500500 B.C.E.), Non-Nok-Tha of Thailand (about 6000 B.C.E.) and Pen-tou-shan of China (7150250 B.C.E.) (Abrol and Gadgil, 1999; Chang, 1998). Great diversity of rice and linguistic evidence support the argument for Southeast Asia as the origin of rice cultivation. It was believed that the early spread of rice was from southern China or northern Vietnam to the Philippines about 2000 B.C.E. and then to Indonesia about 1599 B.C.E. by Deutero-Malayans. Most likely, the techniques of rice cultivation radiated outward from the Yangtze delta of China towards Korea and then Japan. Japan became known as *Mizumono kuni* (the Land of Luxurious Rice Crop) about 30000 B.C.E., but rice did not become the staple Korean dish until the 1930s. Sri Lanka had rice as a crop as early as 1000 B.C.E. Rice was introduced to Greece and the neighboring Mediterranean c. 34424 B.C.E. and then gradually to Europe and Africa (IRRI, 1997). There have been debates about the introduction of Asian rice into Africa. *O. sativa* was believed to have been introduced to Africa primarily from Malayo-Polynesia a few centuries B.C.E. or from Sri Lanka and Indonesia. It was postulated that at different times *O. sativa* was introduced to Egypt from India, to Madagascar from Indonesia as early as 1000 B.C.E., to Mozambique and East Africa from Madagascar, and finally to West Africa by Portuguese spice traders between the fifteenth and seventeenth centuries, or by traders or Muslim missionaries in the ninth or tenth centuries. Lu and Chang (1980) argued that Asian rice entered into the Congo from Mozambique in the nineteenth century. The other possibilities for the introduction of rice into Europe are from Persia, central Asia, or directly from China. Later the Portuguese brought rice to Brazil, and the Spanish introduced rice to Central and South America. The Malays brought rice to Madagascar. The United States might have been first introduced to rice from the Malagasy Republic, Europe, or the Far East. It was often cited that a storm-battered ship from Madagascar had brought rice seeds "Golde Seede Rice" to South Carolina in 1694. Not until 1888 did the first large-scale growing of rice plants occur in Louisiana and Texas, although rice was introduced to Virginia as early as 1609. Later the Gulf Coast grew rice because of the popularization of mechanical farming.

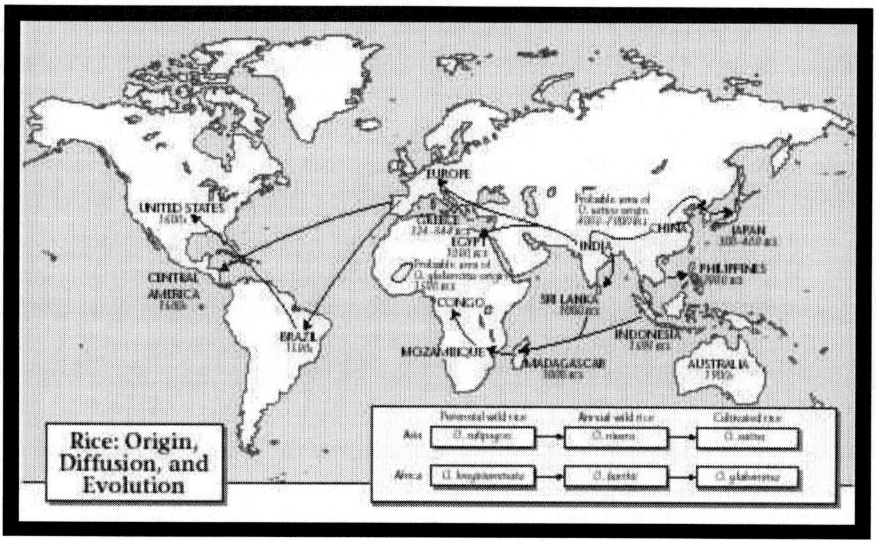

Figure 1. Rice: Origin, Diffusion and Evolution.

The current major rice-producing states in the United States include Arkansas, California, Louisiana, Texas, Mississippi, and Missouri. Rice was first commercialized in Australia in 1924, even though in as early as 1892 experimental planting of rice took place in New South Wales.

1.3. Domestication

Chang (2000) made exhaustive descriptions on the evolution and early spread through several routes for rice, especially Asian rice. In his review the pattern for rice evolution was suggested as from perennial wild to annual wild and then to cultivated. Therefore, the evolution for *O. sativa* is *O. rufipogon* to *O. nivara* to *O. sativa* in Asia. In parallel, the evolution for *O. glaberrima* is *O. longistaminata* to *O. barthii* and then to *O. glaberrima* in Africa. It is believed that rice originated in the marsh areas and spread toward the dry lands and hills. The domestication of rice, including such cultural practices as puddling and transplanting, might have first taken place in China. The ancient Chinese cultural practices for rice then shifted to Southeast Asia and other parts of the world (De Data, 1987). Only in the twentieth century did systematic rice improvement start. For example, IRRI scientists Peter R. Jennings, Te-Tze Chang, and Henry M. Beachell developed the semi-dwarf variety "IR8" in 1966, which initiated the "Green Revolution" (Lang, 1996). China also successfully developed hybrid rice in 1973. These achievements have greatly increased the rice yield so as to feed the increasing world population, especially in Asia.

1.4. Hybrids

Before the 1970s, rice breeders extensively studied the utilization of rice heterosis after they realized the great potential of using heterosis to improve crops such as corn. Generally speaking, hybrid rice has more than 15 to 20 percent yield advantage over the best conventional rice varieties in China, Bangladesh, Brazil, Colombia, Ecuador, India, Indonesia, Malaysia, Myanmar, the Philippines, Sri Lanka, and Vietnam. Unfortunately, no large-scale planting of hybrid rice was successful until China developed a three-line (cytoplasmic male sterile line, male sterile maintainer line, and restorer line) system hybrid rice in 1973 and commercialized this technology in 1976. This was because of the very small reproductive organs and difficulty in finding or developing a cytoplasmic male sterile line in rice. China has greatly increased its rice production by the utilization of hybrid rice technology and therefore has been able to feed its rapidly growing population in the last quarter of the twentieth century. Now China is trying to increase the rice yield by employing the two-line system hybrid rice (male sterile line and restorer line) and intersubspecific heterosis (i.e., the heterosis between *Indica* and *Japonica*) utilization. By following China's success, other countries, including India, Vietnam, Myanmar, and Bangladesh, are currently learning to use the hybrid rice technology to boost the rice yield. FAO is also providing financial support to activities of networks on hybrid rice such as the International Task Force for Hybrid Rice and the Working Group on Hybrid Rice in Latin America and the Caribbean (GRUTHA). Outside China, the planting acreages under hybrid rice in Vietnam, India, and Bangladesh in 1998 were 250,000 ha, 120,000 ha, and 20,000 ha, respectively.

1.5. Outline of the Rice Life History

The rice plant usually takes 3-6 months from germination to maturity, depending on the variety and the environment under which it is grown. During this period, rice completes basically two distinct sequential growth stages: vegetative and reproductive. The reproductive stage is subdivided into preheading and postheading periods. The later is better known as the ripening period. Yield capacity, or the potential size of the crop yield, is primarily determined during preheading. Ultimate yield, which is based on the amount of such that fills spikelets, is largely determined during postheading. Hence, agronomically, it is convenient to regard the life history of rice in terms of three growth stages: Vegetative, reproductive and ripening. The vegetative stage refers to a period form germination to the initiation of panicle primordial; the reproductive stage, from panicle primordial initiation to heading; and the ripening period, form heading to maturity. A 120 days variety, when planted in a tropical environment, spends about 60 days in the vegetative stage, 30 days in the reproductive stage and 30 days in ripening period.

The vegetative stage is characterized by active tillering, gradual increase in plant height, and leaf emergence at regular intervals. All contribute to increasing the leaf area that receives sunlight. Tillering may start when the main clum develops the 5th or 6th leaf. Active tillering refers to a stage when tillering rate - the increase in tiller number per unit of time - is high. The maximum tiller number stage follows active tillering. It is a stage when tiller number per plant or per square meter is maximum – before or after the initiation of panicle primodia, depending on a variety's growth duration. Because tiller number declines after the maximum tiller number stage, there is a period before that stage (often called *the end stage of effective tillering*) when the tiller number becomes numerically equal to panicle number at maturity. That does not necessarily mean that tillers developed at early growth stages normally produced panicle, while those developed later may or may not.

The reproductive growth stage is characterized by culm elongation (which increases plant height), decline in tiller number, emergence of the flag leaf (the last leaf), booting, heading, and flowering. Initiation of panicle primordial usually dates back to about 30 days before heading.

Agronomist often refer to topdressing nitrogen fertilizer at panicle initiation – a stage about 25 days before heading when the panicle has grown about 1 mm long and can be recognized visually or with a magnifying lens. But initiation of panicle primordial can be recognized only under a microscope. Internode elongation usually begins around the initiation of panicle primordia and continues until heading. The top five internodes may be elongated at heading. For this reason, the reproductive growth stage is sometimes called the internode elongation stage.

Heading means panicle exsertion. Spikelet anthesis (or flowering) begins with panicle exsertion, or on the following day. Consequently, heading is considered a synonym for anthesis in terms of calendar days in the life history of rice. It takes 10-14 days for a crop to complete heading because there is variation in panicle exsertion within tillers of the same pant and between plants in the same field. Agronomically, heading is usually defined as the time when 50% of the panicles have exserted. Anthesis normally occurs between 0800 and 1300 tropical environments. Fertilization is completed within 5-6 hours later. Only a very few spikelets have anthesis in afternoon. When the temperature is low, however, anthesis may start late in the morning and last until late afternoon. Within the same panicle it takes 7-10 days for all the spikelets complete anthesis within 5 days.

Ripening follows fertilization, and may be subdivided into milky, dough, yellow-ripe, and maturity stages. These terms are primarily based on the texture and colour of growing grains. Ripening is characterized by leaf senescence and grain growth – increases in grain size and weight and changes in grain colour. During active grain growth, both fresh and dry grain weights increases. Toward maturity, however, dry weight increases slowly but fresh weight decreases as a result of water loss. The length of ripening, largely affected by temperature, ranges from about 30 days in the tropics to 65 days in cool, temperate regions such as Hokkaido, Japan and New South Wales, Australia.

The Basic Processes of the Life History of Rice Can Be Applied to Any Cultivation System, with Some Modification.

First, differences in growth duration are primarily due to differences in the length of the vegetative growth stage. The length of the reproductive stage plus the ripening period may be considered about the same for any variety under a given environment. Early maturing varieties have short vegetative stages. As a consequence, they may initiate panicle primordia before the maximum tiller number stage and heading may be staggered because later tillers may produce panicles. Late maturing varieties have long periods of vegetative stage and may reach the maximum tiller number stage before initiation of panicle primordial. The period from the maximum tiller number stage to initiation of panicle primordial is sometimes referred to as *vegetative-lag phase* (Tanaka et al 1964). When the length of the vegetative stage is adequate, the plant initiates panicle primorida right after the maximum tiller number stage. In the tropics, this is normally attained by a 120 days variety.

Second, direct seeded rice normally starts tillering earlier than transplanted rice because its growth proceeds without the setback caused by growth damage during uprooting. Each direct seeded rice plant, however, usually produces 2-5 tillers while each transplanted rice plant produces 10-30. Thus, tillering is much less important in direct seeded rice.

Third, growth duration of the same variety may be slightly different between the transplanted and direct seeded crops. Transplanted rice usually takes about 1 week more to mature because its growth has been disturbed by uprooting.

1.6. Classification

Rice has 120,000 varieties, the richest gene bank in the plant kingdom. There has been great progress in rice genome sequencing projects recently by using both *Indica* (9311) and *Japonica* (Nipponbare) varieties. This will greatly enhance rice improvements in the near future. From the taxonomy of rice, Asian rice belongs to the grass (Gramineae) family and genus *Oryza.* Wheat, corn, and barley also belong to the grass family. *Oryza* has twenty-three species, which can be classified into four groups. The *O. ridleyi* complex and *O. meyeriana* complex contain species in lowland swamp forests and upland hillside forests, respectively. The *O. officinalis* complex consists of perennial species throughout the tropics. The *O. sativa* complex includes two cultigens and the wild relative of these two cultigens. Only two *Oryza* species, the tetraploid *O. schlechteri* and the diploid *O. brachyantha*, are different from these four groups. Based on another classification method for *Oryza* species, the common wild rice or so-called *O. perennis* complex includes Asian wild rice (*O. rufipogon*), African wild rice (*O. longistaminata* or *O. barthii*), and Oceanian wild rice (*O. meriodinalis*) (Chang, 1976). *O. rufipogon* is the wild relative of *O. sativa* and a noxious weed in rice-growing countries. This wild rice has the characteristic of easy shedding to facilitate easy dispersal and easy crossing with the Asian cultivars, which results in the degradation of the variety and contamination of red-grained plants. The wild rice (or Indian rice or water oats) of North America (*Zizania palustris* or *Z. aquatica L.*, 2n = 30) belongs to a different genera and even a different tribe of the grass family. It was traditionally harvested by Native Americans in the Great Lakes region, and now is commercially produced in Minnesota, Wisconsin, northern California, and Manitoba, Canada. It has nutty and rich flavour and boasts a pleasant chewiness. It is high in protein and B vitamins but low in fat.

Rice has only two cultivated species, *Oryza sativa* Linn. (the Asian cultivated species) and *Oryza glaberrima* Steud. (the West African cultivated species). *O. sativa* is by far the more widely utilized of the two. By contrast, *O. glaberrima* has much less diversity because of a relatively short cultivation history and a narrower dispersal than *O. sativa*. The main differences in botanical morphology between these two cultivated species are the ligule size and glume pubescence. Most of *O. glaberrima* varieties have fewer hairs, short ligules, and fewer or no branches, and also have red-hulled grains on the shattering panicle. They are generally more resistant to flood, alkaline soils, and blast than *O. sativa* varieties. Another difference is that *O. glaberrima* is strictly annual, whereas *O. sativa* is potentially a perennial. It is generally thought that there are two major subspecies in *O. sativa*: *indica* (or *hsien*) and *japonica* (sometimes called *sinica* or *keng*). *Japonica* is generally short, less leafy, and has a strong culm and short grains. *Japonica* varieties are grown in temperate areas such as northern China, Japan, Korea, Spain, Australia, and California. When cooked, *japonica* rice is sticky. *Indica* rice generally has long, slender, and fluffy grains, many tillers, and is tall, leafy, and tolerant of drought. It does not stick when cooked. *Indica* rice is usually grown in hot (tropical or subtropical) climates such as in India, Thailand, Vietnam, and Southern China. The so-called *javanica* can be classified as a tropical *japonica* and is mostly grown in Indonesia and the United States. *Javanica* was originally grown in equatorial areas with abundant water and is generally vulnerable to drought.

2. Rice Morphology

Rice has the following three main developmental phases: the vegetative phase (from germination to panicle initiation), the reproductive phase (from panicle initiation to flowering), and the ripening phase (from flowering to maturity). A typical rice plant during reproductive and ripening stages has the following organs: roots, a main stem and a number of tillers (or side shoots), leaves, panicles, flowers (or florets) when flowering, and grains when maturing. The edible part of a rice plant is in the rice grain, which includes glumes, endosperm, and embryo. Some varieties have awns at the tip of the grain.

Rice has a fibrous root system, that is, it has temporary seminal roots or embryonic roots, then secondary adventitious roots. The fibrous roots only last for a short time after seed germination. An alternative is the classification of rice roots based on the spatial pattern as crown roots (including mat roots) below the soil and nodal roots above the soil. The function of rice roots is to absorb water. It is estimated that it takes 5,000 liters of water to produce 1 Kg of irrigated rice. Owing to the great consumption of water in rice production, future rice cultivation will face the looming water deficiency in some countries.

2.1. Root Characteristics and Its Behavior

The root growth behavior under different water regimes and the different genotypes/variety is shown in Figure 2. It indicated that root growth behavior of the same genotypes/variety varied with the water regimes. Irrespective of genotypes/varieties and water regimes most of the roots were concentrated in the 0-10 cm soil depth (Table 1). Different water regimes showed a distinct variation of distribution of root from top most layer to the bottom layer. AWD water management had the tendency to produce more root in the bottom layer irrespective of variety, which indicated the capacity to uptake more water from the bottom layer.

The vertical distributions along the soil depth for genotypes and water regimes under field conditions appeared to be reduced exponentially (Figures 2, 3 and 4). Beyond 30 cm, the RLD value did not vary a much with respect to genotypes and water managements. The root growth was affected by the growth condition i.e. water regimes (Figure 2). Under water saving conditions root developed more than the anaerobic (CSW) condition. But under CSW root growth was severely inhibited below 20 cm soil depth. The topsoil layer consisted of most of the roots in plants grown under CSW and under water saving condition this amount was less. The RLD was the highest in V4 followed by V3 and the lowest in V1 irrespective in water treatments (Figures 2, 3 and 4).

At extreme soil depth (below 30 cm) V3 was able to maintain better root growth compared to the others. Root-shoot ratio increased significantly with the water saving treatments. More root-shoot ratio were found in 3 DAWD compared to the CSW irrespective of genotypes (Table 2). The highest root-shoot ratio was found in V5 in 3 DAWD followed by V6 and that was in V4 in 5 DAWD followed by V3. But the genotypes V3 consistently maintained the increasing rate in water saving treatments. The water saving treatments considered here was favorable anaerobic to aerobic environment. In contrast the CSW treated plots were kept folded continuously.

From the above data it was indicated that V3 may be the suitable genotypes in all respect for growing in water saving environment. It also revealed that the rooting behavior of the genotypes varied a little under different water regimes.

Table 1. Average cumulative root length below soil surface of different genotypes/ variety under different water regimes during 2008-10

Variety	Cumulative root length (cm)				
	0-10 cm	10-20 cm	20-30 cm	30-40 cm	Total
CSW					
V1	1002.4 (96)	20.6 (2.0)	11.0 (1.1)	10.5 (1.0)	1044.5
V2	1090.4 (89)	89.0 (7.3)	26.3 (2.2)	15.3 (1.3)	1221.0
V3	1614.1 (89)	151.0 (8.3)	35.5 (2.0)	17.7 (1.0)	1818.3
V4	1954.8 (89)	163.9 (7.5)	53.7 (2.5)	11.9 (0.5)	2184.3
V5	1759.4 (91)	149.8 (7.7)	9.6 (0.5)	15.6 (0.8)	1934.4
V6	1049.9 (91)	56.8 (4.9)	25.7 (2.2)	20.5 (1.8)	1152.9
3 DAWD					
V1	1018.6 (90)	63.7 (5.6)	36.1 (3.2)	15.6 (1.4)	1134.0
V2	1145.8 (87)	117.6 (8.9)	28.6 (2.2)	29.0 (2.2)	1321.0
V3	1756.1 (91)	87.7 (4.5)	66.2 (3.4)	28.3 (1.5)	1938.3
V4	2129.5 (93)	129.6 (5.7)	17.4 (0.8)	7.8 (0.3)	2284.3
V5	1765.7 (87)	130.0 (6.4)	91.8 (4.5)	46.9 (2.3)	2034.4
V6	1082.4 (86)	75.3 (6.0)	73.1 (5.8)	22.1 (1.8)	1252.9
5 DAWD					
V1	832.1 (85)	111.4 (11.3)	28.5 (2.9)	12.0 (1.2)	984.0
V2	1098.5 (93)	68.5 (5.8)	4.0 (0.3)	9.0 (0.8)	1180.0
V3	1618.4 (90)	117.6 (6.6)	20.8 (1.2)	31.5 (1.8)	1788.3
V4	1838.3 (88)	135.5 (6.5)	100.1 (4.8)	26.4 (1.3)	2100.3
V5	1616.8 (86)	95.7 (5.1)	121.4 (6.4)	50.5 (2.7)	1884.4
V6	920.5 (83)	81.8 (7.4)	76.9 (7.0)	23.7 (2.1)	1102.9

Maniruzzaman, 2012.

In parenthesis shows the percentage. CSW = Continuous standing water, 3 DAWD = 3 days after water disappearance, 5 DAWD = 5 days after water disappearance, V1 = IR82853-36, V2 = IR82914-11, V3 = IR69515-KKN-4-UBN-4-2-1-1, V4 = IR70179-1-1-1-B, V5 = BRRI dhan28 and V6 = BRRI dhan29.

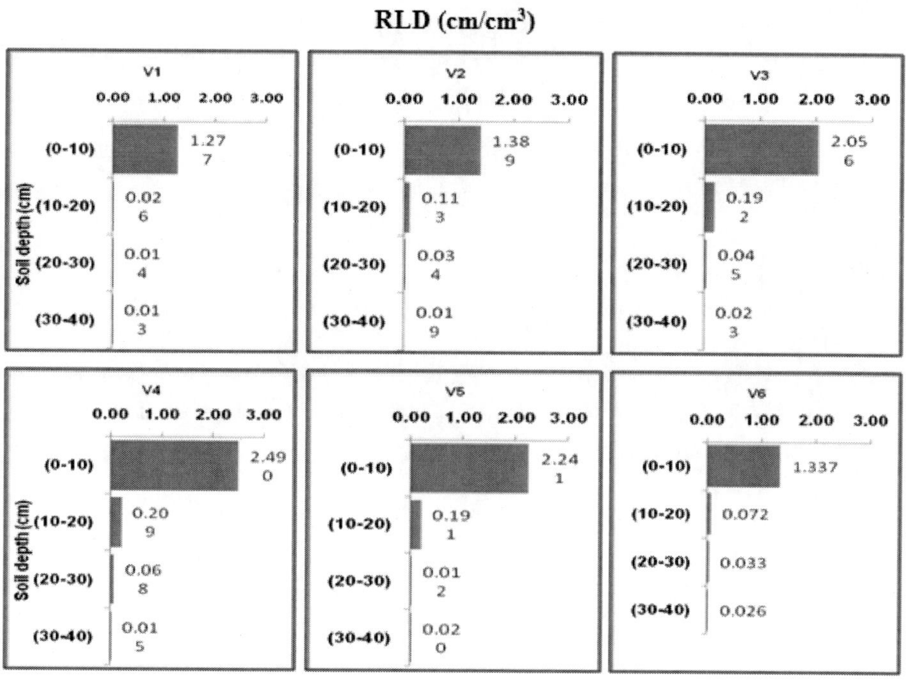

Maniruzzaman, 2012.

Figure 2. Average root length densities of different cultivars in continuous standing water condition of TPR during Boro, 2008-10.

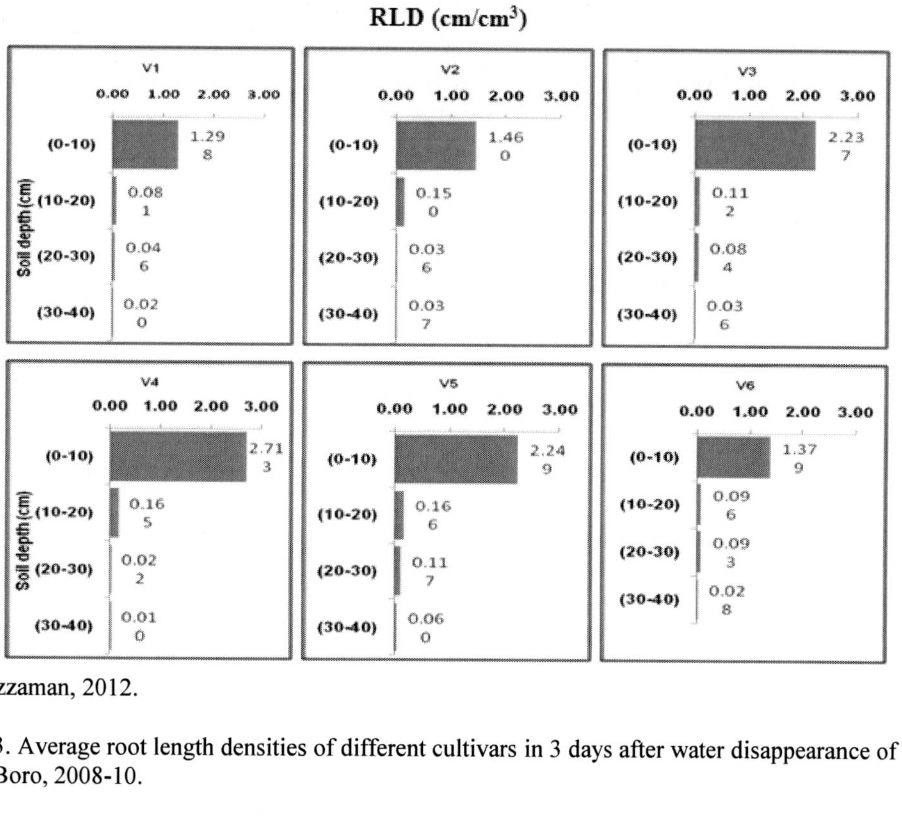

Maniruzzaman, 2012.

Figure 3. Average root length densities of different cultivars in 3 days after water disappearance of TPR during Boro, 2008-10.

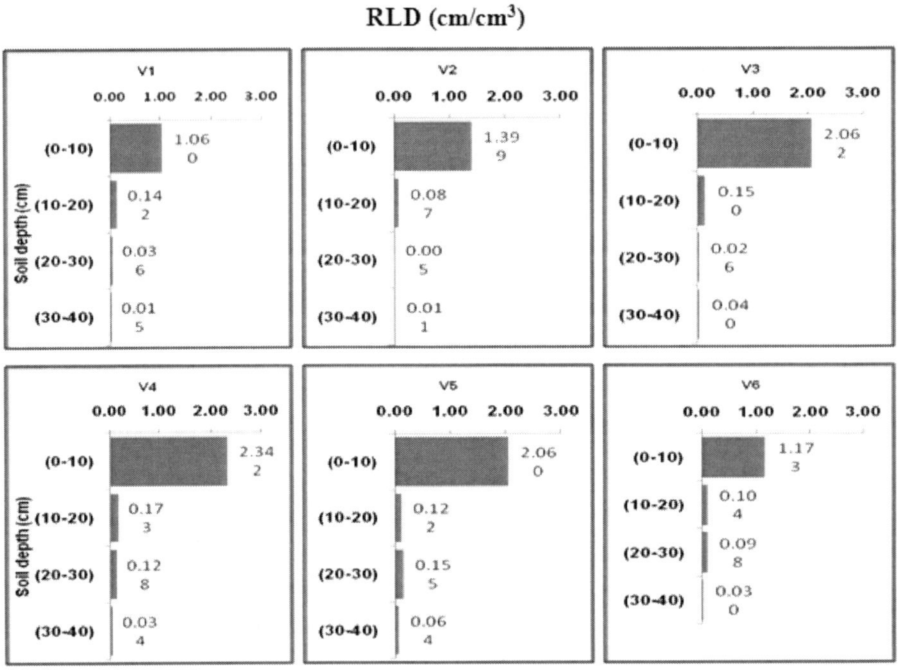

Maniruzzaman, 2012.

Figure 4. Average root length densities of different cultivars in 5 days after water disappearance of TPR during Boro, 2008-10.

Table 2. Average deep root-shoot ratio (below 30 cm depth) of different genotypes/ variety under different water regimes during 2008-10

Variety	Water regimes		
	CSW	3 DAWD	5 DAWD
V1	0.02	0.11	0.02
V2	0.02	0.14	0.18
V3	0.03	0.22	0.37
V4	0.02	0.04	0.43
V5	0.11	0.31	0.24
V6	0.10	0.26	0.19

Maniruzzaman, 2012.

2.2. Water and Nutrient Uptake by Root System

The tip of a single root is relatively impermeable to water and the region some distance away from the tip toward the proximal end also has a low permeability to water due to suberization. The entrance of water and salt is most rapid between the above two low permeable regions of a root. Along the axis of a single rice root, the meristematic region is about 0.25 mm from the tip; branched roots are recognizable at 30-40 mm. Suberization is more intense in regions further away from the tip (Kawata and Lai, 1968). Water absorption is most active between the root tip and the site of branched roots.

Water shortages are frequently a limiting factor for crop growth in upland fields and in rainfed lowland fields. When water is supplied uniformly to all the soil profiles, the shallow layers are depleted of water first. Plants then take water from increasingly greater depths. The rate of water absorption from a given soil volume is proportional to the effective total root length if other conditions are held constant (Gardner, 1960; 1964). The high rate of water absorption results in a more rapid decreases in soil water potential in the shallow layers. Since the rate of water absorption is also proportional to soil water potential if other things are same, the water absorption from the shallow layers will decrease. At this time, more water is available to the roots in the deeper soil profiles, and, hence, more water will be absorbed for these layers. In the mathematical model presented by Gardner (1960), the root was considered an infinitely long cylinder of uniform radius and water-absorbing properties. That is an approximation of the plant's root and may not always apply to the real root. The greater accuracy, the root length should be corrected for root hairs and for degree of suberization. The total root length is also related to the plant's capacity to absorb a nutrient. For corn plants grown in culture solution, the rate of phosphate uptake was proportional to the total root length (Jungk and Barber, 1974). Thus, the capacity of the plant to absorb water and nutrients is closely related to the total length of the root system.

Water and Nutrient Uptake by Old and New Roots

When half of the roots of a rice plant are cut at the base around panicle initiation time and the plant is placed in nutrient solution for above 10 days, new roots will emerge to replace those lost. The capacity of these new roots to absorb water and nutrient can be examined if they are placed in separate nutrient containers (Okajima, 1960). Old roots took up above 60% of the water and nutrients absorbed by the plant. Even if the water and nutrient uptake is expressed on a dry root weight basis, the capacity of the old roots is greater. This higher capacity may be partly attributed to the tissue's low nutrient and high sugar content, which favor active nutrient uptake. The new roots, however, have a higher oxidizing capacity than the old roots.

Root Characteristics Related to Drought Resistance

The rice root system can be studied using several techniques: 1) root box, 2) core sampling and 3) monolith. Since drought occurs when there is an imbalance between water absorption and transpiration, greater root growth can help the plant perform better under a limited water supply. The root-shoot ratio of rice varieties can be studied by the root box technique. The rations range form 0.05 to 0.3 at heading, and average about 0.1. Under drought conditions, the soil starts drying from the surface but the deep soil horizon may remain wet and able to supply water to the plant's roots. Consequently, deep root portions may be more meaningful than shallow root portions when the drought resistance of a variety is to be examined. For this reason, the deep root-shoot ration of rice varieties is defined as the weight, in milligrams, of roots that are deeper than 30 cm/gram shoot. It ranges from less than 10 to greater than 80 mg/g. When shallow-rooted lowland and deep-rooted upland varieties are grown in the field, there is not much difference between their root densities down to 30 cm beneath the soil surface. The differences are greater in the soil horizons below 30 cm. this justifies the use of deep root-shoot ratios in the root box technique. In the field, rice roots grow and extend vertically and laterally. IR20, a shallow-rooted, high tillering variety, has a very high root density around the plant base but the root's growth, both vertically and laterally, is limited. On the other hand, OS4, a deep-rooted, low-tillering variety, ha a well-spaced lateral root distribution and its roots penetrate deeper than those of IR20. Such differences may in part account for the differences in drought resistance between the two varieties.

Plant height can range from approximately 0.4 m to more than 6 m (in floating rice) depending on the rice variety or environmental conditions. The rice culm is round, hollow, jointed, and hairless. A maturing rice plant has a main stem and a number of tillers depending on the variety and cultural conditions. Each stem has a certain number of nodes (generally ranging from 13 to 16 nodes) and internodes under a certain environmental condition for a rice variety.

Rice leaves are rather flat. The first rudimentary leaf or prophyllum is at the base of a tiller and, with only a two-keeled bract, has no blade. The uppermost leaf immediately below the panicle is called the flag leaf. The leaf blade and leaf sheath are attached at the node, where there is an auricle (a pair of claw-like appendages encircling the stem), and the ligule (a thin, upright papery and triangular structure) immediately above the auricle. When young, the major difference between rice and a common grassy weed (*Echinochloa* spp.) is the presence of auricles and ligules.

Rice has a terminate panicle having the following structures: the base, axis, primary, secondary, and sometimes tertiary branches, pedicel, rudimentary glumes and spikelets. Each panicle generally bears from 50 to 300 flowers or spikelets. The rice panicle is a perfect and determinate inflorescence. A rice spikelet has two sterile lemmas, the rachilla and the floret. A rice floret includes six stamens and a pistil having stigmas, styles, and ovary, enclosed by the lemma and palea, sometimes with an awn. Even though rice is pollinated by wind, the natural crossing rate is low (less than 3%) because of the floral characteristics. The stigma (a plumose structure) functions to catch pollen for fertilization. This function is important for the out-crossing rate in hybrid rice seed production.

The rice grain generally refers to rough rice or paddy consisting of brown rice (or caryopsis) and the hull. Brown rice consists of the endosperm, embryo, and several thin layers of differentiated tissueshe pericarp (the ovary wall), the seed coat, and the nucellus. The seed coat consists of six layers of cells, with the aleurone layer the innermost. The rice embryo is small and contains the embryonic leaves (plumule) enclosed by a sheath (coleoptile), embryonic primary root (radicle) ensheathed by the coleorhiza, and the joining part (mesocotyl). Rice endosperm consists mostly of starch granules in a proteinaceous matrix, together with sugar, fats, crude fiber, and inorganic matter. Hull weight is about 20 percent of the total grain weight. The hull of a typical *indica* rice variety has the palea, lemmas, and richilla, but the hull of *japonica* rice usually only includes rudimentary glumes and perhaps a portion of the pedicel. The lemma is usually tough, archmenlike, sometimes awned, and bigger than the palea. Generally, pigmentation in rice does not cause any problems except for the pigmented apiculus or hull, which may stain the endosperm during parboiling, thus affecting the marketing value. Grain ripening stages (155 days) can be subdivided into milky, dough, yellow-ripe, and maturity stages based on the texture and colour of the growing grains.

3. Growth Phases and Stages of Rice Plant

The development of the rice plant may be divided into the following three phases:

1) Vegetative Phase: form seed germination to panicle initiation
2) Reproductive Phase: from panicle initiation to flowering
3) Ripening Phase: from flowering to maturity

The phases may be subsequently divided into different growth stages or period:

Vegetative Phase

During the vegetative phase, the plant undergoes the following stages:

a) *Germination stage:* the vegetative phase begins with the germination of seed. Rice seed germinated by the emergence of radicle or coleoptile form the germinating embryo.
b) *Seeding stage:* the stage follows seed germination and the seedling develops seminal roots. Seedling stage is generally considered form germination until development of the fifth leaf. During this stage, the seedling absorbs most of the endosperm.
c) *Transplanting or recovery period:* only in the transplanting method have this stage. It covers the period from uprooting of the seedling to full recovery. Direct seeded rice pants do not have this stage.
d) *Tillering stage:* this starts with the appearance of the first tiller form the axillary bud in one of the lowermost nodes. The number of tillers increases, at a point more rapidly (active tillering stage) is reached. Then some tillering die and the number of tillers declines and levels off. The plant stops tillering after the tertiary tillers have been produced.
e) *Internode elongation stage:* this stage started with elongation of internode.

The vegetative phase is the most variable among all the growth phases of rice plant. This phase is markedly affected by the prevailing day length and temperature and can be subdivided into the following two component phases:

1) Basic Vegetative Phase (BVP) or Active Vegetative Phase (AVP)
2) Photoperiod Sensitive Phase (PSP) or Lag Vegetative Phase (LVP)

The basic or active vegetative phase is the minimum period of vegetative growth required by the plant before it will initiate panicle primordium. The usual variations of day length have little or no effect o its duration. However, in certain varieties, temperature may either shorten or lengthen its duration.

The photoperiod sensitive or lag vegetative phase is that portion of the vegetative phase in which the flowering date is determined by the day length to which the plant is exposed. It is therefore from the end of the basic vegetative phase to panicle initiation. Only the photoperiod sensitive varieties have this component phase.

Reproductive Stage

a) *Panicle initiation to booting stage:* The reproductive phase begins just before or after the maximum tillering stage, depending on variety and environment. This phase is marked by the initiation of the panicle primordium of microscopic dimension on the growth shoot. As the young panicle develops, it becomes visible to the naked eye in a few days. This stage somewhere included the booting stage also.

Booting is the later part of the panicle development stage. About 16 days after visual panicle initiation, the sheath of the flag leaf swells. This swelling of the flag leaf sheath is called booting. The time of occurrence of internode elongation stage differs among varieties. With late maturing varieties, the accelerated elongation of the upper internodes may begin considerably earlier than the reproductive phase. With early maturity varieties, elongation may begin after panicle initiation.

b) *Heading stage:* This stage is followed by the emergence of panicle out of the flag leaf sheath. Emergence continues until 90-100% of the panicle are out of the sheaths.
c) *Flowering stage:* Flowering (blooming) or anthesis begins with the production of the first dehiscing anther in the terminal spikelets on the panicle branches. Flowering continues successively until all spikelets in the panicle bloom. Pollination and fertilization then follow.

Ripening Phase

The rice grain develops after pollination and fertilization. Grain development is a continuous process and the grain undergoes distinct changes before it fully matures.

a) *Milk stage:* the content of the caryopsis are first watery but later turn milky in consistency.
b) *Dough stage:* The milky caryopsis turns into soft dough and subsequently into hard dough.
c) *Maturity stage:* the individual grain is mature when the caryopsis is fully developed in size and is hard, clear and free from greenish tint. This stage is complete when more than 90% of the grains are fully ripened.

Approximate date of occurrence of some Growth stages of rice

Stage	**Date of occurrence from seedling**	**Date of occurrence before maturity**
Panicle initiation	About 60-70 days after seedling for 130 days non-sensitive varieties and variable in sensitive varieties	60-65 days from date of maturity regardless of variety
Booting	About 75 days after seeding for 130 days non-sensitive varieties and variable in sensitive varieties	About 45 days from date of maturity regardless of variety
Flowering	About 100 days after seeding for 130 days non sensitive varieties and variable in sensitive varieties	25-35 days from date of maturity regardless of variety

Approximate duration of the Growth phases

Phase	Duration
Basic Vegetative Phase Lag Vegetative Phase	25-65 days for most varieties. Varieties greatly according to day length in seasonal varieties
Reproductive stage	About 35 days regardless of variety
Ripening stage	25-35 days regardless of variety

4. RICE GROWING CONDITION: SOIL AND WEATHER

4.1. Soil Types

Rice can be grown a wide range of soil and climatic variability. It can be grown in sandy soil to clay soil in tropical to temperate region of the world based on the water holding capacity of the soil. Though it is a water loving crop, it is widely grown in loam to clay soil, because sandy type soil has a very low water holding capacity. It can be grown in wide range of alkaline to acidic soil, but preferably and mostly grown in neutral soil.

4.2. Effect of Climate

The average rice yields in rice growing countries range from less than 1 to more than 6 t/ha. There are a number of biological, environmental and socioeconomic reasons for the large differences in rice yield. Low yields are associated with upland rice, rainfed lowland rice and the poor socioeconomic conditions in the tropic, whereas high yields are associated with irrigated lowland rice and the good socioeconomic conditions in the temperate regions. Temperature, solar radiation and rainfall influence rice yield by directly affecting the physiological processes involved in grain production, and indirectly through diseases and insects. In the field, these factors are often difficult to separate from one another. From a crop physiologist's point of view, crop period, productivity and stability are important aspects of rice cultivation. Climatic factors affect each of them in different ways. In the temperate regions, irrigated rice cultivation starts when spring temperatures are between 13° and 20° C; the crop is harvested before temperatures drop below 13°C in the autumn. In the tropics where temperature is favorable for rice growth throughout the year and irrigation is not available in most places, cultivation starts with the rainy season. The average dates of the onset and withdrawal of the monsoon are known for particular regions of South Asia. The starting time and duration of rainfed rice cultivation is largely determined by these two dates. In both the tropics and the temperate regions, rice yield per hectare is primarily determined by the level of incident solar radiation. In the tropics, when adequately managed, the dry season crop usually produces higher yields than the wet season crop because it receives more sunlight. Stability of crop yields is affected by several factors.

4.3. Solar Radiation

Most of the radiant energy from the sun has a wavelength between 0.3 and 3 μm, often referred to as short-wave radiation. The earth emits radiations with wavelength ranging from 3 to 50 μm, which is referred to as long-wave radiation. Solar radiation penetrating the earth's atmosphere is reflected, absorbed, and scattered by clouds, gases, and aerosols in the form of soil and salt particles, smoke, insects, and spores. The reflection and scattering occur at all angels, and the downward component is usually referred to as diffuse sky radiation. Thus, incoming solar radiation takes the forms of direct radiation and diffuse sky radiation. The sum of these two is called radiation or simply incident solar radiation. Photosynthesis in green leaves uses solar energy in wavelengths for 0.4 to 0.7 μm, often referred to as photosynthetically active radiation (PAR) or simply light. The ratio of PAR to total solar radiation is close to 0.50 in both tropics and the temperate regions (Monteith 1972), and represents a weighted mean between the fraction for direct radiation and diffuse sky radiation.

4.4. Solar Radiation Requirement at Different Growth Stages

The solar radiation requirements of a rice crop differ from one growth stage to another. Shading during the vegetative stage only slightly affects yield and yield components. Shading during the reproductive stage, however, has a pronounced effect on spikelet number. During ripening it reduces grain yield considerably because of a decrease in the percentage of filled spikelets. Solar radiation at reproductive stage has the greatest effect on grain yield; that at the ripening stage, the next highest effect and that at the vegetative stage, an extremely small overall effect. Solar radiation of 300 cal/cm^2 per day during the reproductive stage makes yields of 5 t/ha possible. Less solar radiation during ripening is required to achieve the same yield level. Thus, the effect of solar radiation is apparent only when grain yield is higher than 5 t/ha. When grain yield is below that, sunlight per se may not have any direct significance.

4.5. Critical Low and High Temperature

Extreme temperatures are destructive to plant growth and, hence, define the environmental under which the life cycle of the rice plant can be completed. The critically low and high temperatures, normally below 20°C and above 30°C, vary from one growth stage to another (Table 3). These critical temperatures differ according to variety, duration of critical temperature, diurnal changes and physiological status of the plant. Subjecting the rice plant to temperatures below 20°C at about the reduction division stage of the pollen mother cells usually induces a high percentage of spikelet sterility (Satake, 1969). Differences among rice varieties in response to low temperatures at this stage have been clearly demonstrated.

Table 3. Response of the rice plant to varying temperature at different growth stages

Growth stage	Critical temperature* (°C)		
	Low	High	Optimum
Germination	10	45	20-35
Seedling emergence and establishment	12-13	35	25-30
Rooting	16	35	25-28
Leaf elongation	7-12	45	31
Tillering	9-16	33	25-31
Initiation of panicle primordial	15	-	-
Panicle differentiation	15-20	38	-
Anthesis	22	35	30-33
Ripening	12-18	30	20-25

* Refers to daily mean temperature except for germination.

4.6. Effect of Water Deficit on Growth and Yield

Water stress at any growth stage may reduce yield. The most common symptoms of water deficit are leaf-rolling, leaf-scorching, impaired tillering, stunting, delayed flowering, spikelet sterility, and incomplete grain filling. The rice plant is most sensitive to water deficit form the reduction division stage to heading. Three days of drought at 11 days and at 3 days before heading reduces yield significantly by causing a high percentage of sterility. Once sterility occurs, there is no way for the plant to compensate. On the other hand, water deficits during the vegetative stage may reduce plant height, tiller number, and leaf area, but yield will not be affected by the retarded growth if water is supplied in sufficient time to permit the plant's recovery before flowering.

4.7. Effect of Water Excess on Growth and Yield

Depending on topography and rainfall patterns, low-lying areas may be subjected to different water depths and to different durations of high water. These areas can be divided into the following categories (Vergara *et al.*, 1976):

a) *Deep-water areas.* The water is between 150 cm and 400 cm deep and usually remains in the field for 3-4 months. Special varieties known as *floating rice* are planted in these areas. Large areas of floating rice are found in Bangladesh, northeast India and Thailand.
b) *Flooded areas.* Flood areas include the margins of deep-water areas, low-lying areas, and tidal swamps. The water is less than 150 cm deep and usually remains in the field for several months. Tall indica varieties, known as *flood rice,* are commonly used.
c) *Submerged areas.* The water level and duration are variable but the plants are usually completely submerged. In some areas submergence occurs almost every year. In

many lowland rice fields submerged occurs only during typhoons or usually heavy, continuous rains.

Attention is currently focused on improving rice varieties adapted to flood areas by incorporating the elongation genes of floating rice into improved semi-dwarf high-yielding varieties. These new varieties, called *deepwater rices*, can remain short when the water is shallow and grow tall in response to increased water depths (Jackson *et al.*, 1972). A large portion of the rainfed lowland rice area falls into the category of submerged areas, where partial submergence is the common problem. When a rice crop is submerged to different degrees at different growth stages, the grain yield decreases in different proportions. For example, with 25% plant height submergence at tillering, yield deceases by 18% in Aman and 25% in Boro; with 75% submergence at ripening, yield decreases by 30% in Aman and 50% in Boro. The yield-decreasing effects of partial submergence could be attributed to impaired tillering and decreased area of photosynthetic leaf surface. To overcome the adverse effects of partial submergence of grain yield, varieties with intermediate plant height (about 110-130 cm, taller than semi-dwarfs and shorter than traditional tall varieties) are replacing semi-dwarf varieties in shallow water areas than are subject to occasional floods.

5. Rice Cultivation Methods

Rice crop establishment includes transplanting and direct seeding. Direct seeding is done either in dry or wet conditions. In wet condition, pregerminated seed are sown. Weed incidence, rainfall pattern, irrigation scope,, field elevation and soil types are the major determinants of crop establishment (Pandey *et al.*, 2002; Balasubramanian and Hill, 2002; Balasubramanian *et al.*, 2003). Moreover, wage rate and availability of family labor are major economic factors influencing the choice of crop establishment method by the farmer.

5.1. Transplanting Method

Transplanting involves raising seedlings in wet or dry nursery beds.. Manual transplanting is labor intensive, requiring more than 30 man-d ha^{-1}. It also requires a large amount of water during land preparation.

5.2. Wet Seeding Method

Direct wet seeding is an alternative to transplanting method, which is commonly practised in irrigated areas. In this method, the fields are puddle, leveled and then sprouted seeds are sown on the wet soil by broadcast manually or by Drum seeder in line. Direct seeding can reduce labor requirement by a significant amount compared to transplanting. The demand for labor is spread out over several months in direct seeded rice (DSR) than transplanting, which allow farmers to make full use of family labor and to avoid labor bottlenecks (Pandey *et al.*, 2002, Singh *et al.*, 1994).

5.3. Dry Seeding Method

Dry seeding is practised in rainfed lowland, upland and flood-prone areas. Direct seeding offers faster and easier planting, reduced labor requirement and less drudgery, earlier crop maturity by 7-10 day (Balasubramania and Hill, 2000; Saleh and Bhuiyan, 1995), more efficient water use and higher tolerance of water deficit and often higher profit in areas with assured water supply. The DSR can be established and harvested earlier, with a reduced risk of drought and opportunity for growing a post-rice crop (Tuong, 1999; Saleh and Bhuiyan, 1995).

6. Water Requirement in Rice Field

The amount of water required for successful growth and yield of a crop without suffering from any water stress is considered as water requirement. Water plays a decisive role in crop production; therefore, improved management of irrigation water is essential for effective implementation of agricultural development programs. Water management for irrigation is the task of proper utilization of the resource based on its availability and quality to increase production without creating any adverse effects.

Current rice production systems consume a high amount of water. It takes about 4000 to 5000 liters of water to produce one kilogram of paddy (BRRI, 2007). This profligate usage of water in irrigated rice production is unsustainable, given the increasing demand for freshwater due to growth in rice demand and growing competition from other sectors (ADB, 2005). With the current rate of water usage, even maintaining productivity in many currently irrigated areas will be difficult unless more water-efficient rice production technologies suitable for irrigated areas are developed.

6.1. General Overview

Chow (1965) reported no yield reduction if water is delivered intermittently, but 20 to 50% water savings can be achieved in this method compared to continuously flooded fields. De Datta *et al.* (1973) found that continuously saturated plots produced similar grain yields compared to continuously flooded plots. However, submergence influenced fertilizer use efficiency and control of weeds and insects. Doorenbos and Kassam (1979) indicated that maintaining moderate submergence during critical periods and keeping soil water at 75% or more in the root depth during rest of the growing period can save water upto 50 percent.

Mowla *et al.* (1992) carried out an experiment in the Ganges-Kobadak (G.K) irrigation project area to determine the amount of water required for rice cultivation in wet season. They supplied 5-7 cm water in each irrigation on rotational basis. They found that water needed for land preparation and growth period were 241 mm and 1453 mm, respectively including 783 mm of rainfall during the growth period. They also mentioned that farmers used more water than the research managed plots. They did not calculate effective rainfall, which is very important because a substantial amount of water is lost through runoff due to heavy rainfall and is not available for the crop. The above studies led to conclude that there is wide variation in water requirement in continuous flooding and alternate wetting and flooding method of rice cultivation. It is site specific and varies from season to season. So, the actual amount of water needed for optimum crop yield is still not known.

6.2. Growth Phase Wise Water Requirement of Rice

The three constituents of water, required to grow lowland rice are evapotranspiration (ET), seepage and percolation (S&P) and surface drainage (SD). Strictly speaking, only ET is a true water requirement for crop growth, but in supplying it, some S&P and SD are unavoidable.

Water Requirement during Land Preparation Period

Land preparation for lowland rice involves land soaking (2-7 days) to saturate and soften the soil, and plowing and harrowing to bring it to a puddled condition. The purpose of this thorough land preparation is to (a) provide weed free, level and soft soil surface for transplanting; (b) incorporate organic matters like weed, rice straw, organic and inorganic manures; (c) level land for better water management and (d) formation of impervious layer for minimizing seepage and percolation (S&P). Kampen (1970) summarized experimental results of water requirements for land preparation in different rice growing countries, such as Sri Lanka (330 mm), India (230–390 mm), Japan (130 mm), Malaysia (176–225 mm), Surinam (420 mm), Taiwan (120–210 mm), Thailand (300-400 mm) and southeast Asia as a region (220–240 mm). Wickham and Sen (1978) suggested that the land preparation requirement can be reduced by shortening the duration of land preparation period.

Studies in the Ganges - Kobadak project (BRRI, 1985 and 1986) in Bangladesh indicated that under researchers management conditions the average water requirement for land preparation during Aus and Aman seasons were 237-248 and 215-261 mm, respectively. These amounts were less than 20% of total water requirement for the growing season. Some people practise dry plowing of field immediately after harvesting of previous crop and flood the field two to three days before final land preparation, which includes puddling. Transplantation of rice seedlings is done after final land preparation, usually on saturated soil, or soil with a thin depth of water. Dry plowing may require less water for land preparation than the traditional way.

Evapotranspiration (ET)

Evapotranspiration is the amount of water used by the plants plus the evaporated from the surface of the ground. Water that moves through the plants is mostly transpired through the leaves but some is retained in the plant tissues during crop growth.

ET is essentially independent of the stage of crop growth. This is not surprising when we recall that ET is made up both of water transpired from the plant surface and of evaporation from the soil/water surface. When the crop is newly planted, plant transpiration is relatively little, but evaporation from the water surface is almost the same as that from a pan. But as the crop grows, transpiration increases while the increased shading of the surface water, reduces direct evaporation form the source. The some of the two effects, ET is thus dependent upon climatic conditions and not on the stage of crop growth.

Seepage and Percolation (S&P)

Seepage and percolation like ET, is also a combination of two effects, one seepage (lateral and surface movement of water) and other percolation (vertical sub-surface movement of water). Seepage normally comes out the soil surface or into streams or rivers, while percolation usually reaches the water table. We consider them, together, they occur simultaneously. S&P rates are higher for light soil than heavy ones. Usually, the S&P rate in rice fields varies from 4 mm/day to 8 mm/day. Water is lost from irrigated rice during the crop season, through transpiration, evaporation and percolation. Water losses through percolation are the most variable. The total water loss ranges from 5.6 to 20.4 mm/day, but most of the observed values for total water loss range from 6 to 10 mm/day. Thus, on the average, about 180-300 mm water/month is needed to produce a reasonably good crop of rice.

In field operation, a total of 1240 mm is an average water requirement for an irrigated rice crop. The same figure is obtained if the assumptions are made that the crop duration is 5 months, monthly water requirement is 200 mm, and land preparation requires 200 mm water. If land Preparation takes more than 30 days, however, as much as 500-600 mm of water may be required.

Several different units are used in discussing the water needs of field crops:

1 liter/ha per second = 8.64 mm/day = 86.4 m^3/ha per day
1 mm/day = 0.116 liter/ha per second = 10.0 m^3/ha per day

Water Balance

The seasonal water balance for a given time span for a rice field, which includes both the crop establishment period and crop growth period, is given by the equation:

$$I + R = E + T + S\&P + D \pm \Delta S \tag{1}$$

where, the inflows are irrigation (I) and rainfall (R) and the outflows are evaporation (E), transpiration (T), seepage and percolation (S&P), drainage (D) and change in storage (ΔS). The water balance components were depicted in Figure 5.

The ΔS include surface storage and change in storage in the root zone. Field water storage was determined from the ponding water depth at transplanting. Soil water storage was determined from the difference of measured saturated water content and moisture content at harvest of the top 25 cm soil layer. It was assumed that the moisture content below that depth was near saturation and the contribution to the seasonal water was small compared to other terms. Alternately, ΔS were the only term that was computed as the closure term. Drainage was zero during the dry seasons because of low rainfall in the crop growing period. In water balance equation, only transpiration is the beneficial outflow component. Therefore, transpiration must not be reduced in order to attain optimum plant growth and achieve high yield. However, other outflows which are non-beneficial losses but are part of the field water requirement may be reduced or minimized in order to reduce the total inflows, especially irrigation water input. Several water saving techniques are to reduce the water use in rice production during (1) land preparation/crop establishment period and (2) crop growth period. But this thesis only discussed the water balance components during crop growth period.

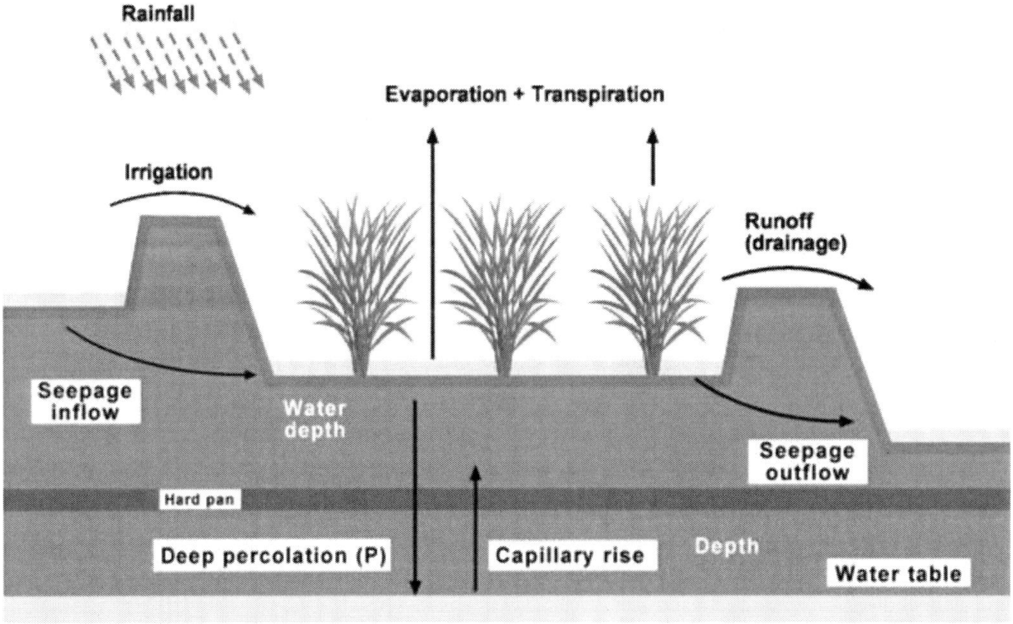

Adapted from IRRI, 2007.

Figure 5. Component of water balance in paddy field.

Table 4. Growth phase wise water requirement (continuous standing water condition)

Variety	Season	Growth phase	Growth duration (days)	Amount of water required (mm) Heavy soil	Amount of water required (mm) Light soil	Seasonal total water requirement (mm)
Medium Duration (BRRI dhan28)	Boro	Vegetative	75	460	600	995 (heavy) 1355 (light)
		Reproductive	35	335	460	
		Ripening	30	200	295	
Long Duration (BRRI dhan29)	Boro	Vegetative	95	650	830	1205 (heavy) 1640 (light)
		Reproductive	35	350	510	
		Ripening	30	205	300	
Medium Duration (BR11)	T. Aman	Vegetative	80	700	815	1255 (heavy) 1460 (light)
		Reproductive	35	395	460	
		Ripening	30	160	185	
Overall average water requirement under continuous standing water condition						1318

Estimation of Water Requirement For Rice

In this estimation, two major constituents of water requirements for rice have been considered: evapotranspiration (ET) and seepage & percolation (S&P). Rice plants require water continuously, but the rate of requirements varies with the atmospheric condition, soil type, growth duration of crop, age of crop, water management status in the plot and also on the topography of land. In addition to the normal water requirement during the growing period, about 200 mm-250 mm of water is required for land preparation in Boro season for heavy and light soils respectively. In T. Aman season these two figures are 150 mm and 200 mm respectively due to higher initial moisture content of soil. A fraction of water (18 to 20 mm of equivalent depth) is also needed for raising seedlings in seed bed. Water should be drain out before 10-15 days of harvest. No irrigation is required during this period. The recommended seedling age of 40-45 days in Boro season and 30-35 days in T. Aman season has been considered for calculating the following growth stage-wise water requirement (Tables 4 and 5).

Table 5. Growth phase wise water requirement (alternate wetting and drying condition)

Variety	Season	Growth phase	Growth duration (days)	Amount of water required (mm)		Seasonal total water requirement (mm)
				Heavy soil	Light soil	
Medium Duration (BRRI dhan28)	Boro	Vegetative	75	380	490	755 (heavy) 1020 (light)
		Reproductive	35	235	325	
		Ripening	30	140	205	
Long Duration (BRRI dhan29)	Boro	Vegetative	95	500	605	910 (heavy) 1175 (light)
		Reproductive	35	245	360	
		Ripening	30	165	210	
Medium Duration (BR11)	T. Aman	Vegetative	80	530	630	925 (heavy) 1085 (light)
		Reproductive	35	280	325	
		Ripening	30	115	130	
Overall average water requirement under continuous standing water condition						978 **(26% saved)**

When to Irrigate and How Much Water to Apply

Grain crops indicate need for water by temporary wilting. The need for water can also be determined by the appearance of the leaves of the crops. But it is better to base the time of irrigation on observation of the moisture status of the soil.

Continuous standing water in rice field is not necessary. A good yield can be obtained if the soil is in saturated condition since transplanting. Number of tillers increased sufficiently if the field is in saturated condition up to 30-45 days after transplanting. Water requirement of rice is maximum at booting and flowering stages. During this period sufficient water should be maintained in the plot to get good yield. For rice a single excessive irrigation should be avoided. Frequent small irrigation can be determined by dividing the depth of irrigation by the water to be depleted for the plot per day. The number of irrigation depends on the depth in each irrigation and duration of crop growth. In alternate wetting and drying condition irrigation is supplied after 3 days of disappearing of standing water from the field. In T. Aman season, about two-third of water requirement is met from rainfall.

7. Irrigation Management Strategies

7.1. Alternate Wetting and Drying (AWD) Method

In *Alternate Wetting and Drying (AWD) method,* irrigation water is applied to obtain flooded conditions after a certain number of days have passed after the disappearance of ponded water. The number of days of nonflooded soil in AWD before irrigation is applied can vary from one day to more than 10 days.

Though some researchers have reported a yield increase using AWD (Wei Zhang and Song, 1989; Stoop et al., 2002), recent work indicates that this is exception rather than the rule (Belder *et al.*, 2004; Cabangon *et al.*, 2002; Tabbal *et al.*, 2002). Experimenting with AWD in lowland rice areas with heavy soils and shallow groundwater table in China and the Philippines, Cabangon *et al.*, 2004; Belder *et al.*, 2004; Lampayan *et al.*, 2005 and Tabbal *et al.*, 2002 reported that total (irrigation and rainfall) water inputs decreased by around 15-30% without a significant impact on yield. In India, in an experiment it is evident that the AWD field had the same yield as continuous flooding, but saved 16-24% in water costs and 20-25% in production costs. In Bangladesh, irrigation and water management division, BRRI reported that on average, for short duration rice variety water saved about 22% and for long duration rice variety water saved about 25% during Boro season (BRRI, 2009). Very little research has been done to quantify the impact of AWD on the different water outflows of rice fields: evaporation, seepage and percolation. The little work done so far suggests that AWD mostly reduces seepage and percolation flow and has only a small effect on evaporation flows. Belder et al., 2007 and Cabangon et al., 2004 calculated that evaporation losses decreased by 2-33% compared with fully flooded conditions. The following potential benefits of AWD have been suggested: improved rooting system, reduced lodging (because of a better control of some diseases such as golden snail).

Practical Implementation of AWD

A practical way to implement AWD is to monitor the water depth in the field using the "field water tube". After an irrigation application, the field water depth will gradually decrease over time. When the water level (as measured in the tube) is 15 cm below the surface of the soil, it is time to irrigate and flood the soil with a depth of around 5-7 cm. Around flowering, form 1 week before to 1 week after the peak of flowering, ponded water should be kept at 5-7 cm depth to avoid any water stress that would result in potentially severe yield loss. The threshold of 15 cm is called "safe AWD" as this will not cause any yield decline since the roots of the rice plants will still be able to take up water form the saturated soil and the perched water in the root zone. The field water tube helps farmers see this "hidden" source of water. In safe AWD, water savings may be relatively small, on the order of 15%, but there is no yield penalty. After creating confidence that safe AWD does not reduce yield, farmers may experiment by following rules should be observed. AWD irrigation can be used form a few days after transplanting till first heading. In the period of first heading to 1 week after flowering, keep the field flooded with 5-7 cm depth. After that, during grain filling and ripening, apply AWD again. When many weeds are present I the early stage of crop growth, the implementation of AWD can be postponed for 2-3 weeks until weeds have been suppressed by the ponded water. Under safe AWD, no special N management regime is needed and local recommendations as for flooded rice can be used (Belder *et al.*, 2004). Apply fertilizer N preferably on the dry soil just before irrigation is applied.

Field Water Tube

The field water tube is used to measure the depth of standing water on the field, be it on top of the surface or just below the surface. Perforate a hollow and bottomless PVC tube of about 10 cm in diameter and 25 cm long with small hole using a drill. The holes should be about 0.5 cm in diameter and spaced 2 cm apart. Install this tube in the field so that the bottom of the tube is buried in the plow sole (about 15 cm deep) and that some 10 cm of the tube is buried in the plow sole (about 15 cm deep) and that some 10 cm of the tube protrudes above the soil surface. Remove the soil form inside the tube down to the bottom of the tube. Water will flow through the holes into the tube, so that the water level inside the tube is same as outside. After irrigation, the level of the water in the tube can be seen going down every day. The tube is easily to record the water depth (no need to walk very deep into the field). Make sure that the spot chosen to install the tube is representative for the whole field (don't place it in a depression or in an elevated patch). The water depth is measured form the top of the tube to the level of the water in the field using a simple scale. Subtract 10 cm form the reading to obtain the depth of ponded water. A negative value means that the water is standing on the field; a positive value means that the water level is above the surface can be measured a few times during the season to check whether it is 10 cm.

AWD method can save 4-5 number of irrigations compared to the farmers' practice during Boro season which save 25% irrigation water and 30% fuel cost. It is also increase yield about 0.5 t/ha. The method is very simple to apply in the field.

7.2. Supplemental Irrigation by Rain Water Harvesting for Drought Mitigation

What is Supplemental Irrigation?

Supplemental irrigation means the application of relatively small quantity of water to a crop for a short period when the crop suffers from water shortage due to the unusual droughts and can be very effective in boosting the production of crops grown under rainfed condition.

In other words, supplemental irrigation is the additional application of the water to the rainfed crop for successful crop growth if the water stress occurs during the growing season.

Role of Supplemental Irrigation

Timely Transplanting

It is very important transplanting of Aman rice on time. Delay transplanting shifts the harvesting time. Consequently, the critical stages of the crop (reproductive and ripening stage of crop) suffer form drought. Thus, drought in the later growth period of T. Aman may cause a large yield reduction. Generally a cumulative of 400 mm rain is needed for land preparation and transplantation in rainfed ecosystem. Form an analysis of 30 years rainfall at Godagari of Rajshahi district, it is seen that the non exceedence probability of occurrence of 400 mm rainfall by mid July is 20%. It means that 2 out of 10 years this amount may not be exceeded by 15 July in other words 8 out of 10 years this amount may be expected by 15 July. But at 80% non-exceedence probability, 400 mm rainfall will occur by mid August. It means that 8 out of 10 years this amount may not be exceeded by 15 August in other words 2 out of ten years this amount may be expected by 15 Augusts. Therefore, farmers of Godagari will need supplemental irrigation 2 out of 10 years if they want to transplant timely.

Early Drought Mitigation

Early drought occurs very often during vegetative stage of T. Aman. This stage has capacity to repair itself form small damage due to short spell drought. But it cannot fully recover large damage caused by long spell drought i.e. very severe drought. Consequence of very severe drought at vegetative stage is yield reduction. It was found form a pot experiment that yield of T. Aman was reduced by 6-50% for 10-15 days drought during vegetative stage. So, early drought cannot be ignored and to mitigate it supplemental irrigation is essential.

Terminal Drought Mitigation

Terminal drought is common scenery in T. Aman season. It occurs during critical stages (reproductive stage and ripening stage) of aman crop. Any severity of drought from low to high during this period cause yield reduction ranges form 40 to 100%. Yield reduction for different severity of drought during different stages of T. Aman crop is shown Figure 6. Therefore, supplemental irrigation is needed must to mitigate terminal drought.

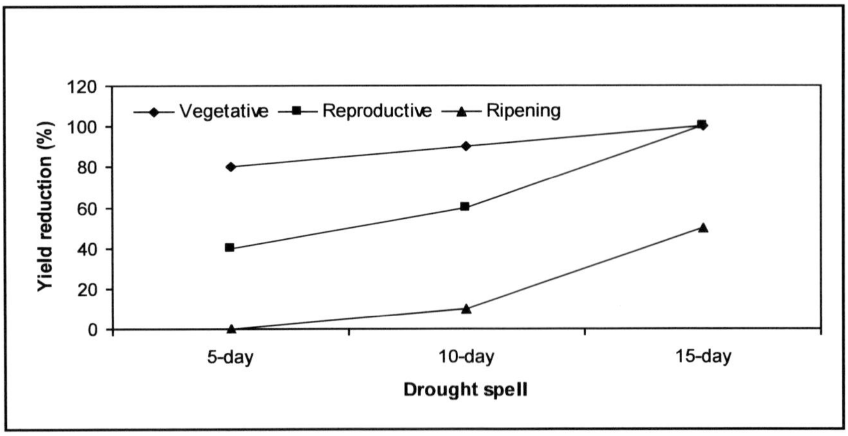

Islam, 2010.

Figure 6. Yield reduction for different drought spell at different growth stages of T. Aman.

Determination of Time for Supplemental Irrigation

Drought is a "creeping phenomenon" the effects of which accumulate slowly over a considerable period of time. Drought may begin at any time, last indefinitely and its effects can be devastating. Farmers neither can understand onset of drought nor can predict the severity (amount and duration) of drought. So they do not apply supplemental irrigation rather wait for rainfall. Meanwhile crop is hampered by drought. Some technique can be adopted to determine time for supplemental irrigation application and they are as follows:

Adoption of Alternate Wetting and Drying (AWD) Method

In this method, a 10 cm diameter and 30 cm long PVC pipe is needed. Its 20 cm perforated is installed below surface and 10 cm imperforated kept above surface of rice plot. After disappearing of standing rain water in the plot, water is appeared inside the PVC pipe. When this water is appeared at the bottom of the pipe i.e. 20 cm below form surface of the plot, supplemental irrigation will have to be applied.

Adoption of Irrigation Three Days after Disappearing Standing Water

When standing rain water is disappeared, supplemental irrigation will have to be applied in the plot after three days.

Forecasting of Rainfall

Time of application for supplemental irrigation can be determined based on weekly forecasting of rainfall. But this method is not yet developed in many countries.

Determination of Supplemental Irrigation Quantity

The amount of supplemental irrigation for each growth stage was determined based on water deficit i.e. drought amount during respective period including seepage and percolation demand. The equation for supplemental irrigation is given by:

$$SI = DA + (S + P)*DD \quad (2)$$

where,

SI = Supplemental irrigation
DA = Drought amount (mm) during particular period
DD = Drought days (day) or duration of drought during that particular period
S = Seepage rate mm/day
P = Percolation rate mm/day

DA is equal to total amount of consumptive use (ET) that remains unsatisfied during drought period (DD).

Sources of Supplemental Irrigation

Surface Water

River, Canal, Khari, Pond, Water body etc can be used as source of supplemental irrigation.

On Farm Rainwater Harvest

Bangladesh Rice Research Institute (BRRI) has studied and developed some techniques on rainwater harvest for supplemental irrigation and those are as follows:

a) *The economic ditch:* A ditch size of 4-5% (around 2 m deep) of the rice plot is required to conserve rainwater for one supplemental irrigation of 4-6 cm. In this study a ditch of 5 m X 5 m X 2 m with a 2:1 side slope was constructed at one corner of the plot.

b) *On Farm reservoir:* OFR system plays a crucial role as source of supplemental irrigation. Harvest surplus rainfall (runoff0 produced in the catchments area and in-situ rainfall, and store the water of subsequent use. OFRs have been in use in rural areas since ancient times. IWM division of BRRI has developed design of farm reservoir for new construction in Barind area which is economically feasible. Area for a new square size OFR is required 7.7% of the service area. The depth and side slope of OFR are 2 m and 1:1.

The OFR can provide 2 supplemental irrigations, each of 5 cm depth during terminal drought. Early harvested water of OFR can be used for land preparation. From the analysis of rainfall and OFR storage data, it was observed that in a dry year the rainfall from the beginning of June to mid July (beginning of T. Aman transplantation) was about 340 mm and the OFR was about 55% full. In a wet year the rainfall during the same period was about 650 mm and the OFR was about 85% full.

Ground Water

Ground water can be used as source of supplemental irrigation. During T. Aman season few farmers are using ground water as source of supplemental irrigation although use of it is economically feasible.

Methods of Application for Supplemental Irrigation

Surface water can be applied by using low lift pump (LLP), swing basket, don etc. If the source of water remains higher elevation than rice plot water can be applied by gravitational flow. Ground water can be applied by using shallow tubewell (STW) and deep tubewell (DTW). Most of the farmers destroy field channel during land preparation of T. Aman. They are not interested to operate STW/DTW for supplemental irrigation. These are the problems of supplemental irrigation. In that case plastic pipe can be used as field channel and awareness about benefit of supplemental irrigation could be developed among farmers so that they become interest for operation STW/DTW.

Impact of Supplemental Irrigation

Generally yield of T. Aman is increased by 30% for applying supplemental irrigation. A study of BRRI conducted from 1978 to 1987 in moderate drought prone area (Gazipur) reveals that yields of supplementary irrigated T. Aman were increased from 2 to 59% over rainfed T. Aman (Table 6). The variation of percent increase in yield depends on effect of drought on rainfed corp.

Economics of Supplemental Irrigation

Benefit of supplemental irrigation is indispensable. Supplemental irrigation is a profitable commodity. But profit in other words benefits-cost ratio (BCR) of it depends on incremental yield. A survey at Godagari of Rajshahi district showed that BCR were 4.1 and 5.88 for applying supplemental irrigation from surface water and ground water respectively (Table 7).

Table 6. Yields of rice under supplemental irrigation and rainfed condition in Aman season, BRRI Farm, Gazipur (Islam, 2010)

Year/Variety	Applied water (mm)	Yield (t/ha)		Yield increased over rainfed (%)
		Supplemental irrigation	Rainfed	
1974/BR4	100	3.66	2.3	57
1979/Pajam	60	3.2	2.7	19
1981/Nizersail	100	2.9	2.4	21
1983/BR11	60	4.6	2.9	59
1984/ BR11	60	5.4	4.9	10
1985/ BR11	60	5.9	5.3	11
s1986/ BR11	60	4.6	4.5	2
1987/ BR11	60	4.5	4.4	2
1987/Nizersail	60	3.7	3.5	6

Table 7. Cost and benefit from incremental yield through supplemental irrigation from surface and ground water during T. Aman in 2006 at Godagari, Rajshahi

Source	Mode	Cost (TK/ha)	Benefit (Tk/ha)	Benefit-Cost ratio (BCR)
Surface water (Pond or Khari)	LLP	6750	27,675	4.10
Ground water	DTW	4550	26,775	5.88

8. Irrigation Application and Drainage System

8.1. Irrigation Application System

Irrigation application systems may be grouped under two broad system types: gravity flow and pressurized systems.

Gravity-Flow Systems

Many irrigation systems rely on gravity to distribute water across the field. Land treatments—such as soil borders and furrows—are used to control lateral water movement and channel water flow down the field. Water is conveyed to the field by means of open ditches, above-ground pipe (including gated pipe), or underground pipe, and released along the upper end of the field through siphon tubes, ditch gates, or pipe valves. Fields are generally rectangular with water runs typically ranging from one-eighth to one-half mile in length. Gravity systems are best suited to medium- and fine-textured soils with higher moisture-holding capacities; field slope should be minimal and fairly uniform to permit controlled water advance.

Pressurized Systems

The decline in gravity-flow acreage has been accompanied by an increase in acreage under pressurized systems. Pressurized systems—including sprinkler and low-flow irrigation systems—use pressure to distribute water. With rare exceptions, the pressure to distribute water involves pumping, which requires energy.

8.2. Irrigation Drainage Systems

The collection and disposal of drainage flows from irrigation and precipitation is an important management consideration in many irrigated areas. Irrigation drainage includes surface runoff and deep percolation from water applied to meet crop consumptive needs. In some areas, periodic flooding of fields may also be required to leach soil salts from the crop root zone, often increasing the need for drainage systems. Irrigation drainage is often collected and reused in irrigated production. Tailwater systems recover drainage flows below the field (or in low-lying areas of the farm), recirculating the water to the top of the field for reuse. Drainage flows may also be used as irrigation supplies downslope, both onfarm and off-farm. In some cases, drainage systems may be used to drain excess water during wet periods as well as "subirrigate" during dry periods by regulating underlying water tables. In many cases, drainage flows of poor quality become a disposal issue. Primary disposal methods include onfarm evaporation ponds, direct discharge to off-farm surface water bodies through drainage canals, and reuse in salt-tolerant crop and tree production.

9. Other Practices Affecting Irrigation

Other practices—while not water-management practices *per se*—can be important components of an irrigated farming system. Such practices, in combination with improved irrigation systems, may enhance returns to irrigated production while reducing offsite environmental impacts.

9.1. Nutrient and Pest Management

Irrigation affects the optimal timing and application rate of chemical applications for nutrient and pest management. Fertilizer use is typically greater for high-valued, high-yielding irrigated production. Weed and pest conditions may also increase under irrigated field conditions, necessitating increased use of pesticides, herbicides, and fungicides. Careful nutrient and pest management increases the effectiveness of water and applied chemicals, while reducing offsite impacts. Chemigation—or the application of fertilizers, pesticides, and other chemicals through irrigation water— ermits controlled applications when used in conjunction with highly efficient irrigation systems. Chemigation can reduce the costs of applying chemicals, while avoiding equipment use and soil compaction. Chemigation is used on all major crops, with the largest treated acreages in orchard crops, hay, and corn—and the greatest concentration of use in potato, rice, and sugarbeet production (USDC, 1996).

9.2. Erosion Control

Soil erosion can be a serious problem for less efficient irrigation systems on sloping fields. Soil erosion creates barriers to even water flow in furrows, reduces long-term field productivity, and contributes to offsite water-quality problems. Irrigation-induced erosion is particularly severe in areas of the Northern Pacific, Southern Pacific, and Mountain regions (USDA, 1992). Measures to improve uniformity of applied irrigation water can help control soil loss. Gravity-flow systems may be modified to reduce flow velocity or field slope in accordance with soil-water infiltration rates. Soil erosion may also be a problem with sprinkler systems, particular on steeply sloping fields and under outer spans of center-pivot systems where water application rates are higher. System adjustments to reduce erosion include reduced water applications per irrigation set, larger pattern sprinkler heads, and booms to increase sprinkler head spacing. Other practices may also limit soil erosion on irrigated fields. Crop residue management to maintain vegetative material on the soil surface increases infiltration while protecting the soil from erosive water flow. In some cases, deep tillage can reduce runoff through increased infiltration. Land treatment measures may be installed to slow runoff and trap sediment on the farm. These include furrow dikes in the field, vegetative filter strips below the field, mini-basins in tailwater ditches, larger sediment ponds constructed in drainage ditches, and tailwater reuse systems. A promising new soil amendment—Polyacrylamide, more commonly known as PAM—may be added to irrigation water to stabilize soil and water-borne sediment. Under experimental field-trial conditions, proper application of PAM with the first irrigation has substantially reduced soil erosion in furrow systems. Potential benefits include reduced topsoil loss, enhanced water infiltration, improved uptake of nutrients and pesticides, reduced furrow-reshaping operations, and reduced sediment-control requirements below the field. An estimated 50,000 irrigated acres were treated with PAM after just 1 year on the market, including 30,000 acres in the Pacific Northwest. Research is underway to determine the best PAM formulations and application techniques (Sojka and Lentz, 1996).

9.3. Water Conservation

Improved irrigation and conveyance technologies may substantially increase on-farm water use efficiency. Whether technology adoption can achieve significant water savings for nonfarm and instream uses, however, will depend on many factors. In general, a given percentage increase in field application efficiency will yield a less-than proportional reduction in applied water. For example, a 50-percent increase in field application efficiency—from 40 percent to 60 percent—may reduce applied water by one-third. Actual quantities of water savings depend in part on the crop irrigated; the more water a crop requires, the greater the potential water savings through improved water management. Water savings also reflect the initial condition of the irrigation system. Improvements in inefficient systems may result in substantial water savings, often at relatively low cost. Under more efficient systems, a comparable increase in efficiency results in lower water savings at a higher cost. For example, an increase from 40 to 60 percent in field application efficiency will yield greater water savings than an increase from 60 to 80 percent for the same crop. The increase from 40 to 60 percent can generally be achieved at lower cost through less expensive system modifications and management adjustments.

9.4. Irrigation Technology and Environmental Benefits

Adoption of improved irrigation technology has been advanced as a means to reduce offsite water quantity and quality problems. The effectiveness of technology in achieving environmental goals has important implications for regional water policy.

10. Factors Affecting Technology Adoption

The choice of irrigation technology is highly site-specific, reflecting locational, technical, and market factors. Field characteristics—such as field size and shape, field gradient, and soil type—are perhaps the most important physical considerations in selecting an irrigation system. Other important factors include technology cost (useful life, financing options); water supply characteristics (cost, quality, reliability, flow rate); crop characteristics (spacing, height); climate (precipitation, temperature, wind velocity); market factors (crop prices; energy cost, labor supply); producer characteristics (farming traditions, management expertise, risk aversion, tenant/owner status, commitment to farming); and regulatory provisions (groundwater pumping restrictions, drainage discharge limits, water transfer provisions). In many cases, current technology choice is limited by fixed investments in existing systems at the site. The 1994 FRIS reports that 38 percent of farms made system improvements from 1990 to 1994, while no improvements were reported on 56 percent of farms. Those farms reporting improvements tended to be larger, accounting for 58 percent of the irrigated acres. Potential benefits of improved irrigation reflect, in part, the rate of technology adoption. FRIS collected information on several key factors affecting technology adoption, including capital requirements, technology information, water-pricing policy, and water-supply considerations.

10.1. Capital Requirements

Improvements in irrigation systems are often highly capital-intensive. FRIS reports that investment in onfarm irrigation equipment, facilities, and land improvements totaled $800 million in 1994, or nearly $10,000 per farm reporting expenditures (USDC, 1996). Capital expenditures included $573 million for irrigation equipment and machinery, $92 mill on for construction and deepening of wells, $82 million for permanent storage and distribution systems, and $51 million for land clearing and leveling. Replacement of existing systems accounted for the largest share of irrigation capital expenditures (64%), followed by irrigation expansion (19 percent) and conservation improvements (17%). While improved irrigation technologies are often economically profitable in a long-run farm plan, high capital outlays may limit their adoption. FRIS reports that nearly 30 percent of respondents indicated that installation of improved practices was either too expensive or could not be financed (USDC,

1996). Smaller farms were less likely to invest in improvements, reflecting more limited financial resources and difficulties in adapting some types of improved systems to smaller fields.

10.2. Technology Information

Lack of information on the availability, use, and profitability of improved irrigation technologies may limit adoption rates. Improved technologies are less familiar and often more sophisticated than traditional practices, requiring additional technical and management expertise. In some cases, improved irrigation systems may necessitate changes in current farming practices and equipment complements. For many producers, the benefits of new technologies are uncertain. Of farmers reporting no system improvements over 1990-94, 74 percent were unaware of improvements that "fit" their operation (explained in part by insufficient information), while 20 percent indicated heightened production risk as a contributing factor (USDC, 1996).

10.3. Water Cost

Limited cost savings for water conservation reduce incentives to adopt improved irrigation practices. Limited cost-savings reflect low purchased-water prices and, in some cases, low energy expenditures for pumping and pressurization. In some cases, the cost of irrigation water is substantially less than both the value of water to producers and the opportunity costs of water in nonfarm uses. (For more discussion of water sources and cost, see chapter 2.1, *Water Use and Pricing*.) Prices paid for off-farm surface-water supplies averaged $16 per acre-foot, or $36/acre, in 1994 (USDC, 1996). Surface-water prices are generally based on operation and maintenance costs of the delivery system. Deliveries are often charged on a fixed rate per irrigated acre, and are not necessarily adjusted for reduced water demand with improved management. Groundwater costs are generally limited to the cost of access—variable and fixed cost of pumping—and vary greatly depending on well yield, pumplift, power source, and other factors. In areas with significant groundwater pumplifts or high-cost surface water, water cost is an incentive to adopt conserving technologies. According to the 1994 FRIS, irrigators recognize the benefits of conservation since only 6 percent of survey respondents reported that water-conserving practices have no economic benefit. Adoption incentives are greatest for producers relying on high-cost water supplies; producers using low-cost ground- and surface-water are less apt to invest in improved technologies (Caswell and Zilberman, 1985; Negri and Brooks, 1990).

10.4. Water Supply

The off-farm water storage and delivery system may limit improvements in irrigation management at the farm-level. High on-farm water-use efficiency depends on adequate and timely supplies of water. This requires a flexible surface-water system with sufficient off-farm storage and conveyance capacity, and effective control facilities and operating policies.

Many older conveyance systems cannot be adapted to delivering water on demand without capital improvements. Limited off-farm water storage may further restrict water deliveries. Coordination is needed between the off-farm conveyance system and on-farm irrigation system to ensure compatible design and water-scheduling procedures. Uncertainty of water supplies is an additional limiting factor. Surface-water supplies for junior water-right holders often vary significantly with water storage conditions and other factors. Producers may apply excessive water during peak-flow periods in an attempt to buffer the effects of potential late-season shortages. Variable water supplies may also restrict investment in more efficient structural system improvements, while favoring the use of portable systems and development of supplemental groundwater supplies. Risk of loss of future water rights further limits incentives to invest in water-conserving technologies. Of those irrigators responding to the question on barriers to adoption almost 20 percent indicated that future water rights were a critical concern (USDC, 1996). Not surprisingly, the greatest concentration of farmers with this concern are in States with growing urban and environmental demands—California, Idaho, Texas, Nebraska, Colorado, Oregon, Washington, Utah, and Florida.

11. Effects of Irrigation Management on Yield and Water Productivity of Inbred, Hybrid and Aerobic Rice Varieties

Irrigated land produces 96% of the annual volume of rice in China (China Agricultural Almanac 2000) and uses more than 50% of all freshwater developed in Asia (Barker et al 1999). Because of continued population growth and economic development, the demand for freshwater to meet industrialization and domestic needs is growing rapidly. It is expected that, in the near future, less water will be available for rice cultivation (Tuong and Bouman 2002). Water savings and "producing more rice with less water" are crucial for food security in China. Several water-saving irrigation techniques for rice have been reported previously (Bouman 2001, Bouman and Tuong 2001). The most widely adopted watersaving practice in China is alternate wetting and drying (AWD) (Mao Zhi 1993, Li 2001, Xu Zhifang 1982). The rice field is allowed to dry for a few days in between irrigation events, including a midseason drainage in which the field is allowed to dry for 7–15 days at the end of the tillering stage. The potential of AWD to reduce water input and its effect on yield and water productivity depend on soil type, groundwater table depth, and climate (Bouman and Tuong 2001).

Tabbal *et al.* (2002) reported reduced water inputs and increased water productivity of rice grown under just-saturated soil conditions, compared with traditional flooded rice. Borrell *et al.* (1997) reported that, in semitropical Queensland, Australia, saturated soil culture with rice grown on raised beds (SSC-RB) reduced the amount of irrigation water by about 32% compared with conventional methods. This system involves growing rice on raised beds with a shallow water table (about 10 cm below the surface of the beds) by maintaining a shallow water depth in the furrows. In a semiarid environment, Thompson (1999) reported no gain in water productivity with SSC-RB since both irrigation input and yield declined by 10%. It has been suggested that rice could be grown aerobically under irrigated conditions just like upland crops, such as wheat or maize (Bouman, 2001). The aerobic condition is maintained by using flush irrigation (FI) or sprinklers so that ponding occurs for only short periods of time just after irrigation or rain, if at all. The potential for water savings is large, but aerobic cultivation using conventional lowland rice varieties almost always leads to yield reduction (De Datta *et al.*, 1973, McCauley 1990, Westcott and Vines 1986). A special type of rice is required to produce high yields under nonflooded conditions in nonpuddled and unsaturated (aerobic) soil. Bouman (2001) named this "aerobic rice"; it is responsive to high inputs, can be rainfed or irrigated, and tolerates occasional flooding.

REFERENCES

Abrol, Yash, P. and Sulochana, Gadgil. (1999). Rice: in a variable climate, APC Publications, New Delhi.

Allen, R.G., Pereira, L.S., Raas, D. and Smith M. (1998). Crop evapotranspiration: guidelines for computing crop water requirements. FAO Irrigation and Drainage Paper 56, Rome (Italy): *Food and Agriculture Organization*: 300.

Allen, Richard, G., Charles, Burt, Clemmens, A.J. and Willardson, L.S. (1996). Water conservation definitions from a hydologic viewpoint. North American Water and Environment Congress, American Society of Civil Engineers.

Arnon D.I. (1949). Copper enzymes in isolated chloroplasts. *Plant Physiol.* 24:1-15.

Barker, R, Dawe, D, Tuong, T.P., Bhuiyan, S.I., Guerra. L.C. (1998). The outlook for water resources in the year 2020: challenges for research on water management in rice production. *In:* Assessment and orientation towards the 21st century, Proceedings of 19th session of the International Rice Commission, Cairo, Egypt, 7-9 September, *Food and Agriculture Organization:* 96-109.

Belder, P., Bouman, B.A.M., Cabangon, R., Lu, G., Quilang, E.J.P., Li, Y., Spiertz, J.H.J. and Tuong, T.P. (2004). Effect of water-saving irrigation on rice yield and water use in typical lowland conditions in Asia. *Agric Water Manage*. 65(3): 193-210.

Belder, P., Spiertz, J.H.J., Bouman, B.A.M., Lu, G. and Tuong, T.P. (2005). Nitrogen economy and water productivity of lowland rice under water-saving irrigation. *Field Crops Res*. 93: 169-185.

Bernardo, Daniel J. and Norman, K. Whittlesey. (1989). Factor Demand in Irrigated Agriculture Under Conditions of Restricted Water Supplies. *U.S. Dept. Agr., Econ. Res. Serv*., TB-1765, July.

Borrell, A., Garside, A. and Shu, Fukai. (1997). Improving efficiency of water for irrigated rice in a semi-arid tropical environment. *Field Crops Res*. 52: 231-248.

Bouman, B.A.M. (2001). Water-efficient management strategies in rice production. *Int. Rice Res. Notes*. 16(2):17-22.

Bouman, B.A.M., Hengsdijk, H., Hardy, B., Bindraban, P.S., Tuong, T.P. and Ladha, J.K. (2002). Waterwise Rice Production. *Proceedings* of the International workshop on Water-wise Rice Production, 8-11 April, International Rice Research Institute: 356.

Bouman, B.A.M., Tuong T.P. (2001). Field water management to save water and increase its productivity in irrigated rice. *Agric. Water Manage*. 49:11-30.

Cabangon, R.J., Castillo, E.G., Bao, L.X., Lu, G., Wang, G.H., Cui, Y.L., Tuong, T.P., Bouman, B.A.M., Li, Y.H., Chen, C.D., Wang, J.Z. (2001). Impact of alternate wetting and drying irrigation on rice growth and resource-use efficiency. *In:* Barker, R., Loeve, R., Li, Y.H., Tuong, T.P. *(editors).* Water-saving irrigation for rice. *Proceedings* of an International Workshop held in Wuhan, China, 13-25 March: 55-80.

Caswell, Margriet, F. and Zilberman, David. (1985). The choices of irrigation technologies in california, *American Journal of Agricultural Economics*. 67(2): 224-234.

Chang, Te-Tze. (2000) II.A.7. Rice, *In*: The Cambridge World History of Food, vol. 1, edited by Kenneth F. Kiple and Kriemhild Coneè Ornelas, Cambridge University Press.

Chang, Te-Tzu and Eliseo, A. Bardenas. (1965). The morphology and varietal characteristics of the rice plant. Technical Bulletin, No 4, International Rice Research Institute.

Chang, Te-Tzu. (1976). The origin, evolution, cultivation, dissemination, and diversification of Asian and African rices. *Euphytica*. 25: 43541.

Chang, Te-Tzu. (1976). The Rice Cultures. Paper presented at Discussion Meeting on the Early History of Agriculture, sponsored by the Royal Society and British Academy, London, April.

China Agricultural Almanac. (2000). China Agricultural Press., pp-619-624.

De Datta, S.K., Abilay, W.P., Kalwar, G.N. (1973). Water stress effects in flooded tropical rice. *In:* Water management in Philippine irrigation systems: research and operations. International Rice Research Institute, Los Banos, Philippines, pp-19-36.

De Datta, S. K. (1987). Principles and Practices of Rice Production, Malabar, Fla.: Robert E. Krieger.

Dong, B., Molden, D., Loeve, R., Li, Y.H., Chen, C.D. and Wang, J.Z. (2004). Farm level practices and water productivity in Zhanghe Irrigation System. *Paddy Water Environ.*, 2: 217-226. 1998.

Grist, D. H. (1986). Rice, 6th ed. London and New York: Longmans.

Hofwegen, P.V., Svendsen, M. (2000). A vision of water for food and rural development, 17 March, The Hague, Netherlands, pp 21.

Howe, C.W. (1985). Economic, legal, and hydrologic dimensions of potential interstate water markets, *American Journal of Agricultural Economics* 67(5): 1226-1230.

Huke, R. E. and Huke, E. H. (1990). Rice: Then and Now, International Rice Research Institute, Philippines.

International Rice Research Institute. (1997). Rice Almanac, 2nd edition, International Rice Research Institute, Philippines.

Islam, M. T. (2010). Supplemental irrigation by rain water harvesting in transplanted aman for drought mitigation, Rice Production Training Manual, Bangladesh Rice Research Institue, pp-257-263.

Islam, M.T. (1999). Plant water relation studies in diverse rice cultivars under Bangladesh climatic conditions, PhD Thesis, Institute of Agronomy and Agriculture Science, Vienna.

Islam, M.T., Salam, M.A. and Kauser, M. (1994). Effect of soil water stress at different growth stages of rice on yield components and yield. *Prog. Agric.*, 5 (20): 151-156.

Kar, S., Varade, S.B., Subramanyam, T.K., Ghidyal, B.P. (1974). Nature and growth pattern of rice root system under submerged and unsaturated conditions. *Riso* 23:173-179.

Kim, C.S., Ronald, Fleming, Richard, M. Adams, Marshall, English and Carmen, Sandretto. (1994). An economic analysis of agricultural practices related to water quality: the Ontario (Oregon) demonstration project. *U.S. Dept. Agr., Econ. Res. Serv.*, AGES94-18, June.

Lang, James. (1996). Feeding a hungry planet: rice, research, and development in Asia and Latin America, chapel hill, The University of North Carolina Press.

Li, Y.H. (2001). Research and practice of water-saving irrigation for rice in China. *In:* Barker R, Li YH, Tuong TP *(editors).* Water-saving irrigation for rice. Proceedings of an International Workshop held in Wuhan, China, 23-25 March, International Water Management Institute, Colombo, Sri Lanka, pp-135-144.

Lu, J. J. and Chang, T. T. (1980). Rice in its temporal and spatial perspectives. *In:* Rice: Production and Utilization, edited by B. S. Luh., Westport, Conn.: AVI Publishing.

Luo, L.J, Zhang, Q.F. (2001). The status and strategy on drought resistance of rice. *Chinese J. Rice Sci.* 15(3):209-214.

Magleby, R., Kasal, J., Walker, D. and Gum, R. (1989). Economics of controlling sediment from irrigation: an Idaho example. *U.S. Dept. Agr., Econ. Res. Serv.*, AGES89-33, September.

Magleby, R., Selley, R. and Zara, P. (1995). Effectiveness and economic efficiency of efforts to reduce nitrate loadings to groundwater: a mid-Nebraska example. Paper presented at Soil and Water Conservation Society meetings, DesMoines, IA, Aug. 6-9.

Maniruzzaman, M. (2012). PhD Thesis. Department of Irrigation and Water Management, Bangladesh Agricultural University, Mymensingh-2202.

Mao Zhi. (1993). Environmental impact of water-saving irrigation for rice. Proceedings of Environmental Sound Water Resource Utilisation, AIT, Bangkok, Thailand, 8-11 Nov.:143-148.

McCauley, G.N. (1990). Sprinkler vs flood-irrigation in traditional rice production regions of southeast *Texas. Agron.* J. 82:677-683.

Moore, Michael R. (1991). The bureau of reclamation's new mandate for irrigation water conservation: purposes and policy alternatives. *Water Resources Research.* 27:145-155

Moore, Michael R. and Dinar, Ariel. (1995). Water and land as quantity-rationed inputs in California agriculture: empirical tests and water policy implications. *Land Economics*, 71: 445-61.

Moore, Michael R., Noel, R., Gollehon and Marc, B. Carey. (1994). Multicrop production decisions in western irrigated agriculture: the role of water price. *American Journal of Agricultural Economics* 76(4): 859-874

National Research Council (1996). Review of the department of interior's national irrigation water quality program: planning and remediation. Water Science and Technology Board,

Negri, Donald H. and John, J. Hanchar. (1989). Water conservation through irrigation technology. *U.S. Dept. Agr., Econ. Res. Serv.*, AIB-576, Nov.

Negri, Donald H., and Douglas H. Brooks. (1990). Determinants of irrigation technology choice, *Western Journal of Agricultural Economics* 15(2): 213-223.

Oka, H. I. (1988). Origin of Cultivated Rice, Japan Scientific Societies Press, New York: Elsevier.

Portères, R. (1956). Taxonomie agrobotanique des riz cultives: *O. sativa* Lin. et *O. glaberrima* Steudel: IV. Journal d'Agriculture Tropicale et de Botanique Appliquée 3: 34184, 54180, 62700, 82156.

Postel, S. (1997). Last Oasis: Facing Water Scarcity, Norton and Company, New York, USA, pp- 239.

Rahman, M.T., M.T., Islam and Islam, M.O. (2002). Effect of water stress at different growth stages on yield and yield contributing characters of transplanted aman rice. *Pak. J. Biol. Sci.*, 5 (2): 169-172.

Schaible, G.D., Gollehon, N.R., Kramer, M.S., Aillery, M.P. and Moore, M.R. (1995). Economic analysis of selected water policy options for the Pacific Northwest. *U.S. Dept. Agr., Econ. Res. Serv.*, AER-720, June.

Schaible, G.D., Kim, C.S. and Whittlesey, N.K. (1991). Water conservation potential from irrigation technology transitions in the Pacific Northwest. *Western Journal of Agricultural Economics* 16:194-206

Shi, Q., Zeng, X., Li, M., Tan, X. and Xu, F. (2002). Effects of different water management practices on rice growth. Proceedings of a thematic workshop on water-wise rice production, 8-11 April, IRRI, Los Banos, Philippines

Siao, T.C. (1973). Plant responses to water stress. *Annual Review of Plant Physiology*, 24: 519-529.

Sojka, Bob and Lentz, Rick. (1996). Polyacrylamide for furrow- irrigation erosion control, *Irrigation Journal* 46(1)

Tabbal, D.F., Bouman, B.A.M., Bhuiyan, S.I., Sibayan, E.B. and Sattar, M.A. (2002). On-farm strategies for reducing water input in irrigated rice; case studies in the Philippines. *Agric. Water Manage.*, 56: 93-112.

Tantawi, B.A. and Ghanem, S.A. (2001). Water use efficiency in rice culture. Agricultural Research Center, Giza (Egypt), *CIHM-Optin Mediterraneennes*, 40: 39-45.

Thompson J. (1999). Methods for increasing rice water use efficiency. Rice Water Use Efficiency Workshop Proceedings. Cooperative Research Centre for Sustainable Rice Production, Leeton, New South Wales, Australia: 45-46.

Ting, Y. (1949). The origin of rice cultivation in china. Agron. Bull. Sun Yatsen Univer. Ser III. 7: 18

Tuong T.P., Bouman, B.A.M. (2002). Rice production in water-scarce environments. Proceedings of the Water Productivity Workshop, 12-14 November, International Water Management Institute, Sri Lanka.

Tuong, T.P., Bouman, B.A.M. and Mortimer, M. (2005). More rice, less water-integrated approaches for increasing water productivity in irrigated rice-based systems in Asia. *Plant Prod. Sci.*, 8(3): 229-239.

U.S. Department of Commerce, Bureau of the Census (1982). *Farm and Ranch Irrigation Survey*, 5(8).

U.S. Department of Commerce, Bureau of the Census (1986). Farm and Ranch Irrigation Survey, *Special Report Series*, AG84-SR-1.

U.S. Department of Commerce, Bureau of the Census (1990). Farm and Ranch Irrigation Survey, Vol. 3, Part 1 - Related Surveys of 1987 *Census of Agriculture*, AC87-RS-1.

U.S. Department of Agriculture, Natural Resources and Conservation Service (1992). National Irrigation Erosion Study.

U.S. Department of Agriculture, Soil Conservation Service (1994). A productive nation in harmony with a quality environment: soil conservation service – strategic initiatives for the 1990's.

U.S. Department of Commerce, Bureau of the Census (1994). 1994 *Census of Agriculture*, Vol 1, Part 1- 51.

U.S. Department of Commerce, Bureau of the Census (1996). Farm and Ranch Irrigation Survey, Vol. 3, Part 1 - Related Surveys of 1992 *Census of Agriculture,* AC92-RS-1.

Vaux, H.J., Jr. (1986). Economic factors shaping western allocation, *American Journal of Agricultural Economics,* 68: 1135-42.

Wadsworth, James I. (1994). Degree of milling, *In*: Rice Science and Technology, edited by Wayne E. Marshall and James I. Wadsworth, New York: M. Dekker.

Wang, Y., Zhou, Y. (2000). Water saving rice culture in north China. Liaoning Science and Technology Press.

Wang, Z.Q., Li, G.S., Yang, J.C. (2000). Responses of the main mid-season japonica cultivars currently used in Jiangsu to nitrogen applications. *Jiangsu Agric. Res.* 21(4): 22-26.

Wang, Z.Y, Xu, Y.C., Li, Z., Guo, Y.X., Ding, Y.P., Wang, Z.Z. (1998). Methane emission from rice fields as affected by water regime and organic manure input. *Acta Agron. Sin.* 24(2):133-138.

Webb, B. D. (1985). Criteria of rice quality in the United States. *In*: Rice Chemistry and Technology, 2nd ed., edited by Bienvenido O. Juliano. St. Paul, Minn.: American Association of Cereal Chemists.

Westcott, M.P., Vines, K.W. (1986). A comparison of sprinkler and flood irrigation for rice. *Agron. J.* 78: 637-640.

Xu, Zhifang. (1982). Irrigation of rice in China. Wuhan, Department of Irrigation and Drainage Engineering, Wuhan Institute of Hydraulic and Electric Engineering, Wuhan, China.

Yang, J.C., Wang, Z.Q., Liu, L.J., Lang, Y.Z., Zhu, Q.S. (2002). Growth and development characteristics and yield formation of dry-cultivated rice. *Acta Agron. Sin.* 28(1):11-17.

Yang, J.C., Wang, Z.Q., Zhu, Q.S. (1995). Drought resistance and its physiological characteristics in rice varieties. *Sci. Agric. Sin.* 28(5):65-72.

Yang, J.C., Wang, Z.Q., Zhu, Q.S. (1996). Effect of nitrogen nutrition on rice yield and its physiological mechanism under different status of soil moisture. *Sci. Agric. Sin.* 29(4):58-66.

Ye, X. D., (2000). Engineering the Provitamin A (Beta-Carotene) biosynthetic pathway into (Carotenoid-free) rice endosperm, *Science* 287: 30305.

Yoshida, S. (1981). Fundamentals of rice crop science, International Rice Research Institute (IRRI), Los Banos, Laguna, Philippines.

Zhang, J.D., Liu, G.P., Shi, Y.L., Gao, F. (1982). Research method for plant physiology. Nanchang (China), Jiangxi People Press.

In: Irrigation Management, Technologies …
Editor: M. Hossain Ali
ISBN: 978-1-62417-862-7

Chapter 2

The Sap Flow Technique: A Precise Means to Estimate Water Consumption of Young Olive Trees (Olea Europaea L.)

***Chiraz Charfi Masmoudi*[1,*], *Moncef Masmoudi*[2], *Jihène Abid Karray*[3] *and Nétij Ben Mechlia*[2]**
[1]Researcher, Olive Tree Institute, Sfax, Tunisia
[2]Professors, National Institute of Agronomy, Tunis, Tunisia
[3]Assistant, High School of Agriculture, Mograne, Tunisia

Abstract

This study was carried out in Northern Tunisia (36.5°N, 10.2°E, Altitude 10 m), in order to estimate water consumption of young olive (Olea europaea L.) trees. The heat dissipation method of Granier (1985) was used. Relationships between sap flux measurements, weather parameters and watering conditions have been investigated. Results showed a satisfactory behavior of the system when working under field conditions. It proved to be robust and to require little maintenance. The alternative 30 minutes heating and cooling regime seems to be correct despite the limited number of observations. Individual sap flow measurements showed typical courses and a consistency between the patterns of hourly measurements. Variability between sensors resulted from the heterogeneity of the conductive system, which was evidenced through observations made on Safranin-red-colored stem sections. Through these results we can conclude that this technique can be used with confidence in young olive trees despite the obvious difficulty of sampling in such a complex ecosystem. It is useful in analyzing between-tree variability of transpiration and has potential to be used as a device to provide a signal from which decisions can be made concerning the irrigation timing.

* Phone and Fax Number: +21671235820. E-mail: masmoudi.chiraz@yahoo.fr.

1. INTRODUCTION

Olive (Olea europaea L.) covers in Tunisia (36.5°N, 10.2°E, Altitude 10 m) one-third of the agricultural area with about 57 millions trees. However, most orchards are cultivated under rain-fed conditions and receive annually no more than 500 mm of rainfall in the north of the country and less than 350 mm in the central and southern olive cultivation areas. In these regions, trees suffer periodically from water or/and thermal stresses. As a result, yields are low and trees are subjected to alternate bearing.

On the other hand, as olive is a strategic sector for both economy and people, and as Tunisia has to provide annually a sufficient amount of olive oil for exportation (60000 tons for EU) and local consumption (40000 tons), management of olive orchards is considered now-a-days as a general concern and even a priority for both, the government and orchardists. For this reason, and in order to achieve good and suitable production, a new strategy was adopted; olive orchards are, therefore, planted at higher densities and irrigated during the dry season. However, because water resources are limited (Hamza, 2009) and fresh water is becoming scarcer, water should be used judiciously, without any wastage.

Irrigation has become in most intensive olive orchards a common practice with more or less regular inputs depending on water availability. However, in most cases, water amounts don't match the crop water needs, making production largely dependent of these inputs. In the irrigated orchards of the north, maximum yield reaches 3.0 tons ha^{-1} for an annual input of about 500 mm, including effective rainfall. This yield is considered too low with regard to the potentialities of the region. In central and southern areas, annual rainfall fluctuates between 350 mm and 80 mm, with minimum values recorded from May to August, *i.e.* during the olive growth period. As a result, yields rarely exceed 1.0 ton ha^{-1} and fruit size is affected. Complementary irrigation is then necessary in both cases to improve yield, stabilize production and reduce alternate bearing. This together, with the increment in irrigated olive area and the increasing national water deficit, makes it crucial to improve the calculation of crop water needs.

The most useful method to estimate water requirements of olive orchards is the climatic method of the F.A.O. (Allen et al., 1998; Rana and Katerji, 2000). It has been used in many studies to determine crop evapotranspiration (ET_c) of adult orchards (Pastor et al., 1998; Fernandez and Moreno, 1999; Michelakis, 2000; Villalobos et al., 2000; Palomo et al., 2002; Bandino and Dettori, 2003; Habaieb and Masmoudi-Charfi, 2003; Testi et al., 2004), but also of young olive plantations (Masmoudi-Charfi et al., 2004a; Masmoudi-Charfi, 2008). This method is based on the reference evapotranspiration (ET_0) estimation and the use of a crop coefficient (K_c), which varies following the season, the growing period and the soil humectation. High K_c values are habitually observed during the rainy period as a result of enhanced soil evaporation due to rainfall (Fernandez and Moreno, 1999). Then, as the season progresses, the probability of rainfall decreases and so did the soil evaporation and consequently the K_c and its variability. Minimum values are recorded during the dry season. The crop coefficient relative to transpiration (K_{cb}) follows approximately the same pattern. This is related to the typical Mediterranean pattern of rainfall in North Africa, with a wet season from October to April and a dry season from May to September. In addition, when the F.A.O. method is used to calculate ET_c for non-standard plantations like olive orchards, values of K_c should be necessarily adjusted by introducing a minoratif parameter (K_r) to take account of the percentage of soil covered by tree biomass (Le Bourdelles, 1977; COI, 1997). This is particularly important in young olive orchards, where soil coverage rarely exceeds 30% (Masmoudi-Charfi, 2004a; Masmoudi-Charfi, 2006).

However, despite being widespread and suitable for many situations, this method has three main disadvantages. First, the published values of K_c are often used in orchards with conditions different from those under which they were obtained. Second, the irrigation amounts are always determined on the basis of long-term average ET_0 values which are calculated for a period of several days; in this case, the daily variations are not considered (for example day with sirocco). In other cases, an average ET_0 value is used for the next irrigation period, so there is a delay in the application of ET_0 data. Third, this method takes into account the soil evaporation amount which is very important in the case of young orchards and low density plantations. The amount of water used for irrigation and deduced from the ET_c estimation may not be used totally by the tree; only a fraction of ET_c is consumed and thus an important amount of water is depleted.

Field measurements used in olive orchards provided also usefull data. In Tunisia, Riou (1978) estimated the actual evapotranspiration of 'Chétoui' adult trees by using a lysimeter, while the soil water balance method was experimented on olive trees of *cv.*, Chemlali, the main olive variety of Tunisia (Abid-Karray, 2006). This work showed through a study of root distribution and monitoring of soil moisture made by a neutron probe, distinguishable soil reservoirs, subjected each to a different process of water use. A mathematical model was then built to characterize lateral water inflows in each of these soil compartments. For young olive trees, Masmoudi-Charfi (2008) found that soil moisture varies consistently following the distance to trunk. This variability makes difficult estimation of crop water consumption when dealing with such orchards. Also, this method is considered as time-consuming and of low temporal resolution (Fernandez et al., 1991; Villagra et al., 1995; Granier et al., 2000), although it provides valuable results when studying water inflows and movements in olive orchards.

The sap flow methods were developed in order to determine precisely the amount of water consumed by trees. Early sap flux studies have been initiated by Huber in 1932 (Swanson, 1994). Since then, several improvements were made and many techniques were developed. Sap flux can be determined as sap (water) flow estimated by the heat-pulse (Green and Clothier, 1988; Fernandez and Moreno, 1999), the heat-balance (Nasr, 1987) or the thermal dissipation (Granier, 1985; Cohen, 1991) techniques. This last one is of the most used technique for measurements of whole-tree water use in eco-physiological and forest hydrological studies. Technical details and studies on the applicability of these techniques in agriculture were described in Granier (1985, 1987a and 1987b), Swanson (1994), Fernandez and Moreno (1999), Fernandez et al., (2003), Do and Rocheteau (2002a, 2002b and 2003) and Pereira et al., (2007).

Sap flow methods have been used to quantify water use by trees and in studies of plant water relations (Moreno et al., 1996; Smith and Allen, 1996; Cabibel and Isbérie, 1997; Fernandez et al., 2001; Fernandez et al., 2003), and becoming increasingly widespread. Results provided useful data which allow calculation of the crop evapotranspiration (ET_c) from independent measurements of its transpiration (T) and the soil evaporation (E) fluxes. But, these methods are suitable only for use on woody stems and restricted to few species. Measurements on tree crops have been less frequent compared with forestry (Granier, 1987a and 1987b). Sap flow was measured in apple and Kiwifruit after developing a compensation heat-pulse velocity (CHPV) system (Green and Clothier, 1988). For olives, this system has been used in Spain on either root to study hydraulic behavior or/and short-term water-use dynamics (Moreno et al., 1996) or trunks to estimate whole-tree water consumption and its relationship with physiological and environmental variables (Fernandez et al., 1991; Fernandez and Moreno, 1999; Giorio and D'Andria, 2002; Giorgio and Giorgio, 2003). In Tunisia, sap flux methods were used for the first time in 1987 (Nasr, 1987) and experimented on olives many years later (Mahjoub, 2004; Masmoudi et al., 2004b; Abid-Karray, 2006; Masmoudi-Charfi, 2008; Masmoudi-Charfi et al., 2011a).

The short-term water-use dynamics of olive tree can be evaluated from sap flow measurements in the trunk or in main roots. Measurements of transpiration by sap flow measurements hold advantages over other techniques. Continuous records of plant water use with high time resolution can be obtained. However, when sap flow measurements are used to estimate transpiration for stands of vegetation, appropriate methods of scaling from plant to unit area of land must be used (Smith and Allen, 1996; Giorgio and Giorgio, 2003). Indeed, after measuring sap flow, it is often necessary to extrapolate water use by sampled plants to an entire stand. Mass or volume flow rates for individual plants must be converted to estimates of transpiration per unit area of land that can be used to address hydrological problems faced by land users and the incomplete soil cover. When variation in tree size and spacing between plants is not uniform, the stand transpiration could be estimated by scaling-up water use on the basis of the ground area occupied by individual trees. Face to these conditions, Smith and Allen (1996) indicated that where members of a stand population vary primarily in size, sap flow for individual plants can be scaled to stand transpiration by determining relationships between sap flow rates and stem diameter, sapwood area or leaf area and then extrapolating transpiration rates to unit area of land on the basis of surveys of tree size or measurements of leaf area index.

So, for sound irrigation of olive orchards, it would be desirable to have a technique or method for irrigation scheduling based on the plant's actual use for water. We have used the heat dissipation technique (Granier, 1985; Do and Rocheteau, 2002a, 2002b, 2003) to carry out a study on young olive trees, for calculating their water consumption over a growing season. The present work is designed to avoid the errors derived from the scarce information on the crop coefficient when the climatic method of the FAO is used. Our objectives are: (1) evaluate the sap flux technique for its applicability with young trees characterized by low soil coverage, (2) examine the spatial variability of sap flows through investigation of the relationships between sap flux measurements, weather variables and soil water status and finally (3) examine the procedure of extrapolation of in situ individual sap flow measurements to whole-tree transpiration for an efficient irrigation management in such orchards. In this paper, experiments on probe calibration and wood conductive section are also presented and discussed.

2. Materials and Methods

2.1. Experimental Site

The study was performed on young olive orchard of about 1.6 ha located at the experimental farm of the *Institut National Agronomique de Tunisie*, about 15 Km South of the capital Tunis, northern Tunisia (Figure 1).

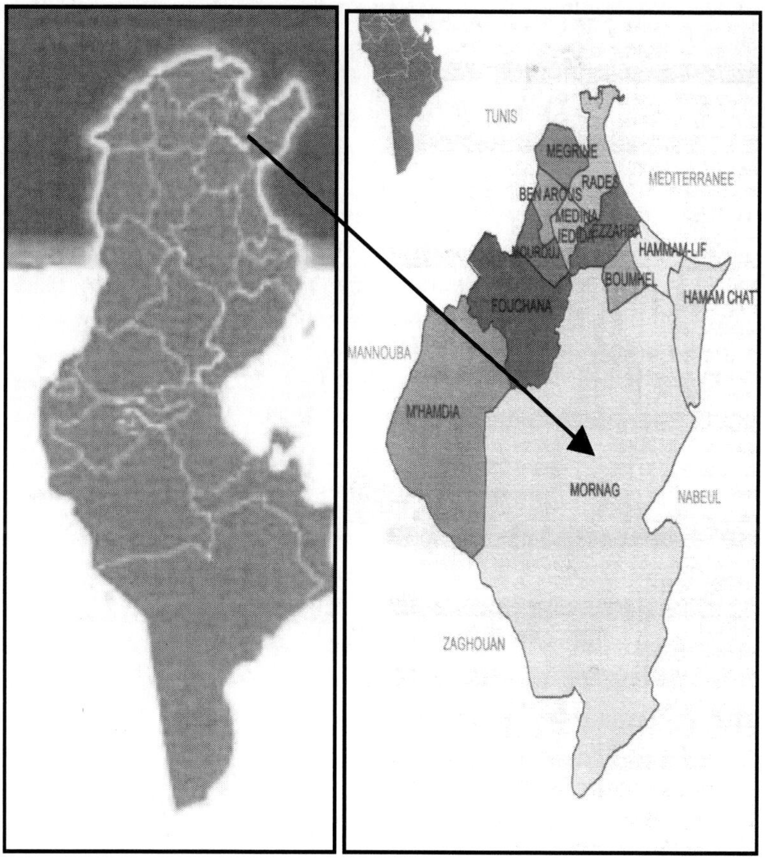

Figure 1. The site of experiment is located in Mornag, northern Tunisia, about 15 Km South of the capital Tunis. The region is alimented by fresh water from Medjerda, the main river of north Tunisia. The region accounts 4500 ha of olive trees.

In this region, climate is Mediterranean with average annual rainfall of 450 mm and reference evapotranspiration (ET_0) of 1200 mm. Soil is clay loam (29%C, 49%L, 23%S), of 2 m depth and has volumetric water contents of 0.50 m^3/m^3 (θ_{cc}) and 0.26 m^3/m^3 (θ_{pFP}), measured at soil matrix potentials of −0.5 MPa and −1.5 MPa, respectively. Weather variables were measured continuously with an automatic weather station located at about 150 m from the olive orchard. Average values recorded from 1979 to 2002 as well as those observed in 2003, are presented in Table 1.

Table 1. Average weather variables recorded from 1979 to 2002 (I.N.M., 2003) and those observed in 2003

Climatic parameter	Average value 1979-2002	2003
Rainfall (mm)	450	790
Reference evapotranspiration (ET_0, mm)	1200	1212
Maximum temperature (°C)	24.2	24.9
Minimum temperature (°C)	13.9	14.9
Relative humidity (%)	30 - 60	-
Wind speed (m/s)	1.4 - 2	-
Global radiation (R_g, Cal/cm^2/day)	180-650	-

Temperature, air relative humidity, wind spread, rainfall and global solar radiation were used to calculate ET_0 following Penman–Monteith equation (Allen et al., 1998). During the experimental year, 2003, ET_0 value, rainfall and effective rainfall (USDA method – F.A.O., 1976) recorded during the irrigation period, from March to September, reached 982 mm, 346 mm and 239 mm, respectively, while annual minimum and maximum average temperatures were 14.9 °C and 24.9 °C, thus exceeding the averages of this region by +3% and +7%, respectively. Rising temperature has been felt during the first three months of the year 2003, resulting in an increase of the growth degree day (GDD) value of about 300 degree-day.

Daily crop water evapotranspiration (ET_c) were determined following Allen et al., (1998) for non-standard conditions by using the average daily ET_0 values as: $ET_c = ET_0 \; x \; K_c \; x \; K_r$ with K_c the crop coefficient, of 0.5 for six-year-old trees and K_r, of 0.75 (C.O.I., 1997) accounting for an average ground cover of 33%. This value was determined experimentally. For this purpose, a large white grilled (10 cm / 10 cm) sheet was placed below the trees and the shade squares were counted and compared to the total number of squares (those lighted by sun and those shaded by leaves). This percentage represents the K_r value. Annual value of ET_c was 454 mm and 368 mm for the irrigation season.

2.2. Olive Orchard and Crop Management Practices

Three olive trees of six-year-old, planted at 6 m×6 m spacing were chosen as representative of the plot. The trial concerned trees of cultivar Chétoui, which is the main oil variety of northern Tunisia. We had intentionally chosen them because of similar shape and grouped within a limited area. They were equipped with sap flow sensors in March 2003. It is important to note here that the study was carried out from 1998 to 2005, but, we have considered for investigation on sap flow, only data recorded from May 2003 to March 2004 because many technical problems appeared after this period and sap flux measurements dropped consistently in comparison with those obtained at the beginning of the experiment.

Crop management practices carried out in the orchard- *i.e.* pruning, fertilizer (Masmoudi-Charfi and Ben Mechlia, 2009) and pest management practices- were similar to those applied in intensive orchards (Masmoudi-Charfi, 2006; Masmoudi-Charfi et al., 2006). After planting, and in order to sustain growth, each tree was provided an amount of N-fertilizer (ammonium–

nitrate, 33%N) in February before the spring growth and in October during the autumn growth period, accounting for an annual increase of 100 g per tree of N-fertilizer for each year of growth (100 g for one year old trees, 200 g for two year old trees…). Trees were pruned considering a standard shape. Irrigation was applied from 15 May to 9 September 2003, with the same irrigation regime for all trees (Table 2). Water was supplied by two drip lines fixed along the row on the soil surface at spacing of about 1.0 m apart and at 0.5 m from trunks, with 4 emitters per tree and a total discharge of 16 l/h. Irrigation amounts ranged between 333 l/tree and 1098 l/tree following to the phenological stage with total seasonal amount of 5449 l/tree as shown in Table 2.

Table 2. Periods of irrigation and amounts of water (l/tree) applied in 2003

Irrigation	1	2	3	4	5	6	7
Period	15/5-20/5	2/6-3/6	30/6-2/7	10/7-15/7	21/7-30/7	5/8-9/8	28/8-5/9
Irrigation amount	823	333	549	843	902	902	1098

2.3. Measurements

Regular measurements of soil water content were performed during the irrigation period around each of the instrumented trees at different distances from trunks and depths (0.30 m - 0.60 m - 0.90 m - 0.120 m). Measurements were carried out with a neutron probe (SOLO 25) and twenty-eight access tubes of 1.5 m long dispatched around the trees. The top 0.3 m layer soil water content was determined gravimetrically. Average values were calculated separately for each position and date. More details are given in Masmoudi-Charfi (2008). Growth measurements were performed on the same trees at the beginning and at the end of the year. Mean tree height and canopy diameter reached values as high as 4.9 m and 4.0 m, respectively (Masmoudi-Charfi and Ben Mechlia, 2007 and 2008). Leaf area was determined on May 2003 by computing the number of leaves on representative branches and the individual foliar area by planimetry as reported by Fernandez and Moreno (1999). Average value was about 14 m^2. Leaf conductance (g_s, cm/s) measurements were made with an ‘AP_4 Delta-T Devices’ porometer on three well-exposed leaves during a cloudless day to determine the plant water status. All these data were used when interpreting results. Sap flow measurements were made on the same trees. Details are presented in the following section.

2.4. Instrumentation and Sap Flux Measurements at Field Level

Sap flux measurements were performed using the heat dissipation method of Granier (1985, 1987a and 1987b). Data were recorded continuously from March 2003 to December 2005 on three olive trees of cultivar Chétoui (Photo 1) aged six years and indexed A_1, A_2 and A_3. Thermal sensors (3/tree) were inserted radially in the trunks of trees at 0.4 m above the ground surface, by drilling holes into the sapwood following North (N), South-East (SE) and

South-West (SW) directions with one probe placed approximately 0.10 m above the other. This was done carefully to prevent excessive damage of vessels.

Photo 1. Sensors were inserted radially in the trunk of three olive trees 'Chétoui' of six-year-old following North, South-East and South-West directions. Sensors were enveloped with a special aluminum paper to reduce the effect of reflected radiation. Trees were drip irrigated.

Sensors are two cylindrical probes (Photo 2) of 3 cm length, equipped with heating resistance (Power = 0.2 Watt) and thermocouple. The upper probe contains a heater element and a thermo-couple junction that is referenced to another junction in the lower probe (Masmoudi et al., 2004). The lower probe was unheated and gives the reference temperature of the wood. Constant power was applied to the heater with an alternative regime of 30 minutes heating and 30 minutes cooling cycles.

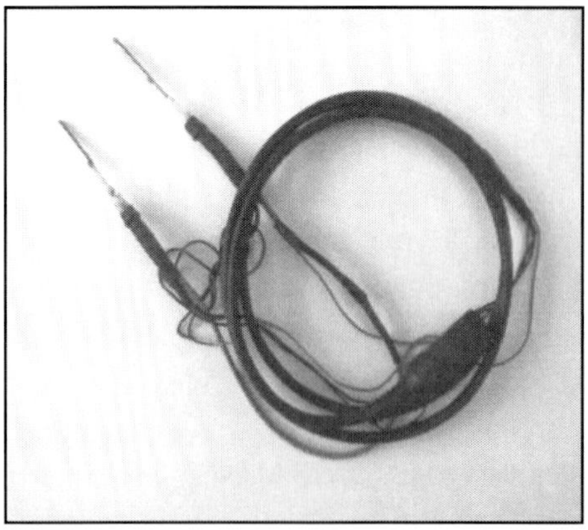

Photo 2. Sap flux sensor is equipped with two cylindrical probes with a heating resistance (red) and thermocouple.

Associated electronic, modules for current regulation and control and worksheets software have been developed locally (Masmoudi et al., 2004). A DL_2 data logger (Delta-T Devices, co.uk) powered by a 12 V battery has been used for continuous monitoring of sensor signals (Photo 3), which were collected every fifteen minutes beginning from March 2003. However, the batteries needed to be recharged every 10–15 days, and with the same frequency, the information in the data logger was downloaded to a portable computer as shown in Photo 3.

Photo 3. Associated electronics and a DL_2 data logger (Delta-T Devices, co.uk) powered by a 12 V battery has been used for continuous monitoring of sensor signals which were collected every fifteen minutes.

The recorded signals were examined before treatment, then sap flow density (F_d, l/h/dm^2) was calculated following Do and Rocheteau (2002a) formulae: $F_d = [a\ K_a /1-K_a]^{0.707}$, where K_a is the sap flux index expressed as follows: $K_a = (\Delta T_{max} - \Delta T_a) / \Delta T_a$. Estimation of F_d relies on the measurements of a single parameter (ΔT_a), the temperature difference. For daily (F_d) calculation, maximum ΔT_a value (ΔT_{max}) was determined for $F_d = 0$ to take into account the night-time flow, if it does occur. Values of ΔT_a were measured at F_d and a was determined during the probe calibration experiment. We have adopted a step of 10 days time for calculating F_d.

2.5. Probe Calibration Experiment

To obtain accurate estimates of the sap flow volume from measurements of sap velocities made at field level, sensors were calibrated. For this purpose, a special high-pressure device was built in the Laboratory to force water through a section of branch of 0.4 m long and 5 cm diameter, obtained from a tree of similar age and variety (Photo 4). Three sets of probes have been implanted considering different directions. The flow rate was adjusted by varying the pressure of air from 0 to 4 bars with the pressure regulator attached to the bottle of compressed air (200 bars). By altering the pressure in the chamber of the aluminum box, we were able to cover the range of sap flows found in the field. Water passing through the branch was collected at the distal side in a flask and weighed. Estimates of flow rate subsequently monitored at regular intervals were compared to the actual flow rates, obtained from the rate of change in the mass of water collected.

The entire system was maintained operational for four successive days. Then, relationship between signals measured at different pressures and heat powers, and the amount of water collected from the stem was established.

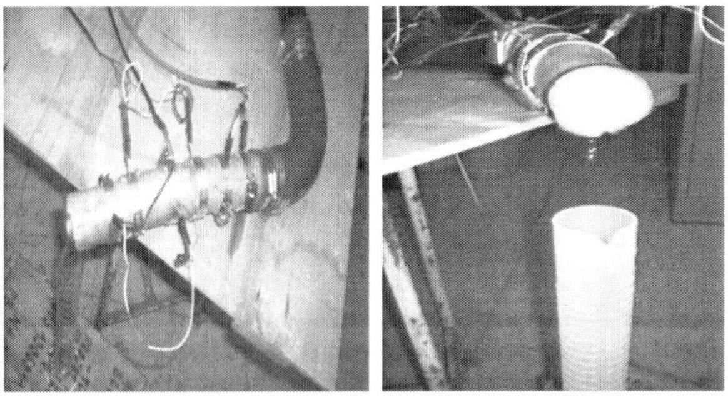

Photo 4. Probe calibration experiment. Three sets of probes have been implanted in a section of branch of 0.4 m long and 5 cm diameter following different directions (Left). The flow rate was adjusted by varying the pressure of air from 0 to 4 bars with the pressure regulator attached to the bottle of compressed air (200 bars). Water passing through the branch was collected at the distal side in a flask and weighed (right).

2.6. Conductive Stem Section Experiment

Total sap flow of the tree (*F*) was calculated as the product of F_d and the conductive stem section (S_c). This area, *i.e.,* the active water-conducting wood, is localized between the cambium and the heartwood and defined as the portion of the cross-section of the sapwood responsible of water conduction. It was determined in the Laboratory by using the *Safranin Orange,* a specific chemical colorant of xylem vessels. We have used this method because it was easy to carry out in the Laboratory. For this purpose, a special device was built (Photo 5), consisting in a butt of a fresh branch, which was placed in a container, previously filled with distilled and filtered (filter pore size of 1/10 mm) red-colored water. The branch was connected to a pump through a pipe. The pump functions continuously and allowed aspiration of the solution, which was absorbed by the branch and colored progressively the conductive wood. The time of coloration depends on the branch thickness and diameter. More details on the experiment are given in Masmoudi-Charfi (2008) and Masmoudi-Charfi et al., (2010).

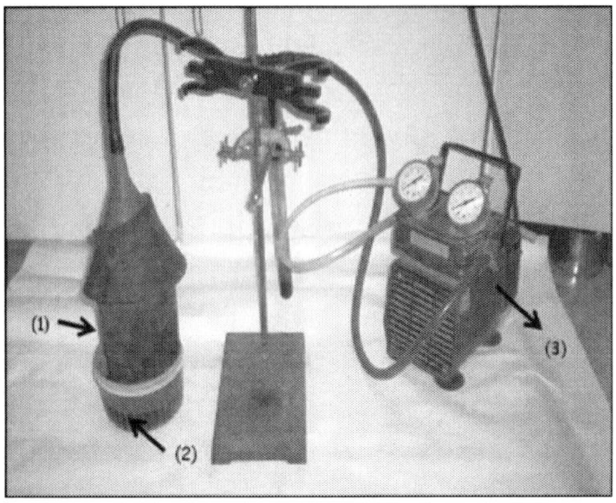

Photo 5. Determination of the water conductive stem section. A fresh portion of a branch was immerged in a container (1), previously filled with distilled and filtered (filter pore size of 1/10 mm) water colored by the red-dye *Safranin* (2). The pump (3) allowed absorption of water by the branch and a progressive coloration of the conductive wood.

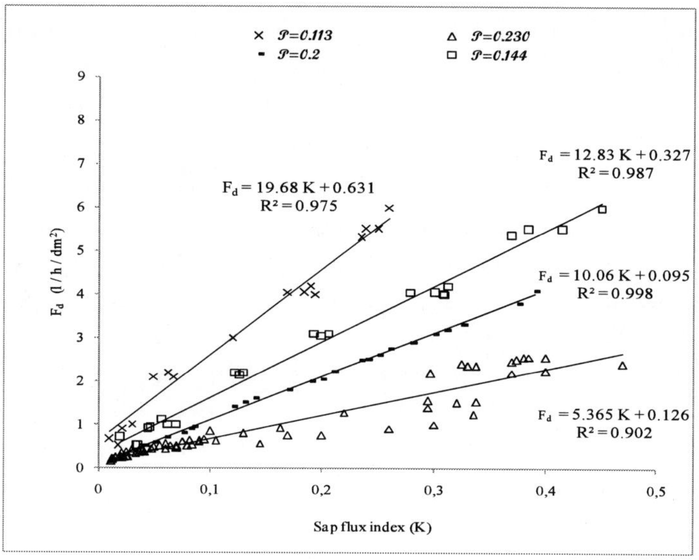

Figure 2. Relationships between the actual sap flux densities (F_d, l/h/dm^2) and the sap flux index (K_a) determined for different sensor's heat powers ($\mathcal{P}$ = 0.200 Watt, $\mathcal{P}$ = 0.144 Watt, $\mathcal{P}$ = 0.230 Watt and $\mathcal{P}$ = 0.113 Watt).

The entire system was maintained operational for three successive days and tested on different stems with diameters ranging between 1 cm and 12 cm. After the device was dismantled, stem sections were cut transversally and photographed. The cross sectional area of flow was determined by the *Arcview* program to estimate the safranin-stained area.

2.7. Calculation of the Whole-Tree Sap Flow

An average value of sap flow (F) was determined for each sensor as the product of F_d and the area of sapwood cross-section (S_c) measured at the level of the heated probe. In order to extrapolate sap flux measurements to the whole plant, daily (F) values recorded for a sensor were correlated to the average value of the tree as suggested by Giorgio and Giorgio (2003). Finally, we normalized the transpired amount of each tree by dividing total sap flow by the tree area (36 m^2).

3. RESULTS

3.1. Probe Calibration

Initially, there was a good agreement between the measured and the calculated values of sap flow (unpublished data). Then, the amounts of water absorbed by the branch decreased and become negligible during the two following days. So, we have considered for investigation only data recorded during the two first days, showing significant relationships between the sap flux index (K_a) and the actual sap flux densities (F_d). But, this positive correlation was highly dependent of the sensor heat power ($\mathcal{P}$). For each value of heat power ($\mathcal{P}$ = 0.200 Watt, $\mathcal{P}$ = 0.144 Watt, $\mathcal{P}$ = 0.230 Watt and $\mathcal{P}$ = 0.113 Watt), a different equation was established as shown in Figure 2. Theses equations demonstrate that as the heat power ($\mathcal{P}$) increases, F_d decreases consistently for the same value of K_a.

Sensors used in the experiment at field level were powered with 0.200 Watt. For this value of $\mathcal{P}$, F_d is expressed by the following equation: $F_d = 10.06\ K_a + 0.095$ with R^2 = 0.998. Heat power was also correlated to (a), as: $\underline{a = 180.74\ e^{-15.23P}}$ with a correlation coefficient $r = 0.969$. As values of heat power (P) increased from 0.113 to 0.230 Watt, those of (a) decreased from 35 to 5 (Figure 3).

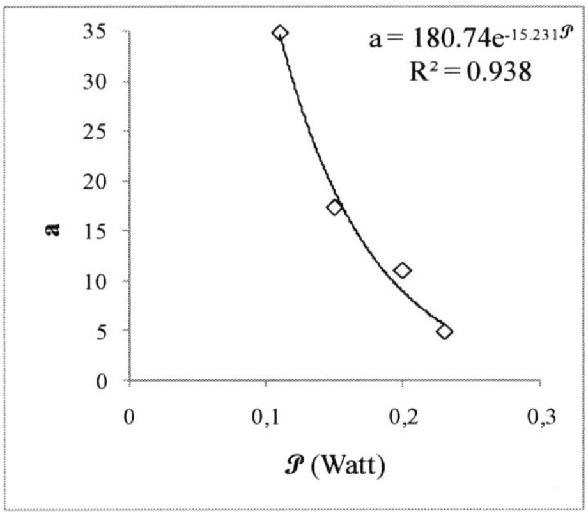

Figure 3. Relationship between the parameter a, as $F_d = (a\ K / 1 - K)^{0.707}$ and the sensor's heat power.

3.2. Water-Conducting Wood Section

Results relative to the estimation of the conductive stem section (S_c) showed positive correlation between the total section of the stem (S_t, cm^2) and that of the sapwood for shoots with diameters ranging between 1cm and 12 cm. This correlation is shown in Figure 4.

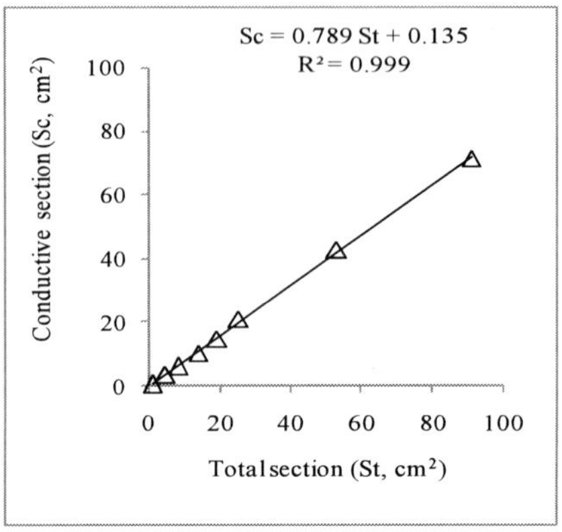

Figure 4. Relationship between the total cross section of the stem (St, cm2) and the conductive cross section (S_c, cm^2) determined for stems with diameter ranging between 1 cm and 12 cm. The number of observations, $n = 8$.

From this correlation we deduced that about 80% of the stem area conduct water. The following photo shows the red-colored cross-sectional area of the sapwood, *i.e.* the active water-conducting wood (Photo 6).

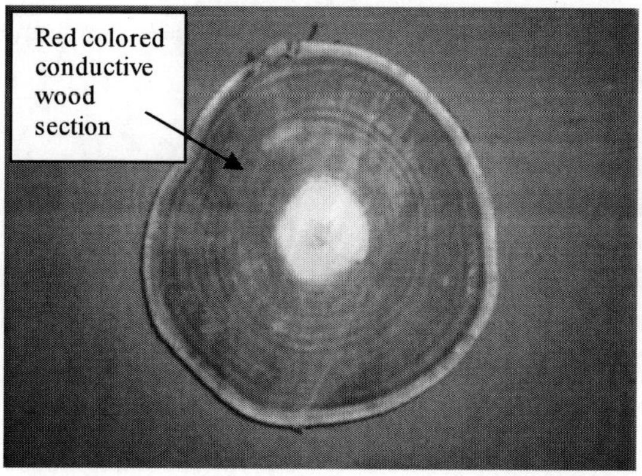

Photo 6. Cross-sectional area of the sapwood (active water-conducting wood) red – colored with Safranin Orange.

3.3. Signal Courses

An example of the signal course recorded on a sunny day (23/3/2003) under well watered conditions (soil moisture of about 33%) is given in Figure 5. This example illustrates the general response of sensors, resulting from variation of temperature between probes following sap flow movements.

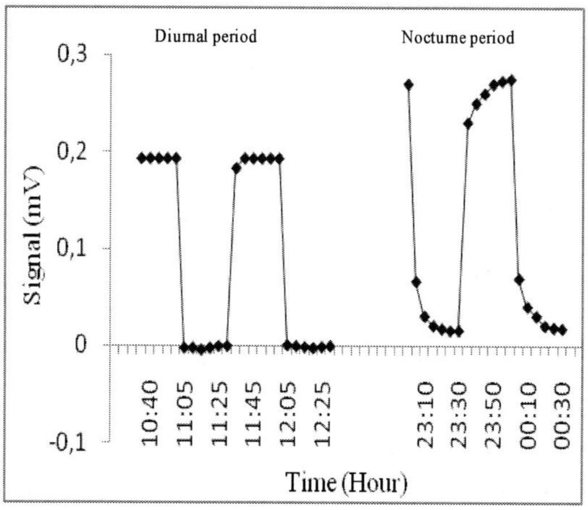

Figure 5. Example of signal course (mV) recorded on 23/3/2003 under well watered conditions.

Diurnal and nocturne signals present different dynamics. A plateau was always observed during the diurnal period with maximum values around 0.20 mV and minimums recorded during the non-heating period, reaching zero or negative voltages. Signals recorded at night were higher with maximum voltage of 0.27 mV. These variations are inherent to the presence or absence of sap flow movement. Indeed, the movement of sap during the diurnal period, enhanced by the atmospheric demand, tends to homogenize the middle by reducing the differences in temperatures between probes, providing lower signals. At night, sap flow is absent. This provided higher differences (signals) between the heated and the reference probes. Transition of signals between night and day was rapid and maximum level was established within the 3rd cycle at dawn or at dusk.

3.4. Sap Flow Density Courses

A typical course of sap flux density is given in Figure 6. It was recorded with sensors FS_4, FS_5 and FS_6 (tree A_2) during a hot and sunny day (26/7/2003) under well watered conditions (soil moisture ranging between 30 and 33% by weight basis). This example illustrates the common behavior of sensors. There was a consistency between the patterns of hourly measurements. Results showed a remarkable concordance between F_d and R_g values in most cases.

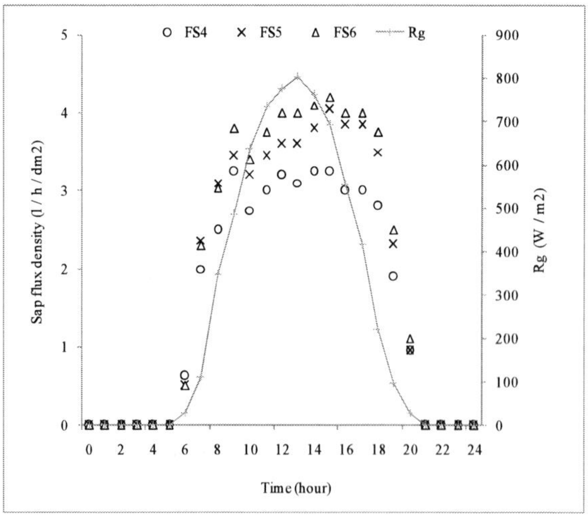

Figure 6. Typical diurnal time course of sap flux density (l/h/dm^2) measured on a hot and sunny day (26/7/2003) under well watered conditions (soil moisture ranging between 30 and 33% by weight basis) with sensors FS_4, FS_5 and FS_6, orientated N, SE and SW, respectively.

3.5. Sap Flux Variations

3.5.1. Sensor Direction

Daily variations of diurnal time courses of sap flux density studied under high and low atmospheric demand (Figure 7), showed marked differences with higher values recorded for the south orientated sensors.

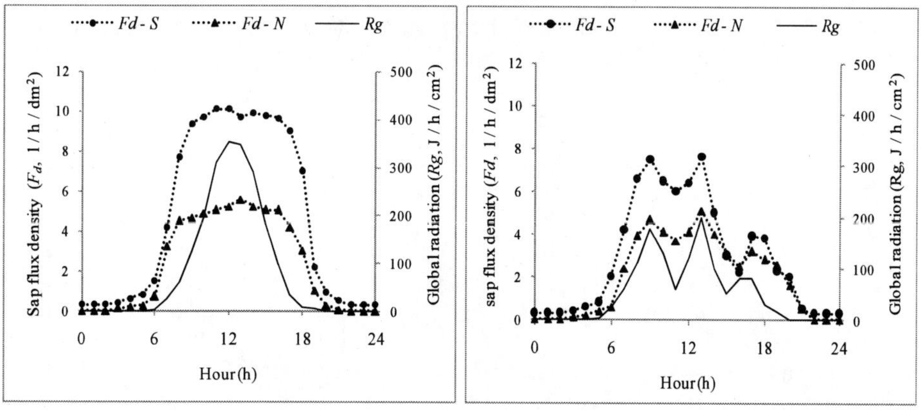

Figure 7. Daily variations of diurnal time courses of sap flux density (l/h/dm^2) measured on individual sensors following North and South directions, under high (left) and low atmospheric demand (right).

The first example is that of a rainy day (Figure 7, right). Rainfall was recorded all over the day. Sap flux density reached maximum values of 7.5 l/h/dm^2 and 5.0 l/h/dm^2 for South and North sensors, respectively, for an atmospheric maximum demand of 200 J/h/cm^2. For both directions, a slim film of water covered the external leaves temporally, hence, interrupting the evaporation process and delaying the response of the tree of about two hours. Afterwards and when water evaporates, higher rates of sap flux density were recorded, reaching values equivalent to those recorded at midday. The response of sensors was enhanced by the presence of water in the rooting-area; soil moisture was in this case high enough (40%) to meet water needs.

The second example (Figure 7, left) was recorded during a hot day (26/7/2003), with higher atmospheric demand condition, *i.e.*, with R_g of about 350 J/h/cm^2. Flux density showed rapid increase during the first hours of the day, reaching maximum values of 10.0 l/h/dm^2 and 4.8 l/h/dm^2 around 8-9:00 am for South and North sensors, respectively. These levels were sustained till 16:00 pm. Values begin to decrease during the afternoon, and cancels after 18:00pm. However, a depression was observed just after midday, at 13:00 pm, resulting from advection phenomenon which provided a partial closure of stomata. Stomata regulation occurred at this time to preserve water and reduce water depletion during these hours of high atmospheric demand. It is important to note here that, on that day, we have recorded very high temperatures and low relative humidity all over the day (conditions of sirocco).

Integration of the velocity profiles (Figure 7) suggests that the 'tree' would have 'transpired' 94.5 l day^{-1} for the North direction and 134.5 l day^{-1} for the south orientated sensor with high values observed between 9:00 am and 18:00 pm. The divergence between responses is due to direction. In looking at the flows of sap over these examples, the South direction seems to be conducting much more water than the North one. Similar result was observed during the rainy day (and others). Heterogeneous soil humectation may be another reason for the divergence. Apparently, roots connected to the North orientated branch may not be wetted enough to give high values, as those obtained for the southern orientated sensor (branch).

In order to characterize variability between sensors, daily values measured on individual sensors were compared to the average sap flux of the tree. This was done considering the procedure used by Giorgio and Giorgio (2003). Daily sap flow densities measured on a sensor in relation to the average value recorded on trees A_1, A_2 and A_3, respectively, are presented in Figure 8.

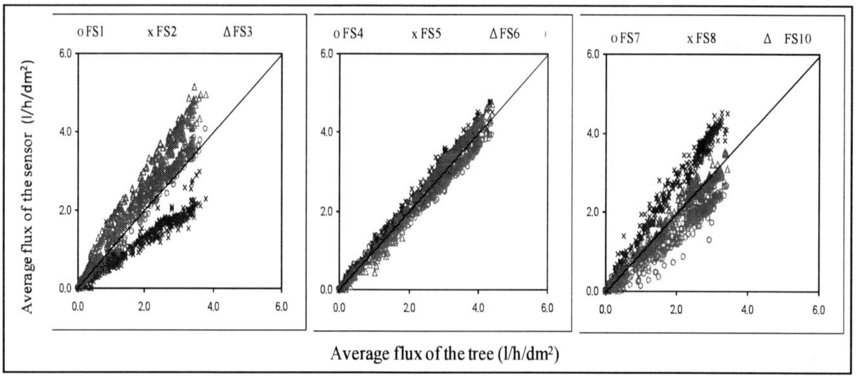

Figure 8. Hourly flux densities (l/h/dm^2) measured on individual sensors compared to the average value recorded on trees A_1, A_2 and A_3, respectively. On tree A_1, sensors FS_1, FS_2 and FS_3 were oriented North, South-East and South-West, on tree A_2, sensors FS_4, FS_5 and FS_6 were oriented North, South-East and South-West and on tree A_3 sensors FS_7, FS_8 and FS_{10} were orientated North, South-East and South-West, respectively.

Comparative results showed also that the average flux given by a sensor was correlated to the average value of the tree. Also, sensors present similar courses for the wide range of 'transpiration' levels. But the response (rates) of a sensor to direction differed following the tree. Indeed, although trees were cultivated under the same conditions and within a limited area, flows of sap showed higher values for the SW orientated sensor (FS_3) inserted in tree A_1 and for the SE orientated sensor (FS_8) implanted in tree A_3. This means that the FS_3 sensor 'overestimates' the flux density. For an average maximum value of 3.8 l/h/dm^2, a maximum density of 5.3 l/h/dm^2 was obtained while the FS_2 sensor, orientated SW peaked at 3.0 l/h/dm^2. Sensors FS_4, FS_5 and FS_6 inserted in tree A_2 presented similar response. For tree A_3, sensor FS_8, orientated SE, gave higher values. For flux densities exceeding 3 l/h/dm^2, values deviate slightly from the rest of the measurements. Lower F_d values were observed by both sensors: FS_2 oriented SE and FS_7 oriented N and implanted in trees A_1 and A_3, respectively. These figures indicate that the relative weight of direction is quite random; there is no influence of the direction of the probe on the average value of sap flux. As shown, a good correlation is observed between mean sap flow of the canopy measured with a set of (n-1) gauges and the sap flow measured by a gauge. It should be possible to estimate the parameters of this linear regression by means of regression analysis and use them to obtain an estimate of the mean sap flow of many trees during other days from values measured with just a single gauge as proposed by Giorgio and Giorgio (2003).

3.5.2. Sensor Type

Comparison between two types of sensors (FS_9 Model Dynamax powered 0.200 Watt and FS_{10} Model INRA-France powered 0.113 Watt) shows marked differences when average sap flux exceeded 2.0 l/h/dm^2 (Figure 9). For an average value of 3.5 l/h/dm^2 measured on sensor FS_9, maximum rate was about 3.0 l/h/dm^2 and exceeds 4.0 l/h/dm^2 for sensor FS_{10} (Dynamax).

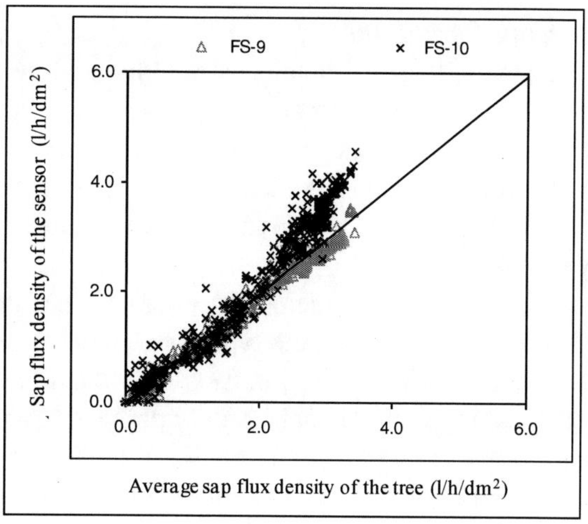

Figure 9. Hourly sap flow density (l/h/dm^2) recorded by two types of sensors: INRA-France (FS_9) and Dynamax (FS_{10}) following the average sap flux of the tree.

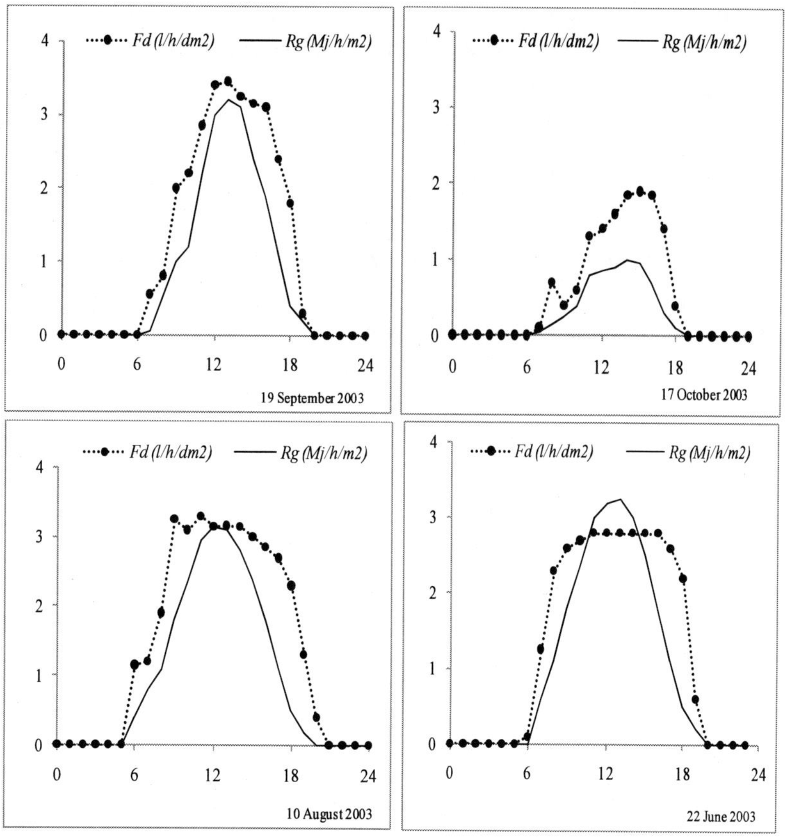

Figure 10. Diurnal time courses (daily variations) of sap flux density (F_d, l/h/dm^2) and global radiation (R_g, Mj/h/m^2) recorded under different watering conditions and atmospheric demand: Moderate climatic conditions (19/9/2003), high evaporative demand with well watering conditions (10/8/2003), water shortage conditions with sirocco (22/6/2003) occurring after a twenty-day long-period without any water supply and low atmospheric demand (17/10/2003) with rain.

3.5.3. Weather and Watering Conditions

Absolute values of sap flux densities obtained by individual sensor vary greatly with environmental conditions. A remarkable concordance between sap flux and global radiation (R_g) was observed in all cases. Daily fluxes followed the same pattern as that of atmospheric demand. This is shown through some typical examples of sap velocity profiles which were considered during the different seasons to highlight the effect of atmospheric demand on sap flux measurements (Figure 10).

Under well watered conditions and moderate evaporative demand, F_d values followed closely R_g variations. This was illustrated by the typical example recorded on 19 September 2003 (ET_0 = 3.7 mm/day, $T_{max.}$ = 28°C and $T_{min.}$= 19°C). Although the day was cloudy with some trace of rain, F_d peaked at high rate of about 3.45 l/hdm^2 for R_g of 3.2 Mj/h/m^2.

During the rainy period, R_g decreased consistently, resulting in significant decrease of F_d. The illustrative example, given in Figure 10 (17 October 2003), showed that for daily ET_0 of 1.4 mm, T_{max} of 25° and T_{min} of 19°C and maximum R_g of 1.0 Mj/h/m^2, maximum flux density was 1.9 l/h/dm^2.

During summer months (June–August), F_d increased as R_g increased. But, increases of F_d were highly dependent of the watering conditions. Indeed, when global radiation exceeded a certain threashold, F_d stopped increasing. The limit value depends on water availability but also on maximum temperature.

Indeed, under these high atmospheric demand and well watered conditions (irrigated period of August), increases of F_d values approximated R_g variations. However, it is important to note here that F_d did not follow systematically water supplies, *i.e.*, when water was applied, we did not observe a systematic increase of F_d. We had observed a punctual increase of F_d during the hour following water application in some cases, only. For example on day 10/8 under high daily evaporative demand (ET_0 = 5.7 mm, T_{max} = 35°C and T_{min} = 24°C). On that day and immediately after irrigation starting, some deviation of F_d was observed around 7:00 am to reach maximum value of about 3.3 l/h/dm^2 around 8:00 am. Then, F_d values dropped. In this case, the amount of water available was suffisant to meet the intense evaporative demand. Beginning from midday, F_d followed closely R_g course.

In June, irrigation was interrupted during 20-day-long period to repair the irrigation system. Unfortunately these days were very hot. Temperature exceeded 35°C, while soil moisture values were reduced consistently during this period of pit hardening. On 22 June (ET_0 = 6.8 mm/day, T_{max} = 35°C and T_{min} = 20°C), and under these water scare conditions, F_d reached soon after sunrise its maximum value (2.8 l/h/dm^2). This rate was maintained at this level for a long period, of about 10-12 hours duration, although R_g continued to grump. During this period of relative stressed conditions (high evaporative demand, high temperature and low water availability), the root system was presumably unable to meet this demand. Soil water content was too low and not sufficient to follow water losses depleted through 'transpiration', resulting in a stomata closure. For this reason, we observed the plateau. Around 18:00 pm, R_g dropped rapidly as well as F_d.

Over the growing season, significant relationships were observed between F_d and ET_0 (Figure 11) with high coefficient of determination (R^2) of 0.887 for the well watered conditions and 0.961 for the water shortage conditions. This indicates that for both situations, F_d is closely dependent of ET_0, which integrate most weather parameters. Deviations between measurements were inherent to the differences in soil water status.

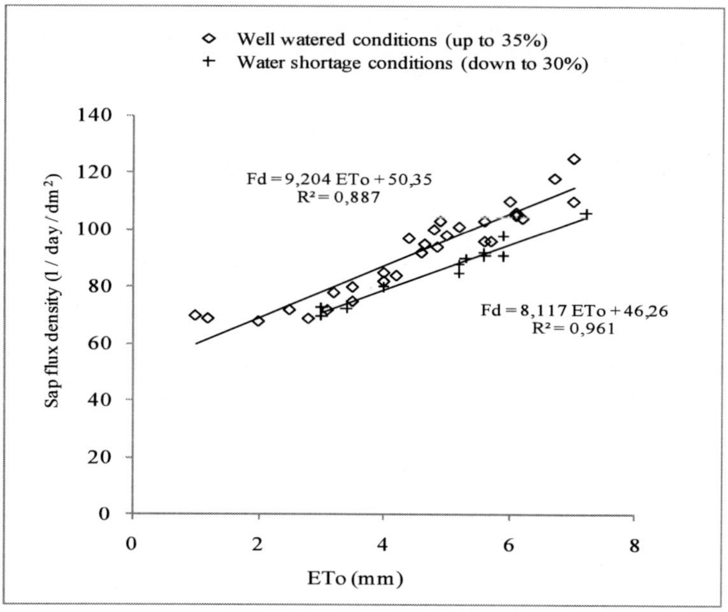

Figure 11. Relationship between sap flux densities (l/day/dm2) and ET0 (mm), determined for 40 day-long-period under water shortage and well watered conditions.

3.5.4. Leaf Conductance

Leaf conductance (g_s) was measured during the irrigation period. Results showed a remarkable concordance between F_d and g_s variations. Maximum leaf conductance occurred when maximum sap flow was recorded. Indeed, sap fluxes given by all sensors increased rapidly between 7:00 am and 9:00 am depending on their position to peak between 11:00 am and 13:00 pm, afterwards, a tendency to decrease was observed for both F_d and g_s. Figure 12 shows the concordance between leaf conductance and sap flow density measurements.

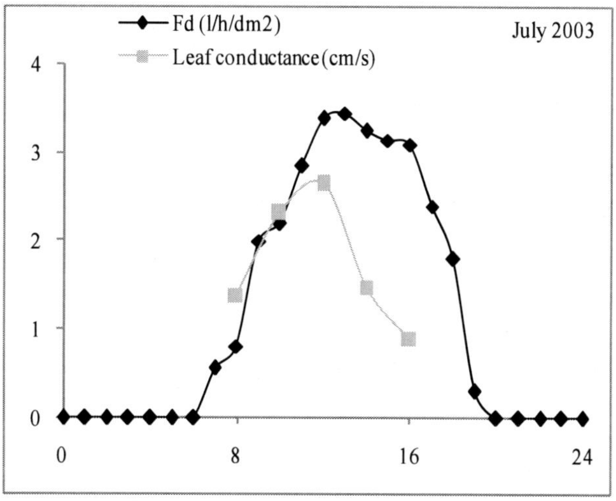

Figure 12. Leaf conductance (cm/s) measured on sunny exposed leaves of six-year-old 'Chétoui' olive trees during the irrigation season.

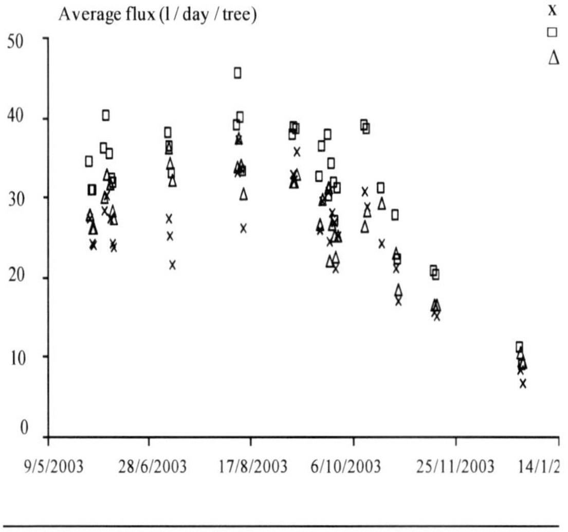

Figure 13. Average annual flux courses (l/day/tree) observed from May 2003 to March 2004 following the tree.

3.6. Annual Sap Flux Course and Transpiration Rates

The annual average course of sap flow measured on each of the three trees during a complete growing season from May 2003 to March 2004 is presented in Figure 13. This figure shows that sap flux varied all over the year, with maximums recorded in tree A_2 and minimums in tree A_1.

Variability between trees results from the heterogeneity of the conductive system characterizing each tree. The flow across the tree was determined by multiplying the sap flux densities by the conductive section as determined previously. Then, transpiration rates were estimated considering daily average sap fluxes, which were ranging between 4.5 l and 41.0 l for an individual tree area of 36 m^2. Daily water consumption ranged between 0.13 mm and 1.13 mm. These values were obtained on February and August, respectively, under the lowest ($ET_0 = 0.5$ mm/day) and the highest ($ET_0 = 8.6$ mm/day) evaporative demands. Maximum monthly consumption reached 36 mm and gave a maximum ratio of T/ET_c of about 0.53; ET_c is the crop evapotranspiration calculated following Allen et al., (1998), which estimates water needs to 67 mm/month for the same period (August) taking into account the transpired and the evaporated amounts (Table 3).

Results given in Table 3 indicate that maximum sap flow, which was recorded in August; represent about 20%ET_0 and 53%ET_c, where ET_0 and ET_c are the reference evapotranspiration and the crop evapotranspiration, respectively, determined following Allen et al. (1998). This result is very important and may have strategic applications to manage irrigation in intensive and irrigated orchards. It shows that the climatic method of the FAO, which is usually used to estimate crop water needs, overestimates water requirements of olive trees. Such information is essential for a better control of the high frequency irrigation systems normally used in olive orchards. Thus, lower amounts of irrigation water could be applied for a better water efficiency.

Table 3. Maximum transpiration rate and crop evapotranspiration value (*ETc*, mm) estimated for six-year-old olive trees for the month of August under the Northern Tunisian conditions

Estimated parameter for August	Value
Maximum value of transpiration	20%ET_o = 36 mm
ET_c (mm/month)	67 mm
T / ET_c	0.53
$I+P_e / ET_o$	0.32

NB: I, is the irrigation amount for August 2003 and P_e, is the effective rainfall for the same period. ET_0 of August is 187 mm.

4. Discussion

Data on sap flow performed on six-year-old drip irrigated olive trees of cultivar Chétoui, were collected regularly and analyzed in their physical form (signal) before treatment. The constant power applied to the heater with the alternative regime of 30 minutes heating and cooling cycles seems to be correct despite the limited number of observations. The problem of the heating system was largely discussed because the natural thermal gradients seem to be inevitable despite the use of appropriate insulation, particularly under some conditions, e.g. the main trunk /stem available for the installation of the sensor is very short; in this case, the daily flux values may be overestimated. To overcome this problem, Granier (1985), first, used a data logger to alternately switch 'on' the power to the heater for 30 minutes and to record temperature differences *(ΔT)* when the system reached a thermal equilibrium and then to switch 'off' the heating for another 30 minutes and to record ΔT when the system again reached a thermal equilibrium. The advantage is that the original calibration can still be used. But, the drawbacks are that under low flow situations (e.g. water stress) the thermal equilibrium may not be achieved within 30 minutes and the measurements may be discontinuous (one reading per hour). Do and Rocheteau (2002a, 2002b, and 2003) further improved this technique of cyclic heating and cooling. They found that the cycle of 15 minutes heating 'on' and 15 minutes heating 'off' was the best option. This new cyclic TDP system is based on the same 'Granier' probe design but with a specific calibration to account for non-steady temperature regimes. It reduces power requirements by half, which may be particularly important for field studies, where power supply is limited and also the influence of the ambient temperature gradients; thus can improve the accuracy of sap flow measurements under conditions where large ambient thermal gradients occur. However, it requires a data logger with a control port and an electronic rely to provide cyclic heating as we do in our experiment, but by using an alternative regime of 30 minutes heating and 30 minutes cooling cycles.

Signals were treated following Do and Rocheteau (2002b), and then values were adjusted taking into account the water conductive section of each tree. The calculation procedure integrate determination of ΔT_{max} which can theoretically be defined as ΔT at $F_d = 0$, to take into account the night-time flow, if it does occur. The non-zero flow occurring during the night time affects the determination of sap flow by underestimating the true ΔT_{max} required in calculation of the sap flux densities thus underestimating sap flux. According to Granier (1987b), value of ΔT_{max} should be determined separately for each sensor, usually at predawn, because ΔT is a compound result of the factors mentioned above that is often probe-specific. To overcome the problem of night-time sap flow, -such as night-time water movement for new growth (vegetative or reproductive), slow restoration of the tree's internal water storage during prolonged drought (Lu et al., 2004) and water loss from the canopy due to high vapor pressure deficit and high wind speed (Snyder et al., 2003)- ΔT_{max} was estimated in our case over a 10 day-period.

The value of ΔT_{max} is influenced by the thermal properties of the wood which is surrounding the heated probe. During the development of a severe soil water deficit there could be a drift in daily values of ΔT_{max}, which could occur during both drying and re-wetting phases. Also, there could be a drift in ΔT_{max} over a long period from several months to years as the contact between probes and surrounding wood deteriorates and the thermal conductivity decreases. This was observed during the 2004–2005 growing season with reduction of signals exceeding 20% in comparison with previous data, resulting from callus formation around sensors. In practice, the probe should be reinstalled when the drift exceeds a certain limit in order to improve the quality of measurements.

Hourly flow values were obtained for each sensor. Daily values were calculated considering each tree individually. The graphical representation of these values showed variations between sensors following the season, phenological stage, direction, atmospheric demand and watering conditions. Results showed also that sap flux values vary consistently with sensor characteristics and particularly their power. High positive correlation between heat power and the response of the sensor was found showing a decrease of F_d as the power increased. This is why we agree with previous works which indicate that, it is desirable that the heating system should be based on constant current rather than on constant voltage to ovoid the effect of the variation in cable resistance. Sap flux varied also with the current regulation and the charge of battery. We have seen that signals decreased when the charge of the battery decreased; for this reason it was recharged regularly every 10-day-period during winter and every two weeks in summer.

About sensor calibration, results indicated that changes in the heat field (heat power) around probes cause deviation in the calibration factors. These results are in agreement with those found by Granier (1985), Green and Clothier (1988) and Smith and Allen (1996) for forest species and Kiwifruit stems. Calibration of sensors was discussed also by Cabibel and Do (1991a and 1991b) and Lu et al., (2004). However, considering the complexity of the calibration procedure and the strict requirement of a section of the stem with homogenous sap flux density, these authors did not recommend carrying out a new calibration for each species, rather, it was suggested that there should be some sort of validation of the method on the species in question for example using gravimetric or 'cut-tree' methods. For 'Dynamax' sensors, 'Dynamax' uses a line-heater instead of a wound heater and a heating power different from the original, but there is no information available as to how they have calibrated their sensors although 'Dynamax' uses Granier's original calibrations for the calculation of the sap flux density. We have calibrated our sensors because we thought that those sensors with modifications in design and/or heating power may not adhere to Granier's original calibration and such modifications may have contributed to underestimate flux density. In addition, deviation from theory may occur because the conditions of experiment were different. We have noted also some difficulties during the calibration experiment which were inherent to the amounts of water consumed. These amounts decreased rapidly and become negligible after the two first days of experiment. This is -we thought- a result of a progressive loss of the hydraulic functioning of the vascular system of the stem. Green and Clothier (1988) and Fernandez et al., (2001) showed similar results in their excision experiments. They reported that it was probably caused by some blockages at the entrance to the xylem vessels at the cut face of the branch.

At field level, sap flux measurements showed a satisfactory behavior of the system. Hourly sap flux densities *(F_d)* recorded all over the growing season showed typical patterns similar to those obtained previously from both continuous heated sensors (Granier et al., 1996) and intermittent heated probes (Abid-Karray, 2006). At dawn, F_d values increased following the increase of radiation, and then oscillated between the minimum and the maximum values following sensor position, atmospheric demand and watering conditions. The morning peak in sap flux density recorded during the first hours refers to a rapid increase to unrealistically high sap flow rates observed soon after sunrise. According to Fernandez et al., (2001) this is commonly observed in isolated trees, e.g. in an orchard where the lower part of the trunk is exposed to direct sun and where a large area of the ground is not shaded. This demonstrates that in an orchard where individual trees, standing between the soil and the atmosphere, have isolated canopies, the type of insulation cover that is commonly used and which just covers the area of the measurements about 20-30 cm around the sensor probes, is inadequate. Full cover of the trunk may be benefit considering the sun flecks and wind which may create thermal gradients along the trunk and such gradients are steeper near the ground and at sunrise and sunset, or near the part of stem on which direct sunlight casts. It could be avoided or alleviated by improving thermal and radiant insulation of the area of measurement (probes, plant and soil surfaces). It also occurs when sap flow is measured very close to the ground surface. Our sensors were inserted at 40 cm above the ground surface

Good agreements and a consistency between sap fluxes and global radiation measurements (courses) were observed, with however some variations depending on water availability and the importance of the atmospheric demand conditions. During cloudy or rainy days, maximum flux density was always reached with some delay. However, when the evaporative demand is intense, sap fluxes reached rapidly after sun rise their maximum value which is generally maintained at the same level for many hours. However, in spite of these variations, significant and positive relationships were found between F_d and ET_0 with high correlative coefficients for both the well watered and water shortage conditions.

In the other hand, analyses of individual measurements showed that after irrigation, flux values do not necessarily translate an increase (unpublished data). They are more or less important following sensor position. This result corroborates Granier et al., (1996) and Abid-Karray, (2006) which observed a great variability of the responses. They are also in agreement with Lu et al., (2004) which observed that soil water supply and drought increase significantly radial variations. But the responses remained in all cases highly dependent of the atmospheric demand. This indicates that watering and the atmospheric demand should be considered together to evaluate the response of a sensor.

Variability between sensors is evoked in most papers dealing with sap flux measurements. In each of these papers, author indicated important azimuthally variations, without however, specifying a preferred orientation. This provided substantial variation in F_d across the entire cross section related to the gradient in sap velocity along an individual heat dissipation probe. It was indicated, firstly, that this variability is 'related' to sensor direction. For example, Giorgio and Giorgio (2003) observed great variability between gauges inserted in the same trunk at different azimuth positions with max-min difference nearly equal to the average sap flow. Similar results were recorded by Granier (1987), Granier et al., (1996), and Abid-Karray (2006). Then, by analyzing the ultra structure of wood, it appears that the azimuthal variability is not due to the effect of exposure but to the variability of sap flow conductivity within the trunk section area, in relation to the naturally irregular nature of sapwood of mature olive trees as it was explained previously by Fernandez and Moreno, (1999). In our case, this was evidenced through observations of transversal sections of representative branches, previously colored with *Safranine orange*. Fernandez et al., (2001) explained also that this variability highlights the heterogeneity of the stream in the timber section, which is enhanced by pruning and leaving many gnarled scars and knots around the circumference of the trunk in the early years after planting. They considered this non-uniformity feature to determine the wood conductive section in their sensor calibration experiment. They reported that measurements of sap flux density using heat pulse probe at different depths into the sap-wood should be able to show the heatwood/sapwood boundary which is responsible of water conduction and which may modify substantially the sensor response. For example, sap flow sensor probes -especially the heated one- may have been inadvertently inserted into an area of less active sapwood, due for example to pathological reasons that are not visible externally. However and despite of these problems, such measurements could be used as preferred non-destructive methods. In our case, for shoots with diameters varying between 1 cm and 12 cm, about 80% of the stem section consists of driver wood water. In other papers, it was indicated that this variability is inherent to the uniformity of irrigation. Abid-Karray (2006) explained that when two sensors are directly inserted into large vessels which are directly connected to an important root, a good response is obtained if the area surrounding this root is well irrigated. We have studied the root-system of the experimented trees and gave a description of root distribution in Masmoudi-Charfi et al., (2011b). Results showed that roots may not be enough wetted even they are developed far from the trunk. Connor and Fereres, (2005) indicated also, that if two sensors are connected to the same root, the ratio of flux densities recorded by these sensors will be low. Consequently, we can conclude, as published previously, that the relative weight of direction is quite random. Furthermore, there is no influence of the direction of the probe on the average value of sap flux because the degree of sap flow variability can change in response to a change in the special pattern of available soil-water within the tree root-zone. This result enhances that we found previously about the response of a sensor to irrigation.

Taking into account the remarkable influence of sensor position, atmospheric demand (reference evapotranspiration) and watering conditions, which seems to be the main sources of F_d variations and which have to be considered when interpreting results and when planning measurements and irrigation, it was suggested that the only way to increase the accuracy of sap flow estimation is to monitor few trees with at least two or if possible four gauges per trunk. Indeed, it was indicated that long term monitoring of the average sap flow of many trees in an olive orchard would be achieved by the continuous monitoring of a single gauge whereas the whole set of gauges on representative trees needs to be monitored only during a very short time. This idea supports our finding. In fact, and as shown, good correlation was found between mean sap flow of the canopy measured with a set of n-1 gauges and the sap flow measured by a gauge. Fernandez et al., (2001) indicated that it should be possible to estimate the parameters of this linear regression by means of regression analysis and use them to obtain an estimate of the mean sap flow of many trees during other days from values measured with just a single gauge. Swanson (1994) suggests also that taken over a sufficiently long time period, the quantity of sap flow upward through the stem must equal transpiration at the leaves because transpiration occurs as a continuum from soil to plant to atmosphere and it may be measured or estimated as moisture loss from the soil, liquid flow through the plant stem (xylem sap flow) or vapor loss to the surrounding atmosphere. Abid-Karray (2006) and Fernandez et al., (2001) concluded that in olive orchards, estimation of the total water consumption of representative trees means of the CHPV technique needs the continuous monitoring of 2-4 gauges per trunk because of the azimuthally heterogeneity of sap hydraulic functioning in this species. According to these authors, it would now seem that two to four probe sets will be needed to provide a robust measure of the tree's transpiration. With increased number of sensors average values will be better reprint the effective sap flow and take into account the heterogeneity of flow within the cross section. Our device seems to be correct and give good estimation of water consumption, which varied in our case experiment between 0.13 mm/day and 1.13 mm/day for six-year-old trees and giving maximum transpiration (August) of about 36 mm and maximum ratio of T/ET_c of 0.53. Our results are concordant with those presented by Fernandez et al., (2001) which found for orchard with LAI of 1.4 and ground cover of 30%, an average ratio of T/ET_c of 0.6. This result is very important and shows that the climatic method of FAO overestimates water needs of olive trees. Such information is essential for a better control of high frequency irrigation systems normally used in olive orchards. Thus, lower amounts of irrigation water could be applied.

Conclusion and Recommendations

The dissipation sap flow technique of 'Granier' was used to determine water consumption of young olive trees. Results show satisfactory behavior of the system when working under field conditions. The technique proved as robust and requires little maintenance. It gave valuable results which are coherent and in agreement with others made under different environments. Continuous records of plant-water use were obtained with high time resolution. They were useful in analyzing between-tree variability of transpiration despite of the obvious difficulty of sampling in such a complex ecosystem. Indeed, uncertainties in estimates of water use can occur as a result of the heterogeneity of the vascular system in individual plants, making difficult the interpretation of data and extrapolation of results to the orchard. In this experiment many constraints and problems were found, amongst, sensor calibration, large variability of sensor responses and determination of the water conductive stem section.

However, this technique, we feel, can be used with confidence in young olive trees. This means that this method has potential to be used as a device to provide a signal from which decisions can be made concerning the timing of irrigation. However, when sap flow technique is adopted to estimate transpiration for stands of vegetation in isolated trees over a long period, appropriate methods of scaling from plant to unit area of land must be used to take into account errors made in scaling transpiration from plant to stand. Comparative results indicate that estimates of olive water requirements cannot be assessed accurately with the crop coefficient method only. An articulate approach estimating soil evaporation (which is very important in young orchards) and transpiration separately is needed.

ACKNOWLEDGMENT

Author wishes to thank *Miss Mahjoub Imène* for help at field level.

DEDICATION

I wish to dedicate this work to my parents Abdelhamid Charfi and Weded Fourati, my husband Nabil and my sons Mehdi and Malek for their help, patience and love. Thank you very much for all nice things you do for me. Chiraz.

REFERENCES

Abid-Karray J., 2006. Bilan hydrique d'un système de cultures intercalaires (olivier-culture maraîchère) en Tunisie centrale: Approche expérimentale et essai de modélisation. Thèse de Doctorat (*PhD*). Université de Montpellier II. p. 172.

Allen R., Pereira L.S., Raes D., Smith M., 1998. Crop evapotranspiration. Guideline for computing crop water requirements. F.A.O., *Irrigation and Drainage Paper*, 56.

Bandino G., Dettori S., 2003. Manuale di olivicultura. Consorzio Interprovinciale per la Frutticoltura Cagliari. Oristano. Nuoro. Regione Autonoma della Sardegna.

Cabibel B., Do F., 1991 a. Mesures thermiques des flux de sève dans les troncs et les racines et fonctionnement hydrique des arbres. I. Analyse théorique des erreurs sur la mesure des flux et validation des mesures en présence de gradients thermiques extérieurs, *Agronomie*, *11*, pp. 669–678.

Cabibel B., Do F., 1991 b. Mesures thermiques des flux de sève et comportement hydrique des arbres. II. Evolution dans le temps des flux de sève et comportement hydrique des arbres en présence ou non d'une irrigation localisée, *Agronomie*, *11*, pp. 757–766.

Cabibel B., Isbérie C., 1997. Flux de sève et alimentation hydrique de cerisiers irrigués ou non en localisation, *Agronomie*, *17*, pp. 97–112.

C.O.I.,1997. L'Olivier. In: Plaza and F.A. Janéf Editors. Encyclopédie Mondiale de l'Olivier. p. 479.

Cohen Y., 1991. Determination of orchard water requirement by a combined trunk sap flow and meteorology approach. *Irrigation Science*, *12,* pp. 93–98.

Connor D.J., Fereres E., 2005. The Physiology of adaptation and yield expression in Olive. Horticultural Review. Volume 31. Edited by Jules Janick.

Do F., Rocheteau A., 2002 a. Influence of natural temperature gradients on measurements of xylem sap flow with thermal dissipation probes.1. Field observations and possible remedies. *Tree Physiology*, 22, pp. 641–648.

Do F., Rocheteau A., 2002 b. Influence of natural temperature gradients on measurements of xylem sap flow with thermal dissipation probes 2. Advantages and calibration of a non continuous heating system. *Tree Physiology*, 22, pp. 654–679.

Do F., Rocheteau A., 2003. Cycle annuel de transpiration d'Acacia raddiana par la mesure des flux de sève brute (Nord-Sénégal). In: Grouzis, M., Le Floc'h, E. (Eds.). Un arbre au désert *Acacia raddiana*. Editeurs Scientifiques/IRD Editions, pp. 119–142.

F.A.O., 1976. Les besoins en eau des cultures. Bulletin F.A.O. d'irrigation et de drainage.

Fernandez J.E., Moreno F., Cabrera F., Arrue J.L., Martin-Aranda J., 1991. Drip irrigation, soil characteristics and the root distribution and root activity of olive trees. *Plant and Soil, 133*, pp. 239–251.

Fernandez J.E. Moreno F. 1999. Water use by the olive tree. *Journal of Crop Production 2*. pp.101–162.

Fernandez J.E., Palomo M.J., Diaz-Espejo A., Clothier B.E., Green S.R., Giron I.F., Moreno F., 2001. Heat-pulse measurements of sap flow in olives for automating irrigation: tests, root flow and diagnosis of water stress. *Agriculture Water Management, 51* (2), pp. 99–123.

Fernandez J.E., Paloma, M.J., Diaz-Espejo, A., Giron, I.F., 2003. Influence of partial soil wetting on water relation parameters of the olive tree. *Agronomie*, 23, pp. 545–552.

Giorgio G., D'Andria R., 2002. Sap flow estimated by compensation heart-pulse velocity technique in olive trees under two irrigation regimes in Southern Italy, *Acta Horticulturae, 586*, pp. 401–404.

Giorgio P., Giorgio G., 2003. Sap flow of several olive trees estimated with the heat-pulse by continuous of a single gauge, *Environmental and Experimental Botany, 49* (1), pp. 9–20.

Granier, A., 1985. Une nouvelle méthode pour la mesure des flux de sève brute dans le tronc des arbres. *Ann. Sci. For* 42 (2), pp. 193–200.

Granier A., 1987 a. Evaluation of transpiration in a Douglas-fir stand by means of sap flow measurements, *Tree Physiology 3*, pp. 309–320.

Granier A., 1987 b. Mesure de flux de sève brute dans le tronc du Douglas par une nouvelle méthode thermique. *Ann. Sci. For, 44* (1), pp. 1–14.

Granier A., Huc R., Barigah S.T., 1996. Transpiration of natural rain forest and its dependence on climatic factors. *Agricultural and Forest Meteorology*, 78, pp. 19–29.

Granier A., Biron P., Lemoine D., 2000. Water balance, transpiration and canopy conductance in tow beech stands. *Agricultural and Forest Meteorology, 100,* pp. 291–308.

Green S.R., Clothier B.E., 1988. Water use of Kiwi fruit vines and apple trees by the heat pulse technique. *J. Exp. Bot, 39*, pp. 115–123.

Habaieb H., Masmoudi-Charfi C., 2003. Calcul des besoins en eau des principales cultures exploitées en Tunisie: Estimation de l'évapotranspiration de référence par différentes formules empiriques. Cas des régions de Tunis, Béja et Bizerte, *Sécheresse, 14* (4), 1-9.

Hamza 2009. Conférence régionale sur la gouvernance de l'eau. Echange d'expériences entre l'OCDE et les pays arabes. CITET – Tunis, 8-9 Juillet 2009. Direction Générale des Ressources en eau. Tunisie.

I.N.M., 2003. Institut National de Météorologie. http://www.meteo.tn/htmlfr/accueil.php.

Le Bourdelles J., 1977. Irrigation par goutte à goutte en Oléiculture. Principe de la méthode, installations et fonctionnement. Olea, Juin (1977).

Lu P., Urban L., Zhao P., 2004. Granier's thermal dissipation probe (TDP) method for measuring sap flow in trees: theory and practice. *Acta Botanica Sinica*, *46,* (6), pp. 631–646.

Mahjoub, I., 2004. Développement d'un dispositif de mesure du flux de sève des arbres. Projet de Fin d'Etude de l'INAT. 48 pages.

Masmoudi-Charfi C., 2006. Gestion de l'irrigation dans les plantations d'olivier. Document Technique n°3. Institut de l'Olivier.13 pages en langue arabe (Technical Paper).

Masmoudi-Charfi C., 2008. Caractérisation et modélisation de la croissance de l'olivier (Olea europaea L.) en rapport avec son alimentation en eau. Doctorat d'Etat en Sciences Agronomiques (*PhD*). *Institut National Agronomique de Tunisie*. p 323.

Masmoudi-Charfi C., Ben Mechlia N., 2007. Characterization of young olive trees growth during the first six years of cultivation. *Advances in Horticultural Sciences*, 21 (2), pp. 116-124.

Masmoudi-Charfi C., Ben Mechlia N., 2008. Changes in olive tree height growth during the first years of cultivation". *Advances in Horticultural Sciences*, 22 (1), pp. 8-12.

Masmoudi-Charfi C., Ben Mechlia N., 2009. Mineral uptakes of macro-nutrients in Tunisian olive orchards (Olea europaea L.) during the first years after plantation. *Advances in Horticultural Sciences*, 23 (4), pp. 211-218.

Masmoudi-Charfi, C., Masmoudi, M.M., Ben Mechlia, N., 2004 a. Irrigation de l'Olivier: Cas des jeunes plantations intensives. *Ezzaitouna*, 10 (1–2), pp. 37–51.

Masmoudi, M.M., Mahjoub, I., Masmoudi-Charfi, C., Karray-Abid, J., Ben Mechlia, N., 2004 b. Mise au point d'un dispositif de mesure du flux de sève xylémique chez l'olivier. In: Actes du Séminaire International 'Aridoculture et Cultures Oasiennes'. Numéro Spécial, Revue de l'Institut des Régions Arides (IRA) - Médenine, Djerba 22–25 Novembre 2004, pp. 242–251.

Masmoudi-Charfi C., Msallem M., Say M.B., 2006. Plantations irriguées d'olivier. Document Technique n°2. Institut de l'Olivier. 7 pages en langue arabe. (Technical Paper).

Masmoudi-Charfi C., Mezghani-Ayachi M., Gouia M., Laabidi F., Ben Reguaya S., Oueled Amor A., Bousnina M., 2010. Water relations of olive trees cultivated under deficit irrigation regimes. *Scientia Horticulturae*, 125, pp. 573–578.

Masmoudi-Charfi C., Masmoudi M.M., Karray-Abid J., Ben Mechlia N., 2011 a. Sap flow Measurements in Young Olive Trees (Olea europaea L.) cv. 'Chétoui' Under Tunisian Conditions. *Scientia Horticulturae*, 129, pp. 520-527.

Masmoudi-Charfi C., Masmoudi M.M., Ben Mechlia N., 2011 b. Root distribution in young Chétoui olive trees (Olea europaea, L.) and agronomic applications. *Advances in Horticultural Sciences*, 25 (2), pp. 112-122.

Nasr, Z., 1987. Une nouvelle technique de pilotage de l'irrigation des arbres: Mesure du flux de la sève par des méthodes thermiques. Mémoire de fin d'études du 2ème cycle de l'Institut National Agronomique de Tunisie, p.70.

Michelakis N., 2000. Water requirements of olive tree on the various vegetative stages, Proceedings of the International Course on Water Management and Irrigation of Olive Orchards. Cyprus. April 2000 (2000).

Moreno F., Fernandez J.E., Clothier B.E., Green S.R., 1996. Transpiration and root water uptake by olive trees. *Plant and Soil*, *184*, pp. 85–96.

Palomo M.J., Moreno F., Fernandez J.E., Diaz-Espejo A., Giron I.F., 2002. Determining water consumptive in olive orchards using the water balance approach. *Agricultural Water Management, 55* (1), pp. 15–35.

Pereira A.R., Green S.R., Nova N.A.V., 2007. Sap flow, leaf area, net radiation and the Priestley–Taylor formula for irrigated orchards and isolated trees. *Agricultural Water Management, 16*, pp. 48–52.

Pastor M., Hidalgo J., Vega V., Castro J., 1998. Irrigation des cultures oléicoles dans la région de la LOMA (Province de Jaén), *Olivae, 71:* 39–49.

Rana G., Katerji, N., 2000. Measurement and estimation of actual evapotranspiration in the field under Mediterranean climate: A review. *Eur. Jour. Agro* 13, 125–153.

Riou C., 1978. Quelques aspects du bilan hydrique d'une plantation d'oliviers. In: Séminaire sur l'olivier et autres plantes oléagineuses cultivées en Tunisie. O.N.H, pp. 101–104.

Smith D.M., Allen S.J., 1996. Measurement of sap flow in plant stems. *Journal of Experimental Botany, 47* (305), pp. 1833–1844.

Swanson R.H., 1994. Significant historical developments in thermal methods for measuring sap flow in trees. *Agricultural and Forest Meteorology, 72*, pp. 113–132.

Testi L., Villalobos F.J., Orgaz F., 2004. Evapotranspiration of a young irrigated olive orchard in southern Spain. *Agricultural and Forest Meteorology, 121*, pp. 1–18.

Villagra M.M., Bacchi O.O.S., Tuon R.L., Reichardt K., 1995. Difficulties of estimating evapotranspiration from the water balance equation, *Agricultural and Forest Meteorology, 72*, pp. 317–325.

Villalobos F.J., Orgaz F., Testi L., Fereres E., 2000. Measurement and modeling of evapotranpiration of olive (Olea europaea L.,) orchards. *European Journal of Agronomy, 449* (13), pp. 155–163.

In: Irrigation Management, Technologies …
Editor: M. Hossain Ali

ISBN: 978-1-62417-862-7

Chapter 3

RESPONSES OF APPLE TO PARTIAL ROOT-ZONE DRYING: A REVIEW

Riccardo Lo Bianco*

Dipartimento DEMETRA, Università degli Studi di Palermo, Palermo, Italy

ABSTRACT

Partial root-zone drying (PRD) is a relatively recent strategy aiming at substantial reduction of irrigation volume. It consists of the alternate irrigation of one portion of the root-zone with volumes reduced to about 40-50% of conventional water. PRD is based on the ability of roots in drying soil to produce hormonal signals, like abscissic acid (ABA), responsible for partial stomatal closure. In turn, partial stomatal closure would improve water use efficiency and increase fruit yield per unit of irrigation water. This technique has been tested on a number of fruit crops, including apple. In addition to climatic conditions, which mainly affect atmospheric water demand, soil properties and water dynamics in the soil seem to play a major role in the outcomes of PRD studies. Irrigation systems and scheduling are also critical for a PRD effect (irrigation placement effect). Apple cultivars with different vigor, period of fruit growth and crop load perform differently in response to PRD and generalizations on the potential use of PRD as a valuable irrigation strategy for apple orchards result in difficulties. Fruit size and number per tree seem to be sensitive to some extent to the reduction of irrigation volumes operated with PRD. On the contrary, vegetative growth is mostly unaffected by PRD and fruit quality may be somewhat improved. These general responses are associated with a relatively good water status of trees under PRD. The hypothesized reductions of tree water loss through stomatal control, however, do not seem to fully explain the lack of water deficit. Higher soil moisture content, in the root-zone of PRD trees compared to trees with similar irrigation volumes but applied to the entire root-zone, has been documented in some studies. This difference, in good part due to diminished soil evaporation, poses some doubts on the real ability of apple trees under PRD to control water loss by root ABA signaling. In this context, longer irrigation events with low delivery rates (i.e., drip systems) should maximize the advantages of PRD in terms of soil

* Email: riccardo.lobianco@unipa.it.

moisture. Also, the use of PRD may be most appropriate for cultivation of late-maturing apple varieties with high crop load, which would suffer from relatively severe dehydration under reduced irrigation volumes.

INTRODUCTION

Fruit crops that are typically grown in cooler climates, like apple, are increasingly being cultivated in Mediterranean regions, where dry summers reduce disease pressure and increase fruit quality. Yet, in Mediterranean areas, the highest crop evapotranspiration rates occur also during the dry season determining a great demand for irrigation water. In those periods, competition by municipalities strongly aggravates water scarcity for agricultural purposes. Under those circumstances of limited water resources, growers should aim at maximizing yield per unit of irrigation water, or irrigation water productivity (IWP), rather than yield per hectare.

Recently, many studies have focused on IWP as being directly related to the increase of plant water use efficiency (WUE) (Kijne et al., 2003). Published research with several crops showed that IWP tends to increase with deficit compared to conventional irrigation (Zwart and Bastiaansen, 2004; Fan et al., 2005).

There are several explanation for this increase of IWP under deficit irrigation, mainly related to dry matter partitioning and maintenance of harvest index (Ali and Talukder, 2008; Fereres and Soriano, 2007). Both evapotranspiration and yield may increase proportionally with irrigation only until a certain point after which a non-linear relationship is established and eventually no more yield response to water volumes is observed. This linear to non-linear transition threshold is not easily and precisely determined and overwatering becomes the most common consequence to avoid yield losses. Water needed to maximize yield depends also on a uniform distribution of irrigation water to minimize deficit areas. Moreover, IWP depends on the yield response function to irrigation. Yield responses to irrigation have been studied for decades, but they cannot be exactly determined in advance as they vary with place and climate, stress parameters, crop, cultivar, etc. For example, sensitivity to water stress varies with crop species and phenological stage, and a deficit irrigation plan must consider this to minimize yield loss. Nevertheless, any yield loss will be relatively less than the decrease of evapotraspiration, determining the mentioned increase of IWP under deficit irrigation. Unfortunately, growers try to maximize profit, not IWP, giving little or no priority to irrigation and environmental sustainability. In any circumstance, knowing yield responses to deficit irrigation for each crop remains fundamental.

Increasing IWP has been more successful in trees than in field crops for several reasons (Fereres et al., 2003). Fruit quality, for example, strongly affects crop value, but is not associated with biomass production and water use. In addition, tree fruit growth may not be sensitive to water deficit in certain periods and developmental stages (Johnson and Handley, 2000). This, in combination with low volume/high frequency irrigation systems, gives the best opportunities to manage fruit trees under controlled water deficit.

REGULATED DEFICIT IRRIGATION

Deficit irrigation is an optimization strategy in which irrigation water is applied during sensitive growth stages of a crop and whereby, outside these periods, irrigation is limited or can even be absent if rainfall guarantees a minimum supply of water. Regulated Deficit Irrigation (RDI) is a system of managing soil water supply to impose periods of predetermined plant or soil water deficit that can result in some economic benefit.

The RDI concept was first proposed by Chalmers et al. (1981) and Mitchell and Chalmers (1982) in the attempt to control vegetative growth in peach. The same Authors soon realized that irrigation water savings could be achieved with no significant yield reduction. Similar results were observed in pear (Mitchell et al., 1989), with different degree of success depending on the environmental conditions (Marsal et al., 2003). Positive results have also been obtained with a number of tree crops like almond (Goldhamer et al., 2006), pistachio (Goldhamer and Beede, 2004), citrus (Domingo et al., 1996; Gonzàlez-Altozano and Castell, 1999; Goldhamer and Salinas, 2000), apricot (Ruiz-Sàncez et al., 2000), grapes (Bravdo and Naor, 1996; McCarty et al., 2002) and olive (Moriana et al., 2003), giving support to the idea that returning full evapotranspiration water to fruit trees may not be the best irrigation strategy in several cases (Fereres and Evans, 2006).

In general, accurate procedures need to be developed for effective irrigation planning. In particular for RDI, soil and plant water status needs to be monitored to minimize yield losses. In fruit trees under micro-irrigation, soil moisture is hard to be monitored and measures of plant water status are most suitable (Jones, 2004). Several options are available to measure plant water status, often accompanied by some limitations. A lot of work is needed to measure stem water potential, while the use of dendrometers seems to produce very variable results (Intrigliolo and Castel, 2004; Naor et al., 2006); infrared thermometers and thermography seem promising to determine water stress, especially in olive (Sepulcre-Cantò et al., 2006). Sap flow measurements represent also a reliable tool for irrigation scheduling in olive, apple, pear and grapevines, although the need of detailed microclimate data and quantification of plant leaf area strongly limits the application of this technique at a farm level (Fernàndez et al., 2008).

The mechanism responsible for no yield loss under RDI has been extensively studied. The best explanation lies in the different sensitivity of various tree organs to water deficit in terms of turgor loss and growth. Those differences, in turn, lead to changes in carbon partitioning, while photosynthesis is not affected by mild water stress. For example, drought often favors root growth (Sharp and Davies, 1979; Hsiao and Xu, 2000), while assimilate partitioning to fruit may not be affected (Gucci and Minchin, 2002).

Water deficit is also known to modify carbon partitioning between vegetative and reproductive sinks depending on each organ's ability to attract photoassimilates. Fruits and seeds, for example, are generally stronger sinks than shoot apexes (Wardlaw 1990), and under water deficit and limited assimilation shoot growth may be preferentially reduced over fruit growth (Higgs and Jones 1991). In stone fruits, drought affects mostly vegetative growth in spring and during pit hardening, whereas it may affect both shoot and fruit growth during final fruit swelling (Flore and Layne 1999). Reduced shoot growth may directly improve light distribution within the canopy and to fruiting shoots, which in turn increases floral induction and return bloom (Caspari et al. 1994), and indirectly facilitates canopy management (less pruning and fertilizers needed).

In general, one period in which water deficit can be applied with little or no risk of yield loss is the window between harvest and leaf fall. This has been showed in peach by imposing relatively severe deficit (Johnson et al., 1992), although continuous application of post-harvest water deficit throughout the years may reduce the productive life of the orchard.

As a result, responses to RDI seem to depend both on timing and intensity of water deficit, but also on crop load. The application of RDI at the farm level is further complicated if we consider that responses change significantly with species and even cultivar. This is mainly associated with genetic differences and genotype-environment interactions determining different sensitivity to water deficit, crop load, but also length, timing and stages of fruit and shoot growth. In apple, for example, fruits and shoots grow concurrently (Forshey et al. 1983), especially in the early to intermediate stages of fruit growth, and water deficit during this period may strongly increase competition for assimilates and cause assimilate starvation in fruit. In many cases, the control and regulation of water deficit in apple is quite complex resulting often in reduced final fruit size and yields, irrespective of the period of deficit imposition (Ebel et al. 1993, 1995; Lötter et al. 1985; Mpelasoka et al. 2001).

Partial Root-Zone Drying

Partial root-zone drying (PRD) is an irrigation technique that was initially developed for grapes (*Vitis vinifera* L.) in Australia, with the primary aim of saving irrigation water (Dry et al., 1995; Dry and Loveys, 1998). With PRD only one half of the root-zone is irrigated, whereas the other half is not, with 50-60% savings over conventional irrigation. The physiological basis for PRD is that roots in drying soil produce abscisic acid (ABA); if there is sufficient sap flow through those roots, ABA is translocated to the shoots (Dodd et al., 2008a) where it signals a developing soil-water deficit (Dry et al., 1995). Accounting for the relative contributions of different parts of the root system to sap flow can also explain shoot xylem ABA concentrations (Dodd et al., 2008b). In the leaves, ABA induces partial stomatal closure, which reduces transpiration and may increase water use efficiency. However, as the other half of the root-zone is kept well watered, the effect on plant water potential is minimal (Gowing et al., 1990) and other metabolic and physiological processes associated to water stress are not affected (Dry et al., 1995; Dry et al., 2000). The technique relies on cyclical wetting and drying of parts of the root-zone in order to stimulate and maintain root derived ABA signals (Dodd et al., 2006). Increased xylem ABA and pH (Davies et al., 2002) (both probably responsible for partial stomatal closure), and reductions of cytokinin levels in roots, shoot apexes, leaves and buds (Kudorayova et al., 2007; Stoll et al., 2000) have been reported in plants under PRD regime. The resulting changes in hormonal balance followed by partial stomatal closure may contribute to reduce vegetative growth and be responsible for higher crop yields than conventional deficit irrigation when the same volumes of water are applied (Dodd, 2009).

Table 1. Indications of soil moisture changes in partial root-zone drying (PRD) studies conducted on apple

Soil water content	Soil water potential	No data shown
Caspari et al., 2004a	Leib et al., 2006	Lombardini et al., 2004
Caspari et al., 2004b	Talluto et al., 2008	O'Connell & Goodwin, 2007
Einhorn & Caspari, 2004	Lo Bianco et al., 2012	Fallahi et al., 2008
Leib et al., 2006	Lo Bianco & Francaviglia, 2012	Zegbe et al., 2011
Talluto et al., 2007		Yang et al., 2011 (potted plants)
Zegbe et al., 2007		Senyigit & Ozdemir, 2011
Zegbe & Behboudian, 2008		
Zegbe et al., 2008		
Einhorn et al., 2009		
Zegbe & Serna-Pérez, 2011		
Zegbe & Serna-Pérez, 2012		

Soil Moisture

Of the 20 studies published on the responses of apple to PRD, 14 reports discussed changes in soil water content and/or soil water potential, while the remaining 6 did not show any soil moisture data (Table 1). Yet, for the efficacy of the PRD technique, a good temporal and spatial separation of wet and dry zones in the soil around the root system is particularly important. For this purpose, use of a drip system, along with a relatively short duration and interval between irrigation events, is generally more effective than other irrigation systems and plans, especially in situations of high tree planting density (Lo Bianco and Francaviglia, 2012). In addition, PRD relies on the production of root-generated signals (presumably ABA), which can be maintained only if roots are kept alive by avoiding permanent or excessive soil drying. This is managed by alternating wet and dry of the root-zone sides. Hence, in field experiments, continuous monitoring of soil moisture change is essential both to verify a good spatial separation between wet and dry sides of the root-zone and to establish the appropriate moment for switching sides. More attention should be posed on this aspect when discussing agronomic responses or physiological mechanisms underlying the functioning of trees subjected to PRD.

To distinguish between volume and placement effects, many PRD studies include a third treatment (DI) delivering the same irrigation volume applied with PRD, but generally on both sides of the root-zone. Although PRD and DI treatments deliver similar irrigation volumes, higher soil moisture has been observed under PRD (average of wet and dry sides) than DI across the irrigation season (Leib et al. 2006; Lo Bianco and Francaviglia, 2012). One explanation for this is that supplying half the irrigation water on both sides of the root-zone (DI) during each irrigation event increases the surface area of wet soil and decreases soil water depth, causing greater evaporative losses from the soil in DI compared to PRD. This would actually convert at least in part the intended placement effect into a volume (soil moisture level) effect, posing some doubts on the real mechanisms underlying PRD, at least in some experiments.

Table 2. Leaf stomatal conductance of apple trees under partial root-zone drying (PRD)

Response	
	Einhorn and Caspari, 2004
PRD = CI	Zegbe et al., 2007
	Zegbe and Behboudian, 2008
	Zegbe and Serna-Pérez, 2012
	Lombardini et al., 2004
	Talluto et al., 2007
	O'Connell and Goodwin, 2007
PRD < CI	Talluto et al., 2008
	Yang et al., 2011 (potted plants)
	Lo Bianco et al., 2012
	Lo Bianco and Francaviglia, 2012
PRD ≠ DI	Lo Bianco et al., 2012
	Lo Bianco and Francaviglia, 2012

CI: irrigation volumes corresponding to about 100% crop evapotranspiration; DI: irrigation volumes similar to PRD but applied to both sides of the root-zone.

Plant Water Status

In PRD but not in RDI, water uptake from the wet side of the root system maintains a favorable plant water status, while roots in the dry side produce the ABA signal responsible for reductions in leaf stomatal conductance (g_s) and a general WUE increase (Gowing et al., 1990; Davies et al., 2002; Costa et al., 2007). Several studies in different environmental conditions have shown that while g_s might be reduced in PRD (the degree of reduction depending on crop drought sensitivity and/or drought severity), photosynthetic rates (Pn) are generally not affected compared to full irrigation (Costa et al., 2007; Sepaskhah and Ahmadi, 2010). In apple, Pn was measured in few PRD studies and results are contrasting (Lombardini et al., 2004; Zegbe et al., 2007; Zegbe and Behboudian, 2008), perhaps due to different types of measurements (single leaf vs full canopy Pn). Also g_s responses to PRD appear inconsistent, with seven studies showing significant g_s reductions and four studies showing no difference with conventional/full irrigation (CI) (Table 2). Such discrepancies may be associated to the specific setup of the experiment, and suggest that a PRD response may not be observed if spatial and/or temporal separation of wet and dry root-zones is not established properly. In any case, g_s measurements are essential to verify a PRD response and discuss any possible effect on yield, fruit quality or IWP.

A g_s reduction under PRD compared to conventional/full irrigation is expected for a tree water-saving mechanism to work and WUE to increase, but it is not sufficient to prove an irrigation placement effect. Reduced g_s levels could be in fact observed in response to a simple reduction of irrigation volume (DI). In this case, for a real placement (or PRD) effect to occur, a difference in g_s level or tree water status between PRD and DI should be detected. In apple, six out of the seven studies showing significant g_s reductions included a DI treatment, but none was able to show lower g_s levels in PRD than in DI. On the contrary, in a few cases, g_s was lower in DI than in PRD (Lo Bianco et al., 2012; Lo Bianco and Francaviglia, 2012), indicating that stomatal control alone may not explain an advantage of PRD over DI.

Table 3. Water status of apple trees under partial root-zone drying (PRD)

Response	Water potential	Relative Water Content
PRD = CI	Einhorn and Caspari, 2004	Talluto et al., 2008
	Zegbe et al., 2007	Lo Bianco et al., 2012
	Zegbe and Behboudian, 2008	Lo Bianco and Francaviglia, 2012
	Zegbe et al., 2011	
	Zegbe and Serna-Pérez, 2012	
PRD < CI	Lombardini et al., 2004	
	O'Connell and Goodwin, 2007	
	Zegbe and Serna-Pérez, 2011	
PRD > DI		Talluto et al., 2008
		Lo Bianco et al., 2012
		Lo Bianco and Francaviglia, 2012

The advantage of PRD over DI is often expressed in terms of tree water status. Of the 11 PRD studies measuring water status in apple trees, eight report leaf water potential, while three report leaf relative water content (Table 3). Among all, only three showed minor reductions of water status in response to PRD (Table 3), suggesting that trees under PRD tend to minimize dehydration. Of the six studies reporting water status data and comparing PRD with DI, only three showed a better water status in PRD than in DI (Table 3), i.e., a real placement effect.

Vegetative Growth

Water deficit is known to generally reduce vegetative growth in plants (Hsiao, 1973), including apple (Irving and Drost, 1987). Growth reductions are often seen as a consequence of water deficit, but they may also be considered as an adaptive mechanism and water saving strategy through reductions of the transpiring surface. In apple, growth reductions are also well documented under RDI (Mills et al., 1996), whereas contrasting expectations and results are reported for trees under PRD.

PRD never affected trunk growth, in one study (and only one year) it reduced shoot length, while in three out of four studies it reduced total leaf area or dry mass (Table 4). A summary of all these responses suggests that PRD induces only minor reductions of vegetative growth, which can be statistically detected only when growth is measured in a very accurate way or in more sensitive (fast growing) tree organs. Typically, in a tree, leaves grow faster (and are more drought sensitive) than trunk, with shoots growing at variable intermediate rates depending on type and size. In addition, leaf shedding is a common response to drought, increasing the chances of detecting a water deficit effect when total leaf area or dry mass is measured. In any case, it is not particularly surprising that vegetative growth is not reduced by PRD if we consider that a good hydration level is generally maintained in PRD trees (Table 3), at least in part by reducing leaf g_s (Table 2).

Table 4. Vegetative growth responses of apple trees under partial root-zone drying (PRD)

Response	Trunk growth	Shoot length	Leaf area/dry mass
PRD = CI	Fallahi et al., 2008	Lombardini et al., 2004	Talluto et al., 2007
	O'Connell & Goodwin, 2007	Talluto et al., 2007	
	Zegbe et al., 2007	Talluto et al., 2008	
	Lo Bianco & Francaviglia, 2012	Zegbe et al., 2007	
	Zegbe & Behboudian, 2008	Zegbe et al., 2011	
	Zegbe & Serna-Pérez, 2012	Lo Bianco & Francaviglia, 2012	
		Zegbe & Serna-Pérez, 2012	
PRD < CI		O'Connell & Goodwin, 2007	Talluto et al., 2008
			Yang et al., 2011
			Lo Bianco et al., 2012

Table 5. Yield responses of apple trees under partial root-zone drying (PRD)

Response	Crop load	Fruit weight	Yield
	Caspari et al., 2004a	Caspari et al., 2004a	Caspari et al., 2004a
	Lombardini et al., 2004	Einhorn & Caspari, 2004	Caspari et al., 2004b
	Caspari et al., 2004b	Leib et al., 2006	Einhorn & Caspari, 2004
	Einhorn & Caspari, 2004	Talluto et al., 2007	Leib et al., 2006
PRD = CI	Leib et al., 2006	Talluto et al., 2008	Talluto et al., 2007
	Talluto et al., 2007	Fallahi et al., 2008	Talluto et al., 2008
	O'Connell & Goodwin, 2007	Zegbe et al., 2007	Fallahi et al., 2008
	Zegbe et al., 2007	Senyigit & Ozdemir, 2011	Zegbe & Behboudian, 2008
	Zegbe & Serna-Pérez, 2012	Zegbe & Behboudian, 2008	Zegbe & Serna-Pérez, 2012
		Zegbe & Serna-Pérez, 2012	
	Talluto et al., 2008	Caspari et al., 2004b	Lombardini et al., 2004
PRD < CI	Senyigit & Ozdemir, 2011	Lombardini et al., 2004	O'Connell & Goodwin, 2007
		O'Connell & Goodwin, 2007	Senyigit & Ozdemir, 2011
	Caspari et al., 2004a	Caspari et al., 2004a	Caspari et al., 2004a
	Senyigit & Ozdemir, 2011	Caspari et al., 2004b	Caspari et al., 2004b
DI < CI		Talluto et al., 2008	Leib et al., 2006
		Fallahi et al., 2008	Girona et al., 2010
		Girona et al., 2010	Senyigit & Ozdemir, 2011

Yields

Yield expressed as fruit mass per tree is the result of number of fruit per tree (a measure of crop load) multiplied by the average fruit weight. Hence, yield reductions may be due to fewer and/or smaller fruit. In several cases, PRD did not reduce yield, or at least not consistently and significantly (Table 5). Final fruit size seems to be more markedly affected by irrigation than crop load as it is shown in the few cases where yield was reduced by PRD. Yet, fruit size reductions are predominantly associated with reduced irrigation volumes as indicated by responses of DI trees in PRD trials (Table 5) as well as by RDI studies in apple (Ebel et al. 1993, 1995; Lötter et al. 1985; Mpelasoka et al. 2001). This suggests that in those studies where PRD reduced fruit size, PRD settings may not have been entirely effective and a mere volume effect was generated by the PRD treatment.

Moreover, reductions of crop load under PRD are not necessarily due to wrong thinning or pruning as late fruit drop in response to PRD has been observed in 'Pink Lady' (Talluto et al., 2008). On the contrary, late fruit drop may be a specific regulatory response of trees under PRD, possibly due to changes in hormonal balance (Bangerth, 2000; Masia et al., 1998) specifically mediated by root-generated signals. Alternatively, fruit drop may become an issue when deficit occurs during the early stages of fruit growth, i.e. late ripening varieties like 'Pink Lady' (Talluto et al., 2008) or dry season starting very early (Senyigit and Ozdemir, 2011).

Fruit Quality

Apple fruit quality is generally not affected by PRD irrigation (Table 6). Sporadic increases of flesh firmness and soluble solids in response to PRD or DI have been reported for cultivars grown in dry environments, whereas peel color is never really improved by deficit irrigation. The presence or absence of response may be due, just like for yield parameters, to the phenology of the cultivar and/or period of deficit imposed in the various experiments. Soluble solids are well known as important factors of internal fruit quality, and increased soluble solids are easily related to higher sugars and economic value also in apple. Also higher flesh firmness may be considered a positive feature for handling and storage, especially when, like in this case, it is not attributable to differences in fruit maturation.

It is interesting to notice that DI irrigation does not improve apple fruit quality (especially in terms of soluble solids) any more than PRD; actually only two studies show significant soluble solid increases in response to DI (Table 6). Several factors like light interception, temperature, and nutrient concentrations may affect peel color, but the absence of peel color improvements is generally consistent with no significant reduction of vegetative growth under PRD (Table 4). A couple of studies also showed some advantages of PRD over DI in terms of fruit distribution into more valuable size categories (Leib et al., 2006; Talluto et al., 2008) and the same seems to be the case for 'Gala' and 'Fuji' apples grown in central Sicily (Lo Bianco, unpublished data).

Table 6. Fruit quality responses of apple trees under partial root-zone drying (PRD)

Response	Firmness	Soluble solids	Peel color
	Leib et al., 2006	Talluto et al., 2007	Talluto et al., 2007
PRD = CI	Talluto et al., 2007	Talluto et al., 2008	Talluto et al., 2008
	O'Connell & Goodwin, 2007	Fallahi et al., 2008	Fallahi et al., 2008
	Zegbe et al., 2007	Zegbe et al., 2007	O'Connell & Goodwin, 2007
	Zegbe & Behboudian, 2008	Senyigit & Ozdemir, 2011	Zegbe et al., 2007
	Zegbe & Serna-Pérez, 2011	Zegbe & Behboudian, 2008	Zegbe & Behboudian, 2008
		Zegbe & Serna-Pérez, 2011	
	Leib et al., 2006	Leib et al., 2006	
PRD > CI	Talluto et al., 2008	O'Connell & Goodwin, 2007	
	Senyigit & Ozdemir, 2011	Senyigit & Ozdemir, 2011	
	Leib et al., 2006	Leib et al., 2006	
DI > CI	Senyigit & Ozdemir, 2011	Senyigit & Ozdemir, 2011	

Conclusion

Overall, differences in stomatal control alone, or together with minor reductions of leaf growth, are not sufficient to explain a similar water status in conventionally and PRD irrigated apple trees or a better hydration in PRD than DI trees (placement effect). Alternatively, lower soil moisture due to increased evaporation from soil surface may play an important role in determining more severe leaf water deficit in DI trees compared to PRD trees. This actually converts, at least in part, the expected placement effect primarily based on root-derived hormonal signals into a simple irrigation volume effect, posing some doubts on the real mechanisms underlying PRD.

Regardless of the actual mechanism (if mainly due to adjustments of hormonal balance and/or hydraulic components), the good water status of PRD trees translates into adequate apple yields and fruit quality in association with significant water savings. Actually, fruit quality improvements, especially in terms of soluble solids and flesh firmness, seem to be possible with PRD. Finally, the increased crop value, along with the mentioned water savings, contribute to a positive evaluation of PRD for apple orchards under limited water availability. Generally, a more detailed economic evaluation of indirect savings associated with a reduction of water, energy and pruning costs, as well as an estimate of long-term environmental benefits, is still needed to prove full sustainability of the PRD irrigation strategy for apple orchards.

REFERENCES

Ali, M.H. and Talukder, M.S.U. (2008). Increasing water productivity in crop production - A synthesis. *Agric. Water Manage.* 95: 1201-1213.

Bangerth, F. (2000). Abscission and thinning of young fruit and their regulation by plant hormones and bioregulators. *Plant Growth Regul.* 31: 43-59.

Bravdo, B. and Naor, A. (1996). Effect of water regime on productivity ad quality of fruit and wine. *Acta Hort.* 427: 15-26.

Caspari, H.W., Behboudian, M.H. and Chalmers, D.J. (1994). Water use, growth, and fruit yield of 'Hosui' Asian pears under deficit irrigation. *J. Amer. Soc. Hort. Sci.* 119: 383-388.

Caspari, H.W., Einhorn, T.C., Leib, B.G., Redulla, C.A., Andrews, P.K., Lombardini, L., Auvil, T. and McFerson, J.R. (2004a). Progress in the development of partial rootzone drying of apple trees. *Acta Hort.* 664: 125-132.

Caspari, H.W., Neal, S. and Alspach P. (2004b). Partial rootzone drying. A new deficit irrigation strategy for apple? *Acta Hort.* 646: 93-100.

Chalmers, D.J., Mitchell, P.D. and van Heek, L.A.G. (1981). Control of peach tree growth and productivity by regulated water supply, tree density, and summer pruning. *J. Amer. Soc. Hort. Sci.* 106:3 07-312.

Costa, J.M., Ortuno, M.F. and Chaves, M.M. (2007). Deficit irrigation as a strategy to save water: physiology and potential application to horticulture. *J. Integr. Plant Biol.* 49: 1421-1434.

Davies, W.J., Wilkinson, S. and Loveys, B. (2002). Stomatal control by chemical signalling and the exploitation of this mechanism to increase water use efficiency in agriculture. *New Phytol.* 153: 449-460.

Dodd, I.C. (2009). Rhizosphere manipulations to maximize 'crop per drop' during deficit irrigation. *J. Exp. Bot. 60*: 2454-2459.

Dodd, I.C., Egea, G. and Davies, W.J. (2008a). ABA signalling when soil moisture is heterogeneous: decreased photoperiod sap flow from drying roots limits ABA export to the shoots. *Plant Cell Environ.* 31: 1263-1274.

Dodd, I.C., Egea, G. and Davies, W.J. (2008b). Accounting for sap flow from different parts of the root system improves the prediction of xylem ABA concentration in plants grown with heterogeneous soil moisture. *J. Exp. Bot.* 59: 4083-4093.

Dodd, I.C., Theobald, J.C., Bacon, M.A. and Davies, W.J. (2006). Alternation of wet and dry sides during partial rootzone drying irrigation alters root-to-shoot signalling of abscisic acid. *Funct. Plant Biol.* 33: 1081-1089.

Domingo, R., Ruiz-Sanchez, M.C., Sanchez-Blanco, N.J. and Torrecillas, A. (1996). Water relations, growth and yields of Fino lemon trees under regulated deficit irrigation. *Irr. Sci.* 16: 115-123.

Dry, P.R. and Loveys, B.R. (1998). Factors influencing grapevine vigour and the potential for control with partial rootzone drying. *Aust. J. Grape Wine Res.* 4: 140-148.

Dry, P.R., Loveys, B.R., Botting, D.G. and Düring, H. (1995). Effects of partial rootzone drying on grapevine vigour, yield, composition of fruit and use of water. *Proc. Ninth Aust. Wine Ind. Tech. Conf.* 128-131.

Dry, P.R., Loveys, B.R. and Düring, H. (2000). Partial drying of the rootzone of grape. I. Transient changes in shoot growth and gas exchange. *Vitis* 39: 3-7.

Ebel, R.C., Proebsting, E.L. and Evans, R.G. (1995). Deficit irrigation to control vegetative growth in apple and monitoring fruit growth to schedule irrigation. HortSci. 30: 1229-1232.

Ebel, R.C., Proebsting, E.L. and Patterson, M.E. (1993). Regulated deficit irrigation may alter apple maturity, quality, and storage life. *HortSci.* 28: 141-143.

Einhorn, T. and Caspari, H.W. (2004). Partial rootzone drying and deficit irrigation of 'Gala' apples in a semi-arid climate. *Acta Hort.* 664: 197-204.

Einhorn, T., Green, S. and Caspari, H.W. (2009). Estimation of containerized single-stem and split-rooted, non-fruiting apple tree water use using miniaturized heat pulse probes. *Acta Hort.* 846: 285-291.

Fallahi, E., Fallahi, B. and Shafii, B. (2008). Effect of irrigation systems and rootstocks on water use, tree growth, fruit quality and mineral nutrients in apples during the third and fourth year after planting. *Acta Hort.* 772: 33-39.

Fan, T., Stewart, B.A., Payne, W.A., Wang, Y., Song, S., Luo, J. and Robinson, C.A. (2005). Supplemental irrigation and water: yield relationships for plasticulture crops in the loess plateau of China. *Agr. Jour.* 97: 177-188.

Fereres, E. and Evans, R.G. (2006). Irrigation of fruit trees and vines: an introduction. *Irrig. Sci.* 24: 55-57.

Fereres, E., Goldhamer, D.A. and Parson, L.R. (2003). Irrigation water management for horticultural crops. Historical review compiled for the American Society of Horticultural Science's 100th Anniversary. *HortSci.* 38: 1036-1042.

Fereres, E. and Soriano, M.A. (2007). Deficit irrigation for reducing agricultural water use. *J. Exp. Bot.* 58: 147-159.

Fernández, J.E., Green, S.R., Caspari, H.W., Diaz-Espejo, A. and Cuevas, M.V. (2008). The use of sap flow measurements for scheduling irrigation in olive, apple and Asian pear trees and in grapevines. *Plant Soil* 305: 91-104.

Flore, J.A. and Layne, D.R. (1999). Photoassimilate production and distribution in cherry. *HortSci.* 34: 1015-1019.

Forshey, C.G., Weires, R.W., Stanley, B.H. and Seem, R.C. (1983). Dry weight partitioning of 'McIntosh' apple trees. *J Amer. Soc. Hort. Sci.* 108: 149-154.

Girona, J., Mata, M., Goldhamer, D.A., Johnson, R.S. and DeJong, T.J. (1993). Patterns of soil and tree water status and leaf funcioning during regulated deficit irrigation scheduling in peach. *J. Amer. Soc. Hort. Sci.* 118: 580-587.

Girona, J., Behboudian, M.H., Mata, M., Del Campo, J. and Marsal, J. (2010). Exploring six reduced irrigation options under water shortage for 'Golden Smoothee' apple: Responses of yield components over three years. *Agric. Water Manag.* 98: 370-375.

Goldhamer, D.A. and Beede, R.H. (2004). Regulated deficit irrigation effects on yield, nut quality and water-use efficiency of mature pistachio trees. *J. Hort. Sci. Biotech.* 79: 538-545.

Goldhamer, D.A. and Salinas, M. (2000). Evaluation of regulated deficit irrigation on mature orange trees grown under high evaporative demand. Proc. Intl. Soc. *Citruc. IX Cong.* 227-231.

Goldhamer, D.A., Viveros, M. and Salinas, M. (2006). Regulated deficit irrigation in almonds: effects of variations in applied water and stress timing on yield and yield components. *Irrig. Sci.* 24: 101-114.

González-Altozano, P. and Castel, J.R. (1999). Regulated deficit irrigation in 'Clementina de Nules' citrus trees. I. Yield and fruit quality effects. *J. Hort. Sci. Biotech.* 74: 706-713.

Gowing, D.J.G., Davies, W.J. and Jones, H.G. (1990). A positive root-sourced signal as an indicator of soil drying in apple, *Malus* x *domestica* Borkh. *J. Exp. Bot.* 41: 1535-1540.

Gucci, R. and Minchin, P.E.H. (2002). Translocation of newly-assimilated carbon in the vegetative shoot of olive. *Acta Hort.* 586: 461-463.

Higgs, K.H. and Jones, H.G. (1991). Water relations and cropping of apple cultivars on a dwarfing rootstock in response to imposed drought. *J. Hort. Sci. Biotech.* 66: 367-379.

Hsaio, T.C. (1973). Plant responses to water stress. Ann. Rev. Plant Physiol. 24: 519-570.

Hsiao, T.C. and XU, L.K. (2000). Sensitivity of growth of roots versus leaves to water stress: biophysical analysis and relation to water transport. *J. Exp. Bot.* 51: 1595-1616.

Intrigliolo, D.S. and Castel, J.R. (2004). Continuous measurement of plant and soil water status for irrigation scheduling in plum. *Irrig. Sci.* 23: 93-102.

Irving, D.E. and Drost, J.H. (1987). Effects of water deficit on vegetative growth, fruit growth and fruit quality in Cox's Orange Pippin apple. *J. Hort.* Sci. 62: 427-432.

Johnson, R.S., Handley, D.F. and DeJong, T.M. (1992). Long-term response of early maturing peach trees to postharvest water deficits. *J. Amer. Soc. Hort. Sci.* 117: 881-886.

Johnson, R.S. and Handley, D.F. (2000). Using water stress to control vegetative growth and productivity of temperate fruit trees. *HortSci.* 35: 1048-1050.

Jones, G.H. (2004). Irrigation scheduling: advantages and pitfalls of plant based methods. *J. Exp. Bot.* 44: 2427-2436.

Kijne, J.W., Barker, R. and Molden, D.J. (2003). Water productivity in agriculture: limits and opportunities for improvement. Wallingford, UK: CABI, IWMI.

Kudoyarova, G.R., Vysotskaya, L.B., Cherkozyanova, A. and Dodd, I.C. (2007). Effect of partial rootzone drying on the concentration of zeatin-type cytokinins in tomato (*Solanum lycopersicum* L.) xylem sap and leaves. *J. Exp. Bot.* 58: 161-168.

Leib, B.G., Caspari, H.W., Redulla, C.A., Andrews, P.K. and Jabro, J.J. (2006). Partial rootzone drying and deficit irrigation of 'Fuji' apples in semi-arid climate. *Irrig. Sci.* 24: 85-99.

Lo Bianco, R. and Francaviglia, D. (2012). Comparative responses of 'Gala' and 'Fuji' apple trees to deficit irrigation: Placement versus volume effects. *Plant Soil* DOI 10.1007/s11104-012-1149-z.

Lo Bianco, R., Talluto, G. and Farina, V. (2012). Effects of partial rootzone drying and rootstock vigour on dry matter partitioning of apple trees (*Malus domestica* cvar Pink Lady). *J. Agric. Sci. Cambridge* 150: 75-86.

Lombardini, L., Caspari, H.W., Elfving, D.C., Auvil, T.D. and McFerson, J.R. (2004). Gas exchange and water relations in 'Fuji' apple trees grown under deficit irrigation. *Acta Hort.* 636: 43-50.

Lötter, J. De V., Beukes, D.J. and Weber, H.W. (1985). Growth and quality of apples as affected by different irrigation treatments. *J. Hort. Sci.* 60: 181-192.

Marsal, J., Mata, M., Arbonés, A., Rufat, J. and Girona, J. (2003). Regulated deficit irrigation and rectification of irrigation scheduling in young pear trees: an evaluation based on vegetative and productive response. *Europ. J. Agron*. 17: 111-122.

Masia, A., Ventura, M., Gemma, H. and Sansavini, S. (1998). Effect of some plant growth regulator treatments on apple fruit ripening. *Plant Growth Regul*. 25: 127-134.

McCarthy, M.G., Loveys, B.R., Dry, P.R. and Stoll, M. (2002). Regulated deficit irrigation and partial rootzone drying as irrigation management techniques for grapevines. In: Deficit irrigation practices. FAO Water reports 22: 79-87.

Mills, T.M., Behboudian, M.H., Clothier, B.E. (1996). Water relations, growth, and the composition of 'Braeburn' apple fruit under deficit irrigation. *J. Amer. Soc. Hort. Sci*. 121: 286-291.

Mitchell, P.D. and Chalmers, D.J. (1982). The effect of reduced water supply on peach tree growth and yields. *J. Amer. Soc. Hort. Sci*. 107: 853-856.

Mitchell, P.D., Van Den Ende, B., Jerie, P.H. and Chalmers, D.J. (1989). Responses of 'Bartlett' pear to withholding irrigation, regulated deficit irrigation, and tree spacing. *J. Amer. Soc. Hort. Sci*. 114: 15-19.

Moriana, A., Orgaz, F., Pastor, M. and Fereres, E. (2003). Yield responses of mature olive orchard to water deficits. *J. Amer. Soc. Hort. Sci*. 123: 425-431.

Mpelasoka, B.S., Behboudian, M.H. and Green, S.R. (2001). Water use, yield and fruit quality of lysimeter-grown apple trees, responses to deficit irrigation and to crop load. *Irrig. Sci*. 20: 107-113.

Naor, A., Gal, Y. and Peres, M. (2006). The inherent variability of water stress indicators in apple, nectarine and pear orchards, and the validity of leaf-selection procedure for water potential mesurements. *Irr. Sci*. 24: 129-135.

O'Connell, M.G. and Goodwin, I. (2007). Responses of 'Pink Lady' apple to deficit irrigation and partial rootzone drying: physiology, growth, yield, and fruit quality. *Aust. J. Agric. Res*. 58: 1068-1076.

Ruiz-Sànchez, M.C., Torrecillas, A., Perez-Pastor, A. and Domingo, R. (2000). Regulated deficit irrigation in apricot trees. *Acta Hort*. 537: 759-766.

Sepaskhah, A.R. and Ahmadi, S.H. (2010). A review on partial root-zone drying irrigation. *Int. J. Plant Prod*. 4: 241-258.

Sepulcre-Canto, G., Zarco-Tejada, P.J., Jimenez-Munoz, J.C., Sorbino, J.A., De Miguel, E. and Villalobos, F.J. (2006). Detection of water stress in an olive orchard with thermal remote sensing imagery. *Agr. Forest Meteor*. 136: 31-44.

Senyigit, U. and Ozdemir, F.O. (2011). Effects of partial rootzone drying and conventional deficit irrigation on yield and quality parameters of "Williams Pride" apple cultivar drafted on M9 rootstock. *Sci. Res. Essays* 6: 3776-3783.

Sharp, R.E. and Davies, W.J. (1979). Solute regulation and growth by roots and shoots in water-stressed maize plants. *Planta* 147: 43-49.

Stoll, M., Loveys, B. and Dry, P. (2000). Hormonal changes induced by partial rootzone drying of irrigated grapevine. *J. Exp. Bot*. 51: 1627-1634.

Talluto, G., Farina, V. and Lo Bianco, R. (2007). Growth, fruit yield and quality of 'Golden Delicious' apple trees under fixed partial rootzone drying. *J. Appl. Hortic*. 9: 50-55.

Talluto, G., Farina, V., Volpe, G. and Lo Bianco, R. (2008). Effects of partial rootzone drying and rootstock vigour on growth and fruit quality of 'Pink Lady' apple trees in Mediterranean environments. *Aust. J. Agric. Res*. 59: 785-794.

Wardlaw, I.F. (1990). The control of carbon partitioning in plants. *New Phytol.* 116: 341-381.

Yang, Q., Zhang, F. and Li, F. (2011). Effect of different drip irrigation methods and fertilization on growth, physiology and water use of young apple tree. *Sci. Hort.* 129: 119-126.

Zegbe, J.A. and Behboudian, M.H. (2008). Plant water status, CO_2 assimilation, yield, and fruit quality of 'Pacific Rose^{TM}' apple under partial rootzone drying. *Adv. Hort. Sci.* 22: 27-32.

Zegbe, J.A., Behboudian, M.H., Clothier, B.E. (2007). Reduced irrigation maintains photosynthesis, growth, yield, and fruit quality in 'Pacific Rose^{TM}' apple. *J. Sust. Agric.* 30: 125-136.

Zegbe, J.A., Behboudian, M.H., Clothier, B.E. and Lang, A. (2008). Postharvest performance of cv. 'Pacific Rose^{TM}' apple grown under partial rootzone drying. *HortSci.* 43: 952-954.

Zegbe, J.A. and Serna-Pérez, A. (2011). Partial rootzone drying maintains fruit quality of 'Golden Delicious' apples at harvest and postharvest. *Sci. Hort.* 127: 455-459.

Zegbe, J.A. and Serna-Pérez, A. (2012). Partial rootzone drying to save water while growing apples in a semi-arid region. *Irrig. Drain.* 61: 251-259.

Zegbe, J.A., Serna-Pérez, A. and González-Fuentes, J.A. (2011). Nutrient status of apple leaves not affected by three years of irrigation using partial rootzone drying. *J. Plant Nutr. Soil Sci.* 174: 459-464.

Zwart, S.J., Bastaanssen, W.G.M. (2004). Review of measured crop water productivity values fro irrigated wheat, rice, cotton and maize. *Agric. Water Manag.* 69: 115-133.

In: Irrigation Management, Technologies ...
Editor: M. Hossain Ali

ISBN: 978-1-62417-862-7

Chapter 4

FIELD STUDIES ON THE USE OF SALINE WATER FOR DEFICIT IRRIGATION IN ARID TUNISIA

Kamel Nagaz[1,*], Mohamed Moncef Masmoudi[2] and Netij Ben Mechlia[2]
[1]Institut des Régions Arides, Tunisia
[2]National Institute of Agronomy of Tunisia (INAT), Tunisia

ABSTRACT

In areas with limited water resources, better management strategies of irrigation water is required for farmers seeking a viable means to maximize crop production. Options to improve water productivity and reduce salt input to soils could be sought through the practice of drip irrigation, precise scheduling, and deficit irrigation. Building on available knowledge, participatory investigations have been conducted on deficit irrigation practices with saline waters in the arid areas of Medenine, located in south-eastern Tunisia. Research concerned the combined impacts of cultural practices and irrigation methods on selected crops: potato, carrot, barley and millet. Scheduling consisted in the total and/or partial replacement of accumulated crop evapotranspiration (ETc), as derived from meteorological data and dual crop coefficient. Results give quantitative assessments of the impact of decreasing amounts of irrigation waters on crop yield and soil salinity, over various seasons with waters having a salinity of 3 and 9 dS/m. Drastic improvements in yield obtained under actual farming conditions support the use of the soil water balance strategy for irrigation. Under the very dry conditions of southern Tunisia, optimum supply seems to correspond to a replacement of 100 to 80% of ETc, because of a lower salinity build up in the root zone. Under these treatments, natural salt leaching is facilitated by more humid soil profile. Yield and soil salinity increase almost linearly with decreasing irrigation water amounts for potato, carrot and Millet. Barley showed a very high tolerance to water salinity. High water use efficiencies are obtained with appropriate scheduling and drip irrigation. Moreover, yield improvements associated with conversion to subsurface drip irrigation system has been substantial. Interactive effects of salinity and fertilizers are found to be significant, insinuating that the effect of salinity could be mitigated by increasing nitrogen fertilizer. Future work should focus on

[*] Email address: Nagaz.Kameleddine@ira.rnrt.tn.

the integration of data on water salinity and rainfall for better selection of the cultivar and management of fertilizers.

1. INTRODUCTION

While on a global scale water resources are still ample, serious water shortages are developing in many regions of the world as existing water resources reach full exploitation. Water scarcity is expected to be intensified by the global climate change and with increasing demand for food. The great challenge for the coming decades will be the task of increasing crop production with less water and lower irrigation water quality. This is especially the case of countries which are subject to frequent droughts and to chronic deficiencies in food production.

Limited supplies of quality water for irrigation and increasing demand from other users are forcing farmers to use saline waters. In Tunisia, irrigation of a wide range of crops is expanding around shallow wells having a salinity ranging from 3 to 9 dS/m. Water is typically applied on a routine basis without scheduling and without drainage. Local practices give to vegetables and cereals an important place in the annual crop successions, but productivity is usually low (Nagaz et al., 2009). Current practices carry the danger of a rapid soil degradation. Farmers who are attempting to reduce salt input turn towards crops with short cycle and try to make the cropping cycle coincide with the rainy season. Various species and high economic value crops are planted over contrasting seasons in order to optimize the use of water. Although there is increased awareness about the necessity to reduce soil salinization, traditional irrigation scheduling techniques remain empirical, in terms of timing and irrigation water quantities. The "fixed amount approach" adopted by the majority of farmers leads to water losses during periods with low water needs and water deficits when the needs are important.

Many experiments carried out around the world on water use strategies, indicate that sustainable water and salinity management is possible when water is supplied according to soil characteristics and crop water requirements (Bastiaanssen et al. 1996). Irrigation practices considered by Shalhevet (1984) as important in the management of saline waters are: irrigation scheduling, deficit irrigation, irrigation method and management of multi-source irrigation water of different qualities.

The term irrigation scheduling includes the crop irrigation requirements using the appropriate irrigation interval. Information on consumptive use of many crops is available for irrigation with non-saline water. The question arises whether this information is also applicable for saline water irrigation. The answer to this question is of importance. Irrigation scheduling is complicated under saline conditions, mainly due to the fact that information of consumptive use of many crops under saline conditions is not available and the leaching fraction must be calculated and included in the crop water requirements.

Deficit irrigation (DI) provides means of reducing water consumption while minimizing adverse effects on yield. DI is one way of maximizing water use efficiency (WUE) for higher yields per unit of irrigation water applied (English and Nakamura, 1989; English and Raja, 1996; Mugabe and Nyakatawa, 2000; Ghinassi and Trucchi, 2001; Kirda, 2002; Mao et al., 2003; Panda et al., 2003; Zhang et al., 2004). With this method, the crop is exposed to a certain level of stress either during a particular period or throughout the whole growing season. In water quality-limiting situations, the expectation is that any yield reduction, will be compensated by reduced degradation of the soil and water resources.

Irrigation methods are becoming increasingly important for efficient use of water for crop production in regions with limited water resources. Reports by Ayers et al. (1986), Saggu and Kaushal (1991), Goldberg and Shamueli (1970), Bernstein and Francois (1973) and Fereres et al. (1985) show that saline water can be efficiently used through drip irrigation even on saline soils. It results in considerable saving in irrigation water (Yohannes and Tadesse, 1998; Cetin and Bilgel, 2002) thus reducing the risks of salinization. More frequent and localized water applications to only part of the crop's potential root zone reduces adverse effects of over-irrigation, thus improving crop yield and quality of production. However the potential of drip irrigation are not yet fully known by small farmers in arid regions.

Water management strategies based on the use of multi-source irrigation water of different qualities could further enhance the feasibility of using saline waters for irrigation. Many studies indicate that waters previously thought unsuitable for irrigation can often be used successfully to grow crops without hazardous long-term consequences to crops or soils. One approach blends saline water with good quality water to an acceptable salinity as characterized by electrical conductivity (EC) and then uses this water to irrigate crops (Rains et al., 1987; Roltson et al., 1988). Choice of EC is based upon the salt tolerance of the specific crop and availability of the water supplies. A second method uses a cyclic strategy for waters of various salinities separately, either during one season or in a crop rotation, and also in alternate irrigation as a function of crop salt tolerance (Rhoades 1984, 1987).

Under severe water constraints, good irrigation management practices are required in order to improve water use, particularly in the absence of sufficient rainfall events used elsewhere for natural leaching. Irrigation management should take into consideration the effect of irrigation on the crop yield and at the same time on the environment, particularly the risk of soil salinization. This work is reporting on experiments, carried out in southern Tunisia, that aimed to derive irrigation strategies that save water, reduce salt input and improve crop productivity. All field works were conducted with full farmers' participation. It concerned crops having different tolerance to salinity: potato, carrot, barley and millet. Most of the investigations reported on the effect of different irrigation water supplies on yield, water productivity and soil salinization. Used layouts normally include distinct water treatments with decreasing amounts in relation to maximum crop evapotranspiration (ETc):

- Full irrigation (I100) is the treatment that receives 100 % of accumulated ETc.
- Deficit treatments were irrigated at the same frequency as the control, but with lower quantities for instance equal to 80, 60 and 40% of accumulated ETc (I80, I60 and I40).
- Other treatments with regulated deficit amounts were used, as specified in the text.

The readily available water (RAW) represents normally 35-60 % of total available water (TAW), in the root zone. ETc was estimated for daily time step by using reference evapotranspiration (ETo) combined with dual crop coefficients, following the FAO-56 method given in Allen et al. (1998). Weather data used in this experiment were collected daily from a nearby meteorological station located at Institut des Regions Arides (IRA-Medenine). For irrigation scheduling, the water balance method was used, by means of a spreadsheet program for Excel. The program calculates the soil water depletion on daily basis and projects the next irrigation event based on the target depletion. The soil depth of the effective root zone is increased with the irrigation scheduling program from a minimum depth of about 0.15 m at planting to a maximum of 0.60-0.80 m in direct proportion to the increase in the crop coefficient. Once the maximum root depth is reached, it is held constant.

2. Deficit Irrigation of the Potato Crop

The Potato crop (Solanum tuberosum L.), although considered relatively susceptible to salinity and normally not suited for stressful conditions, is becoming a widely used crop in the arid region of Tunisia. Its cultivation, as a cash crop, has been gaining popularity because temperature and radiation conditions allow for cropping over the spring, fall, and winter seasons. However, inadequate water management is an important limiting factor to yield, in spite of a widespread adoption of drip irrigation. Farmers generally have lack of knowledge on aspects of soil-water-plant relationships and apply water to the crops regardless of plant's need. They tend to relate irrigation occurrences to days after planting with fixed intervals and water amounts rather than to crop growth stages progress. In areas irrigated with well waters, accurate scheduling is manageable and requested by farmers to achieve tangible productivity improvements. Based on farmers' interviews it was hypothesized that simplified irrigation scheduling methods could have an important impact on improving crop productivity in the region.

Irrigation-scheduling experiments were carried out during the years 2000-2004 in commercial farms to determine the effect of irrigation regimes with saline water (3.2 dS/m) on soil salinity, yield and water use efficiency of Potato over three distinct growing periods: spring, autumn and winter seasons. The cultivar Spunta was grown on a sandy soil, following standard cultural practices as reported by Nagaz et al. (2004). The rainfall received during the three cropping seasons was respectively 26, 59 and 72 mm. Irrigation water, obtained from a well, was applied by means of polyethylene dripper lines with an emission rate of 4 l/h and passed through a water meter, gate valve, before passing through laterals placed in every potato row. The layout included five distinct water treatments with decreasing amounts in relation to maximum crop evapotranspiration (ETc). Full irrigation (I100) is the treatment that received 100 % of accumulated ETc. Deficit treatments were irrigated at the same frequency as the control, but with quantities equal to 80, 60 and 40% of accumulated ETc (I80, I60 and I40). The fifth treatment (I100-D) was irrigated daily with amounts equal to 100% of the previous day's water consumption. The readily available water RAW represents 35 % of total available water (TAW), in the root zone.

The soil depth of the effective root zone is increased with the irrigation scheduling program from a minimum depth of 0.15 m at planting to a maximum of 0.60 m in direct proportion to the increase in the crop coefficient (Figure 1).

Results concerning salt distribution in the soil profile indicate that the most irrigated treatments (I100-D and I100) decreased the soil salinity beneath the emitter as the zone of salt accumulation moved from the emitter (Figure 2). Salts were concentrated midway between the emitters and towards the wetting front.

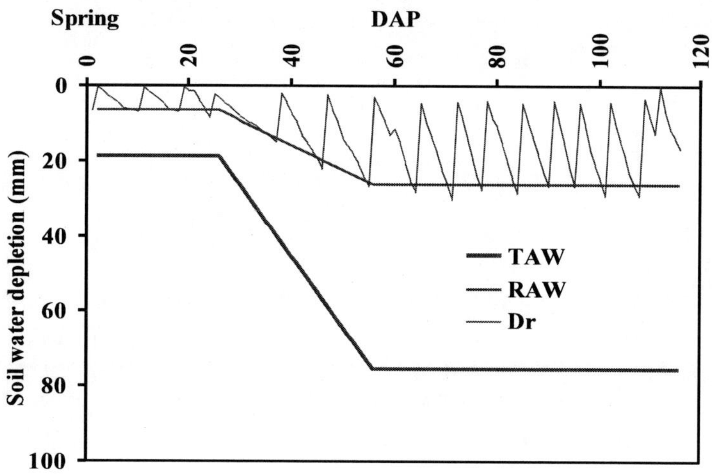

Figure 1. Daily soil water depletion for potato under full irrigation treatment (I100) during the spring cropping season (after Nagaz et al., 2007).

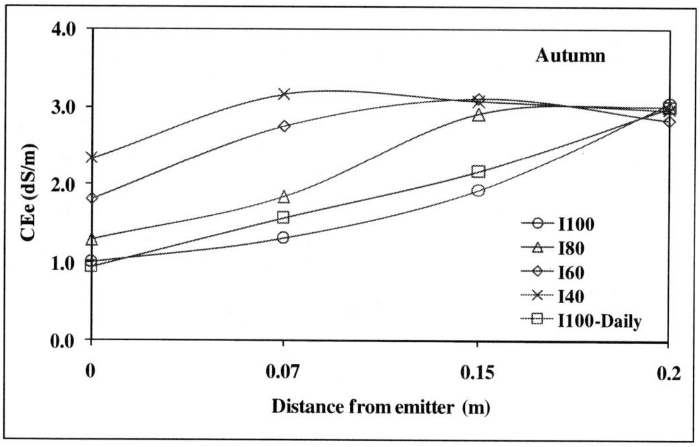

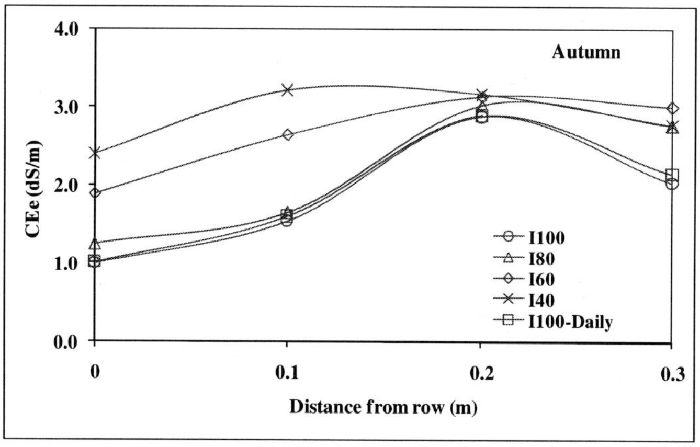

Figure 2. (Continued)

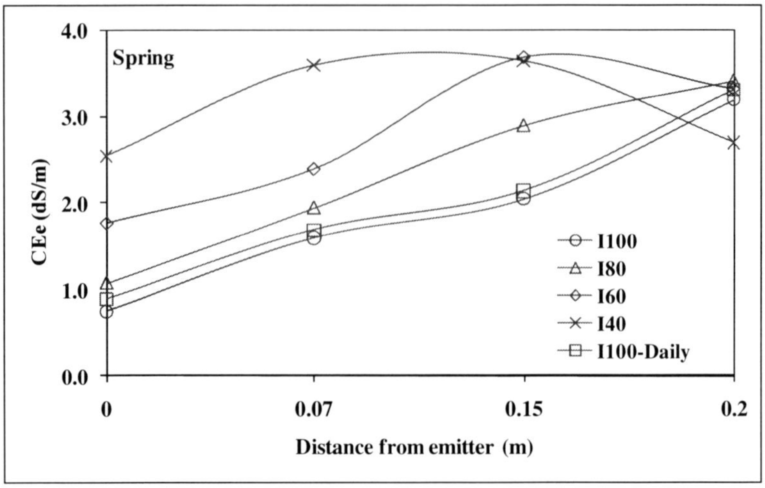
Spring
CEe (dS/m)
4.0
3.0
2.0
1.0
0.0
0
0.07
0.15
0.2
Distance from emitter (m)
I100
I80
I60
I40
I100-Daily

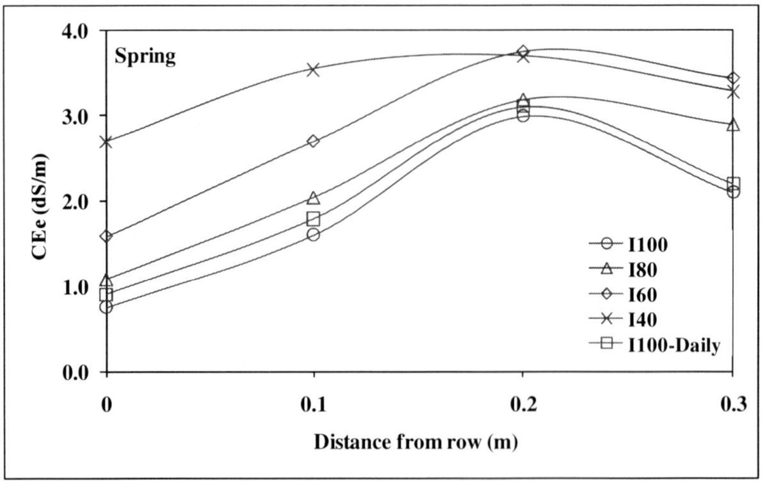
Spring
CEe (dS/m)
4.0
3.0
2.0
1.0
0.0
0
0.1
0.2
0.3
Distance from row (m)
I100
I80
I60
I40
I100-Daily

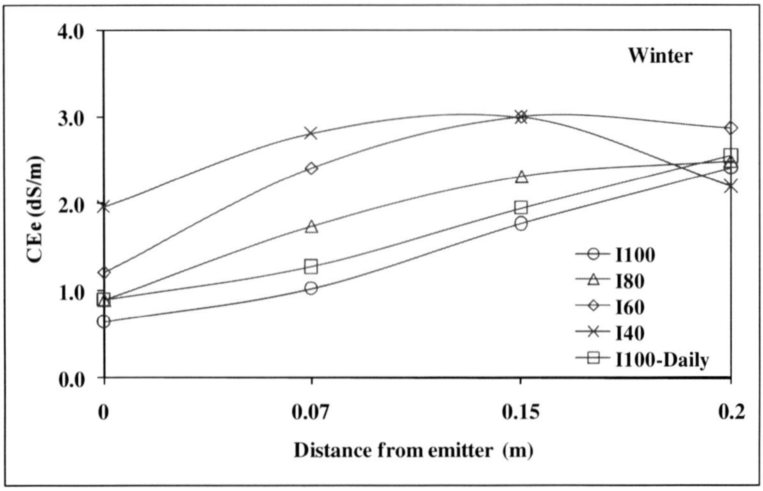
Winter
CEe (dS/m)
4.0
3.0
2.0
1.0
0.0
0
0.07
0.15
0.2
Distance from emitter (m)
I100
I80
I60
I40
I100-Daily

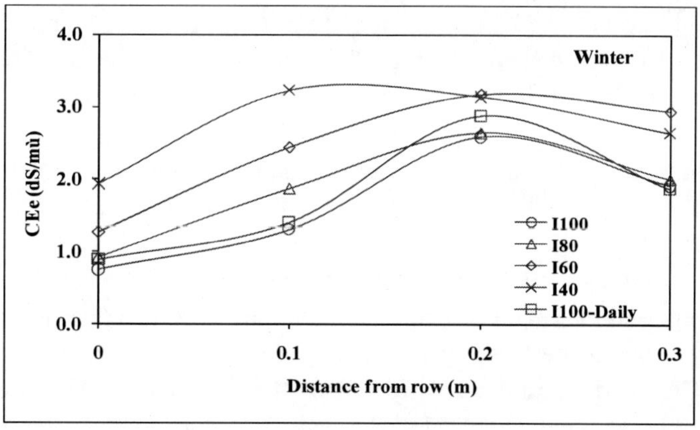

Figure 2. Soil salinity (ECe, dS/m) under different irrigation treatments along the row and across row (after Nagaz et al., 2007).

Relatively high values of soil salinity were observed beneath the emitter for I40 and I60 treatments; whereas, the highest soil salinity occurred at a distance of 7 and 15 cm from the emitter, and 10 and 20 cm from the drip line.

ECe data, presented in Figure 3, show the importance of the cropping season to benefit from the leaching capacity of rainfall. For the spring period, a clear increase in the soil salinity was observed, from 1.3 dS/m at plantation in February to approximately 2 dS/m after harvest in May.

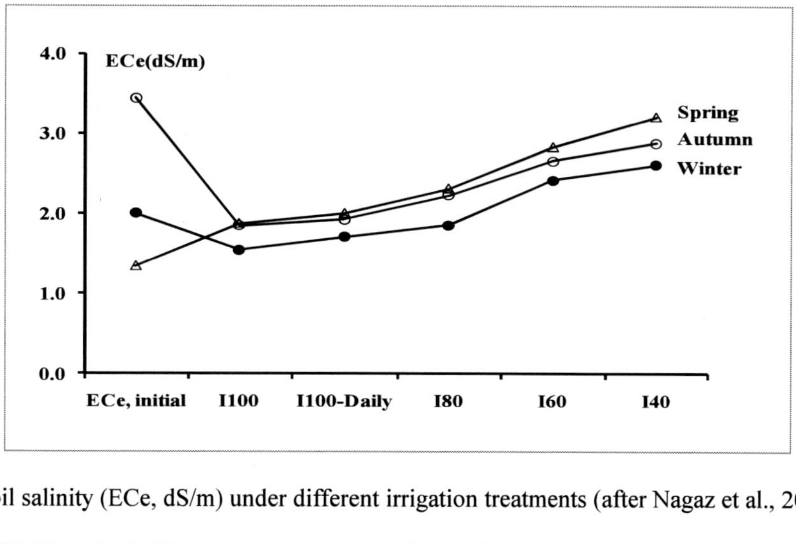

Figure 3. Soil salinity (ECe, dS/m) under different irrigation treatments (after Nagaz et al., 2004).

Table 1. Yield and yield components under full irrigation treatments (I100 and 100-D) obtained in three different growing seasons (after Nagaz et al., 2007)

	Spring		Autumn		Winter	
	I100	I100-D	I100	I100-D	I100	I100-D
Yield (t ha^{-1})	39.6	36.5	30.4	28.9	22.7	20.5
Tubers m^{-2}	36.0	34.0	32.2	29.7	25.0	23.5
Tuber weight (g)	110.3	107.2	100.6	97.2	90.6	87.3

For the winter crop, which takes place from November to February, ECe remains approximately at about 1.5 dS/m. However, for the autumn period, the ECe falls from 3.4 dS/m in September to about 2 dS/m at harvest in January. The application of deficit irrigation with reductions of 20, 40 and 60% of ETc increased salt accumulation in the root zone over the three cropping periods, although to a lesser degree for autumn and winter seasons than for spring (Figure 3). The ECe values under the different irrigation treatments were lower than the ECi of the irrigation water used, due to the leaching of soluble salts with the received rainfall. Singh and Bhumbla (1968) observed that the extent of salt accumulation depended on soil texture and reported that in soils containing less than 10% clay the ECe values remained lower than ECiw. Among the three seasons, the maximum tuber yield was obtained in the I100 treatment. The yields in the 100-D treatment were not statistically different from those obtained with I100 treatment (Table 1). However a higher salinity in the rooting zone associated with deficit irrigation caused important reductions in tuber yield. Figure 4 gives a quantitative indication of the yield loss caused by the decrease in water applications. For the spring period, a clear increase in the soil salinity was observed, from 1.3 dS/m at plantation in February to approximately 2 dS/m after harvest in May. For the winter crop, the reduction in tuber yield is mainly attributed to reduction in tubers number and weight as a consequence of water supply shortage during tubers initiation and development. These results, obtained under actual farming conditions, support the practicality of using the appropriate scheduling methodology to implement an efficient use of saline water for potato production.

Under severe water shortage, irrigation could be reduced voluntarily, with known impact on soil salinization within the root zone. A deficit of 20% seems to constitute a good compromise to maintain low ECe values with limited impact on yield.

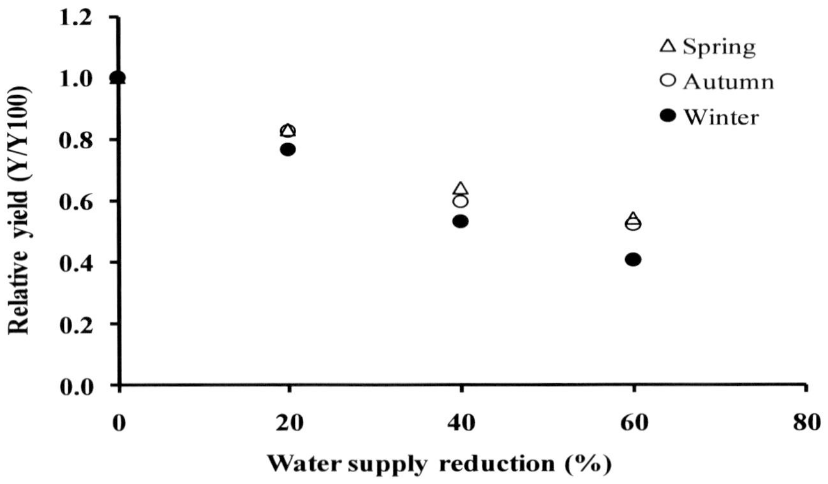

Figure 4. Relative yield (Y/Y100) decrease in relation to water supply deficit for different production seasons in Southern Tunisia (after Nagaz et al., 2007).

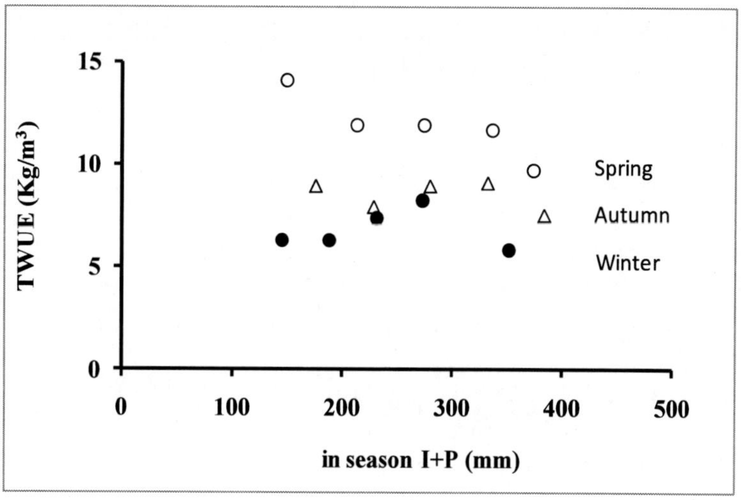

Figure 5. In season water supply (mm) from planting to harvest and water use efficiency (WUE, (kg/m^3) for different irrigation treatments and cropping seasons (redrawn from Nagaz et al., 2007). An irrigation of 75mm supplied just before planting is not included in the totals.

Water use efficiency (WUE) reflected yield differences between seasons (Figure 5). WUE varied typically around 8-9, 6-8, and 11-14 kg/m^3 respectively for autumn, winter and spring. These values are comparable with those obtained in other field studies (Fabeiro et al., 2001, Ferreira et al., 1999, Singh et al., 1977) and were affected by irrigation treatments. In the autumn season, the WUE for fresh tuber yield was lowest under I100-Daily treatment and highest under I100. The WUE with I100 treatment was not considerably different from those obtained with I80 and I40 treatments but different from those obtained with I100-Daily and I60 treatments.

In the spring season, WUE with I40 was significantly higher than those obtained with I100, I80, I60 and I100-Daily treatments. The difference was also significant between 100-Daily treatment and the I100, I80 and I60 treatments. For winter season, the WUE was the lowest under I100-Daily and highest under I100 treatment. The WUEs with I100 were not significantly different from those obtained with I80, however the values obtained with I100-Daily, I60 and I40 were considerably. The relatively high yields and water use efficiencies noted in I80 treatments in three seasons indicate an acceptable response of potato to mild water deficit. Full irrigation with daily application resulted in the lowest WUE values, most likely because of higher evaporative losses.

From an economical point of view, the 100 % of ETc irrigation treatments (I100-D andI100) produced the highest fresh tuber yields for the three seasons. Treatment I80 gave also a good yield. Deficit irrigation treatments gave lower yield, due to a large reduction in tubers number and weight. Lower yields and reduction in marketable tuber size (tuber weight) under I40 and I60 cannot favor the use of these strategies for scheduling. I100 appears to be a good irrigation strategy. In case of situations where water supply is limited, irrigation of potato could be scheduled using 80% of ETc.

3. Combined Effects of Salinity and Nitrogen on Potato Production

In addition to soil and crop appropriate practices, fertilizer management has been emphasized by many researchers to be a practical approach in alleviating the effects of salinity on agricultural crops (Bernstein et al., 1974; Feigin, 1985). The performances of different potato cultivars under different levels of N have been reported by Goffart et al. (2002), Feibert et al. (1998), Ünlü et al. (2006), Halitligil et al. (2002) and Waddel et al. (1999). These studies brought out the positive role of nitrogen fertilizer for growth, yield and quality of potato.

However the response of crop plants to irrigation with different water qualities and nitrogen fertilizer hasn't been the subject of numerous studies. Hamdi and Samti (1995) noted that fertilized potato irrigated with saline water (6 dS/m), produced higher yield than that obtained with fresh water (0.8 dS/m) but without nitrogen fertilizer.

Sandy soils used for potato cultivation are very low in nitrogen, and little information is available concerning optimum doses for deficit conditions. To fill the data gap, a two-year study was carried out during spring season to determine a suitable level of nitrogen for potato cultivation under the arid conditions of Tunisia (Nagaz et al., 2011). Treatments, consisting of two irrigation water qualities: canal (3.2 dS/m) and well (5.2 dS/m) and four levels of nitrogen as ammonium nitrate (0, 100, 200 and 300 kg/ha) were used for both years.

The level of available nitrogen in the soil was very low (less than 0.007%) and thus was not taken into consideration while calculating the fertilizer amounts for treatments. Before planting, soil was spread with 17 t/ha of organic manure.

The P and K fertilizers were applied as basal dose before planting at rates of 300 and 200 kg/ha, respectively. Nitrogen was divided and delivered with the irrigation water during early vegetative growth. Irrigations were scheduled to meet crop water requirements for drip-irrigated potato based on previous study at the experimental area by Nagaz et al. (2007). The rainfall received during the experimental period was taken into account in calculating the irrigation water applied by subtracting it from the total amount required for that particular day. In order to differentiate between the treatments, the irrigation water amount was kept the same in all plots. Results show that the use of saline water increased the ECe.

However, average ECe were respectively lower than EC values of irrigation waters. Salinization was maintained relatively low due to the leaching of soluble salts with the received rainfall (35 and 62 mm) and also due to the low initial soil salinity (1.9 and 1.5 dS/m) (Figure 6).

For both experiments potato production was significantly affected by water quality. It decreased when water salinity changed from 3.2 to 5.2 dS/m (Figure 7). The crop was found to be responsive to increasing nitrogen amounts. Differences in tubers yield were highly significant between the various nitrogen rates except for 100 and 200 kg N/ha doses. For both water qualities, maximum fresh tubers yield was obtained with 300 kg N/ha and decreased as nitrogen supply is reduced.

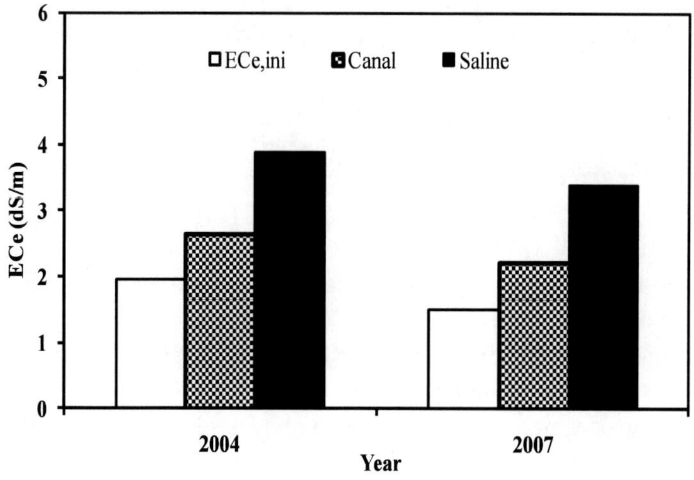

Figure 6. Mean soil salinity values under irrigation with two water qualities (after Nagaz et al., 2011).

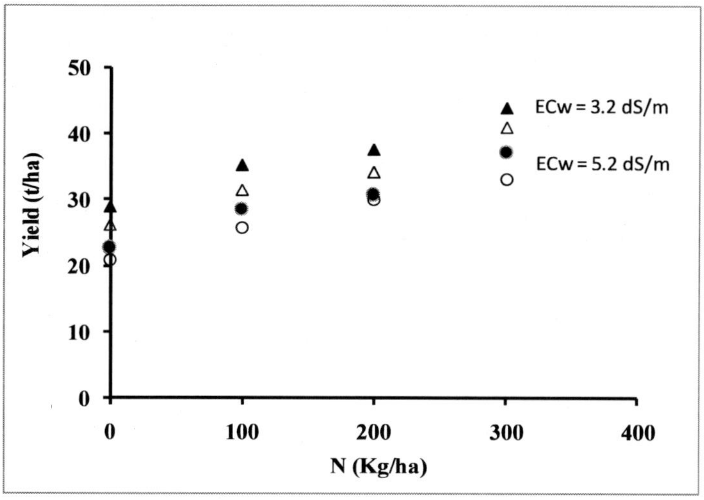

Figure 7. Effect of water quality and nitrogen interaction on potato fresh tubers yield (t/ha) (after Nagaz et al., 2011).

Due to improved growth, yield increased with N application under both water qualities, however the response to nitrogen is more pronounced under irrigation with canal water than with saline water. There are also systematic differences between results obtained during the two experimental years. Yields were higher in 2007 because of the lower initial soil salinity.

The reduction in fresh tubers yields because of higher salinity is almost the same in 2004 and 2007 and corresponded respectively to 19 and 18 % decrease. For both years, the effect of increasing salinity was counteracted by the nitrogen fertilizer. In 2004, the tubers yield obtained with canal water and 0 kg N/ha (26.1 t/ha) was the same as the yield obtained with saline water and 100 kg N/ha. Yield attained 30.0 t/ha when the nitrogen level increased to 200 kg/ha under irrigation with saline water. That yield is not significantly different from the one obtained with canal water and 100 kg N/ha. Yield produced with saline water and 300 kg N/ha (33.1 t/ha) is similar to the yield obtained with canal water and 200 kg N/ha (34.2 t/ha).

Similar trends are observed for the year 2007, where the highest value (44.1 t/ha) was obtained with canal water and 300 kg N/ha and the lowest value with saline water and 0 kg N/ha (22.7 t/ ha). It could be also noted for this second year that yield obtained under irrigation with saline water and 300 kg N/ha (37.3 t/ha) was similar to that obtained with canal water and 200 kg N/ha (37.7 t/ha). Thus, the reduction in yield due to increasing irrigation water salinity could be compensated by increasing nitrogen application. The findings of Hamdy and Samti (1995), and Al-Tahir et al. (1997) confirm these results.

Water use efficiency (WUE) calculations refer to ratios between fresh tubers yield and in-season irrigation and total water supply. Considered irrigation quantities are 315 and 321mm and rainfall 35 and 62mm respectively for 2004 and 2007 and didn't 75mm of water supplied just before planting. Estimated efficiencies for irrigation (IWUE) and total water (TWUE) increased in response to nitrogen supply due to a considerable increase in yields.

Interaction effects of water quality x nitrogen showed that significantly higher WUE for fresh tubers production were observed with nitrogen rate of 300 kg/ha under both canal and saline waters (Figure 8). WUE decreased significantly as the rate of nitrogen fertilizer decreased and for each nitrogen level, it decreased as the salinity of irrigation water increased. However, the values of WUE, based on fresh tubers yield, with canal water irrigation and nitrogen rates of 0 kg/ha were considerably lower than the WUE obtained with saline water and nitrogen levels of 200 and 300 kg/ha. It could be also noticed that the WUE values obtained with nitrogen rates of 100, 200 and 300 kg/ha under irrigation with saline water were not significantly different from those obtained with canal water and nitrogen rates of 0, 100 and 200 kg/ha.

Use of irrigation water of higher salinity increased ECe and caused further reductions in fresh tubers yield of potato. However, the effect of increasing levels of nitrogen fertilizer was substantial. The reduction in potato yields due to increasing irrigation water salinity is in a way compensated by increasing nitrogen rate.

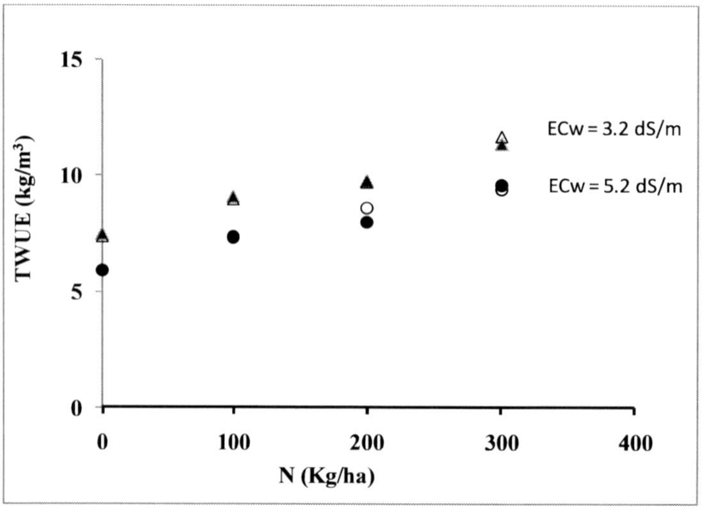

Figure 8. IWUE and TWUE for fresh tubers yield under different water qualities and nitrogen applications (after Nagaz et al., 2011).

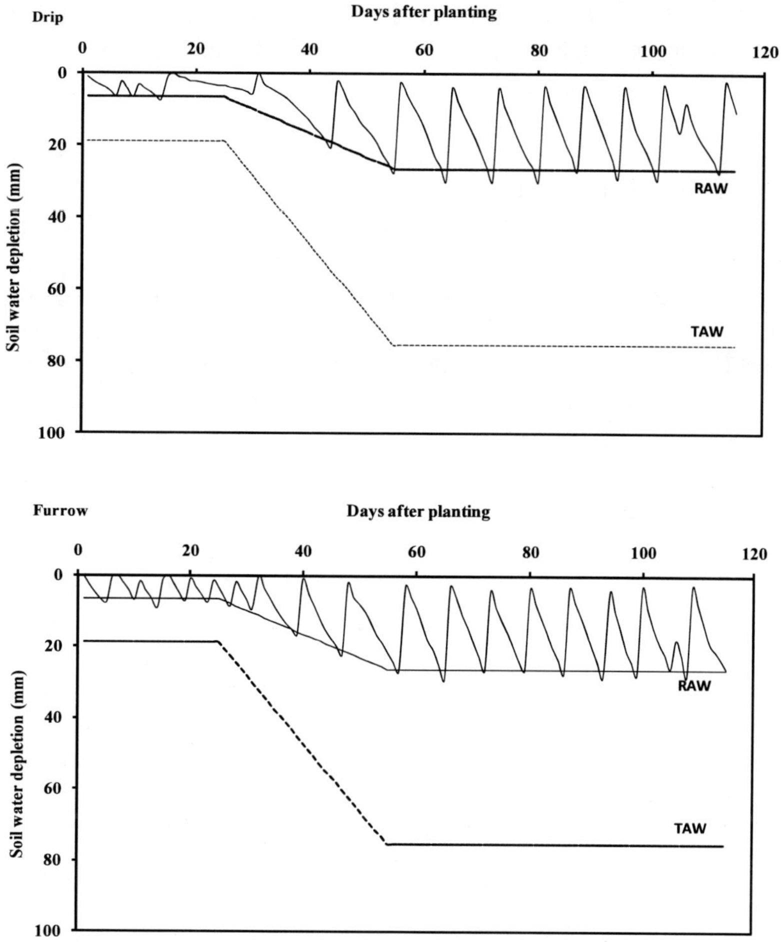

Figure 9. Daily soil water depletion for potato under drip and furrow irrigation methods during the cropping season (after Nagaz et al., 2008).

This observation is consistent with the experimental results of Hamdy and Samti (1995) and Al-Tahir et al. (1997). Significant interactions observed between water quality and nitrogen suggest that fertilizer amounts needed by the crop for optimum growth are not constant. They should be, in each case, increased or reduced according to the salinity of irrigation water.

4. Use of Drip and Furrow Irrigation Systems

Furrow irrigation, used with low amounts on sandy textured soils, is difficult to manage since the infiltration rate of these soils exceeds 10 cm/h. The conversion of this technique towards the localized irrigation must be among the strategic orientations; considering its efficiency and capacity of water saving in particular for vegetable crops. An evaluation of on-farm drip irrigation system was conducted to determine crop response and water savings as compared to conventional techniques under deficit irrigation strategies (Nagaz et al., 2008).

The field experiment was conducted on a sandy soil during spring of 2005 for evaluating the effects of drip and furrow irrigation methods on soil salinity, yield and water use efficiency of potato. For both irrigation methods, irrigations were scheduled when soil water content in the root zone was depleted by the crop to 35% of TAW, and plants in these treatments received 100% of accumulated crop ET (Figure 9). Water with an ECi of 3.2 dS/m was used for irrigation. For drip irrigation system, each dripper had a 4 l/h flow rate. Water for each drip-irrigated treatment passed through a water meter, gate valve, before passing through laterals placed in every potato row. Irrigation water quantity applied through drip irrigation was measured by water meter. For furrow irrigation, plots were made to small basins, which was furrowed and each furrow was fed individually. Measured amounts of water were delivered to the furrows using a hosepipe and water meters. Results show that higher soil salinity was developed in case of furrow irrigation compared to drip irrigation (Figure 10). The reason may be evaporation of ridge soil during the periods between the respective irrigations. Since irrigation was applied through furrows, little leaching of the ridge soil is expected.

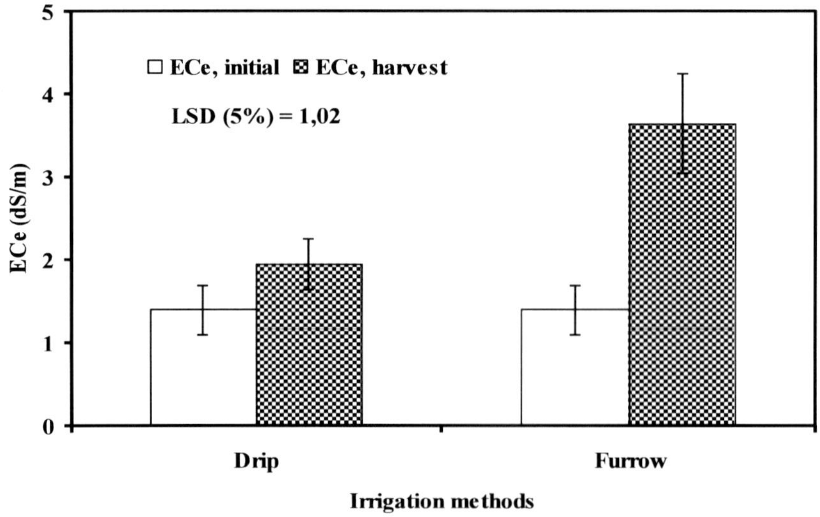

Figure 10. Soil salinity (ECe, dS/m) under furrow and drip irrigation methods (after Nagaz et al., 2008).

Table 2. Yield, water supply and water use efficiency (WUE) for drip and furrow irrigated potato (after Nagaz et al., 2008)

Yield components	Drip	Furrow	LSD (5%)
Fresh tubers yield (t/ha)	36.4	32.6	3.4
in-season irrigation (I, mm)*	326	412	-
Precipitation (P, mm)	42	42	-
Irrigation WUE (kg/m^3)	11.1	7.9	1.5
Total WUE (kg/m^3)	9.9	7.1	1.0

* An irrigation of 75mm supplied just before planting is not included in these totals.

However, the application of water in the vicinity of roots through drippers provides better soil moisture in the root zone and thus lowering the salinity under drip irrigation. Low values of ECe under the prevailing climatic conditions were due to the leaching of soluble salts with the received rainfall (42 mm). Higher potato yield and water use efficiency were obtained with drip irrigation than with furrow irrigation. As expected, the drip-irrigated treatment required less water than the furrow-irrigated treatment and resulted in saving 21% of the irrigation water. As the salinity increased, there was an important reduction in crop yield and WUE under furrow irrigation. The tuber yield was increased by 11% and WUE by 29% with drip irrigation compared to furrow technique, following the same irrigation scheduling rule for both (Table 2). This is obviously due to higher yield accompanied by saving of irrigation water with the drip method.

5. Surface Drip vs Subsurface Drip Irrigation Methods

The continuous use of saline water with conventional on-surface drip irrigation (SDI) might result in salt accumulation close to soil surface (DeMalach and Pasternak, 1993; Oron et al., 1995) due to increased evaporation. However, salt accumulation in the root zone under conventional on-surface drip irrigation can be partially avoided by using subsurface drip irrigation (SSDI) (Oron et al., 1990, 1991, 1992; Phene 1993). Previous research shows that crop marketable yields and quality and water use efficiency have been improved through the use of SSDI (Phene et al., 1991; Camp, 1998; Ayars et al., 2001; Alexiou et al., 2003; Hanson and May, 2004; Patel and Rajput, 2008; Patel and Pandey, 2008; Hassanli et al., 2009). SSDI, while maintaining yields and improving water use efficiency compared to surface irrigation, can be used successfully in commercial fields.

This work is concerned with the use of on-farm SDI and SSDI for deficit irrigation with saline waters. The field experiments were conducted on a sandy soil during spring of 2009 and autumn of 2010 in Medenine for evaluating the effects of these two drip irrigation methods and three irrigation regimes on soil moisture, soil salinity, yield, and water use efficiency of potato (Nagaz et al., 2012). Irrigation regimes consisted in replacement of cumulated ETc when readily available water is depleted with levels of 100% (I100), 60% (I60) and 30% (I30). Well water with an ECi of 7.0 dS/m was used for irrigation. The subsurface drip lines with 70 cm spacing were buried at a depth of 15 cm below the soil surface (Camp, 1998) in each row. The surface drip lines with 70 cm spacing were laid on the soil surface beside the plant rows. The drip laterals for both the SSDI and SDI systems were 16 mm diameter polyethylene pipes with in-line emitters 40 cm apart. The drippers had 4 l/h flow rate in both methods. A drip line for each plant row and an emitter for each potato plant were used in the experiment. Fertilizers were supplied for the cropping seasons in the same amounts (Nagaz et al., 2007). For both irrigation methods, irrigations were scheduled when soil water content in the root zone was depleted by the crop at 35% of TAW.

Results show that average soil moisture values were directly related to the amount of water applied at full or deficit-irrigation treatments and irrigation methods (Figure 11). Higher moisture content in the soil profile initially in all the treatments was due to the irrigation amount applied before planting to replenish the soil profile to field capacity. For all irrigation treatments, significant differences were observed between the soil moisture content of the subsurface irrigated plots and those irrigated with the SDI system during the development, mid-season and harvest periods. The reduction of the soil moisture content with regard to the field capacity (18%) was more important in the case of SDI than that of SSDI. This is dependent on reduced evaporation from soil surface by setting drip line under soil surface. With the SSDI method the surface soil layer is not completely wetted as in the case of SDI. Therefore, with SSDI the first 10-15 cm below the soil surface remains relatively dry reducing thus the direct soil evaporation as compared to surface drip irrigation. Phene et al. (1983) showed that with the SSDI the upper soil layer has a lower humidity resulting in a reduction of the direct soil evaporation. Soil moisture content under I100 was significantly higher than that of deficit treatments for both irrigation systems ($p<0.05$) (Figure 11). In both cropping seasons, the fluctuation in moisture content in the soil profile for SDI and SSDI was in the order I100 > I60 > I30, which can be attributed to irrigation amounts. Soil salinity exhibits trends similar to that of the soil moisture content (Figure 12). The ECe values are lower in case of SSDI than SDI for all irrigation treatments. The difference observed between irrigation methods is mainly due to the difference in soil moisture content since the irrigation water supplies were similar. According to the soil moisture content values for each irrigation method, we can conclude that the SSDI keeps higher soil water content in the root zone which may help to maintain a continuous leaching of accumulated salts and thus reduce the soil salinity. Oron et al. (2002) reported that high moisture content in the root zone with SSDI could increase the leaching process of accumulated salts; whereas the conventional SDI facilitated sufficient leaching just below the emitter in the top soil layer, contributing to extra accumulation of salts in the active root zone of the crop and the soil salinity level remained high under the SDI system. For both irrigation methods, I100 resulted in a significantly lower ECe values than with deficit irrigation treatments (Figure 12). Higher soil salinity levels observed for I60 and I30 is attributed to the absence of substantial leaching under deficit irrigation conditions. Schoups et al. (2005), Kaman et al. (2006) and Geerts et al. (2008) reported that one consequence of reducing irrigation water use by deficit irrigation is the greater risk of increased soil salinity due to reduced leaching. During the fall period, a decrease in ECe values measured at development, mid-season and harvest is observed under all irrigation treatments and methods, compared to initial soil salinity (5.9 dS/m). The decrease of ECe is attributed to the leaching of soluble salts by the 61 mm of rainfall (Figure 13) and concomitant decrease in evaporation in autumn. However, in the spring period, ECe values during development, mid-season and harvest were higher than initial ECe (2.7 dS/m) for all irrigation treatments and methods despite spring rains (46 mm). This increase can be explained by the sampling date which corresponded to the spring season, a period of high evaporation.

ECe values under the different irrigation treatments for both SDI and SSDI systems were lower than the EC of irrigation water used (7 dS/m), due to the natural leaching of soluble salts by rainfall. Thus, under actual farming conditions, the use of high saline waters for irrigation of short-cycle crop during the rainy season seems to have relatively moderate impact on soil salinization as salts added by irrigation are removed from the root zone by natural leaching.

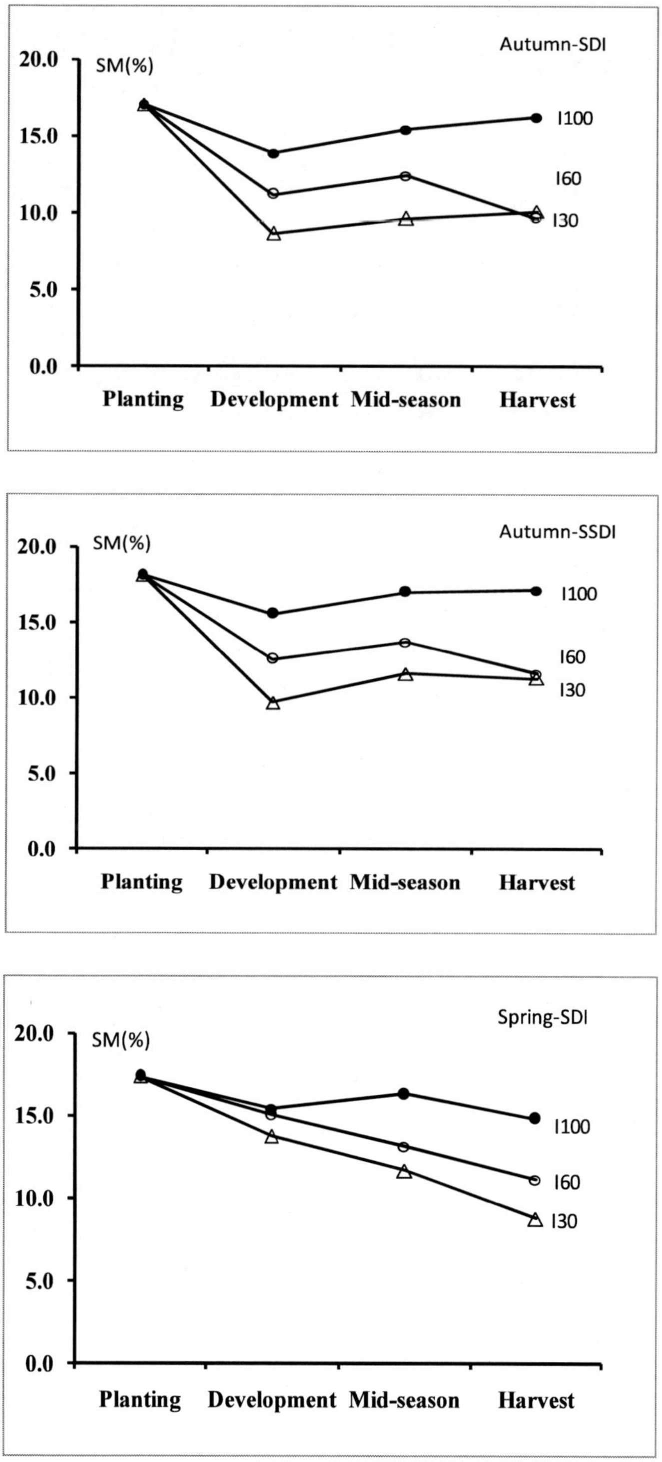

Figure 11. (Continued)

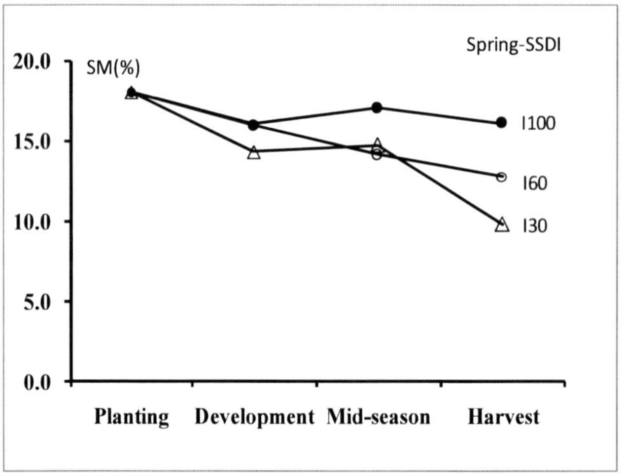

Figure 11. Soil moisture content (SM %) in the root zone of potato under different irrigation treatments (I100, I60 and I30) for surface (SDI) and subsurface drip irrigation (SSDI) methods (redrawn from Nagaz et al., 2012).

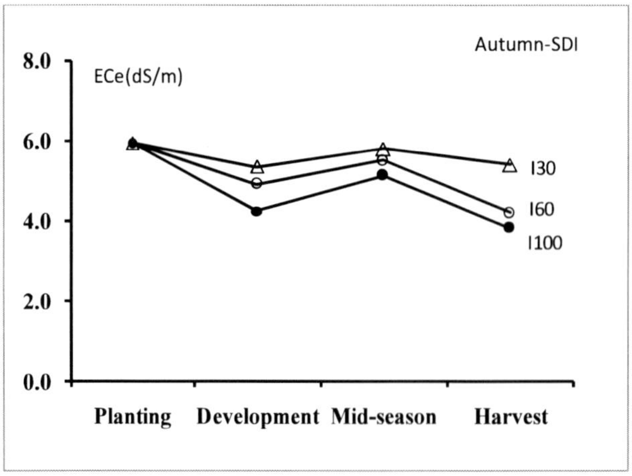

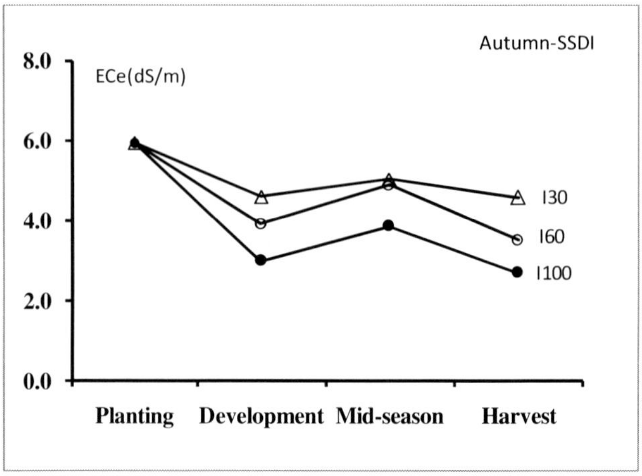

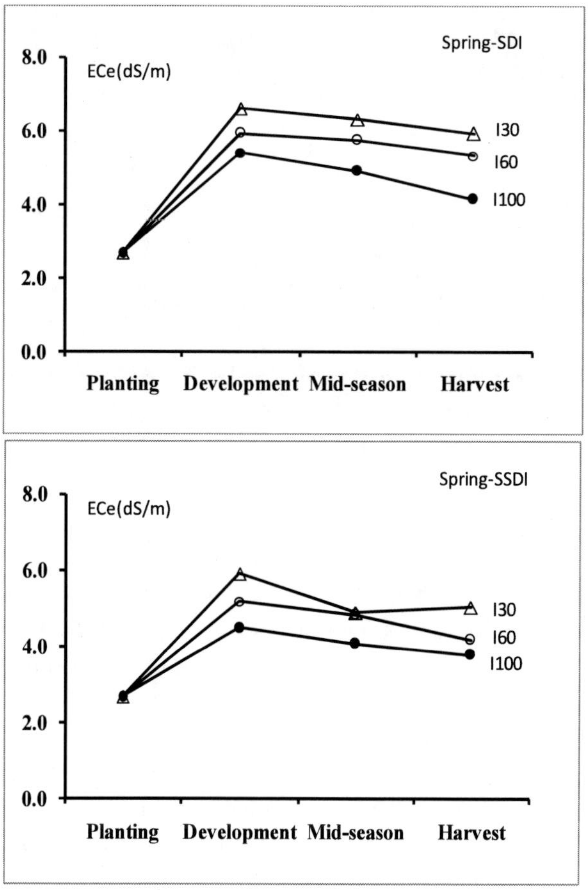

Figure 12. Soil salinity (ECe, dS/m) under different irrigation treatments (I100, I60 and I30) for surface (SDI) and subsurface drip irrigation (SSDI) (redrawn from Nagaz et al., 2012).

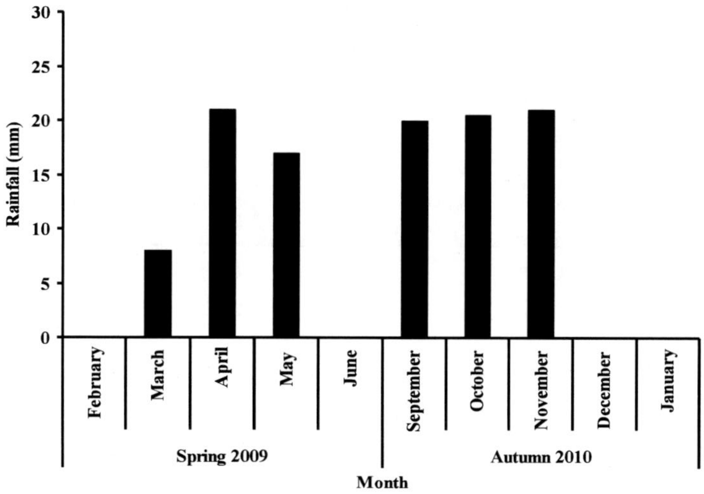

Figure 13. Monthly values of rainfall received during the potato cropping periods for 2009 and 2010 (after Nagaz et al., 2012).

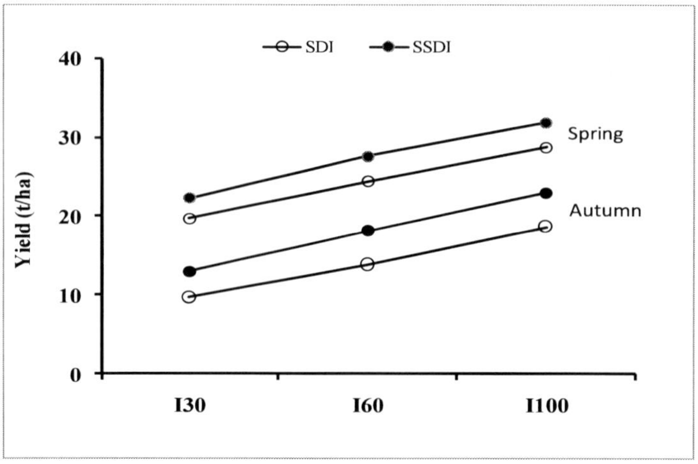

Figure 14. Potato yields under different irrigation methods (SDI and SSDI) and treatments (I100, I60 and I30) during the two cropping periods of potato (after Nagaz et al., 2012).

The highest potato yield for both SDI and SSDI methods was observed in the I100 treatment, for both seasons (Figure 14). Yield values were 28.8 and 31.9 t/ha in spring season and 18.7 and 23.0 t/ha in fall under SDI and SSDI methods, respectively. Although yields observed for I100 under both irrigation methods are numerically higher than I60, the difference was not statistically significant during the two cropping seasons. However, yields decreased significantly for I30 treatment in comparison to I100 under both irrigation methods. Potato yields of I60 and I30 treatments were also significantly different ($p<0.05$). The reduction in potato yield was mainly attributed to reduction in tubers number, weight and size as a consequence of water supply shortage and higher soil salinity associated with the deficit irrigation strategies during tubers initiation and development stages. Previous studies have shown that adequate irrigation water supply before and during tubers initiation increases the number of tubers per plant (Cappaert et al., 1992); whereas, after tubers initiation, it increases their individual sizes (Shock et al., 1998). In both seasons, SSDI increased yield compared to SDI but the difference between both irrigation methods was not found to be statistically significant. Phene (1995), Weatherhead and Knox (1998) also reported no significant differences between surface and buried tape methods on potato yields. However, analyzing the data regarding soil-moisture content and soil salinity during the growing season revealed significant differences among the irrigation methods (Figures 11 and 12). These results are in agreement with those of Oron et al. (1999) who reported that soil moisture content values under SSDI were higher and the ECe values were lower than those obtained with SDI. Thus, the observed yield variation, although not highly significant, seems to be due to the combined effect of the moisture content and salinity in the soil. Under both irrigation methods, maximum yield was obtained with treatment I100. However, yield response to irrigation water is more pronounced under SSDI than that with SDI (Figure 14). Potato yield was enhanced under SSDI compared with SDI for each irrigation treatment. Thus, the yield reduction noted under SDI for all irrigation treatments can be overcome by applying saline water through an SSDI system (Figure 14). The findings of Oron et al. (2002) also confirm these results. The high yields obtained with SSDI under the prevailing climatic conditions indicate its high potential to manage the irrigation of potato with saline water.

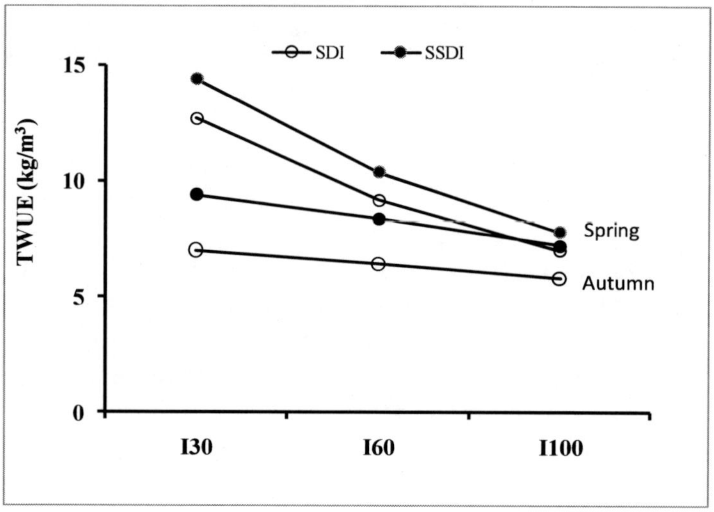

Figure 15. Irrigation and total water use efficiencies (TWUE) under different irrigation methods (SDI and SSDI) and treatments (I100, I60 and I30) (redrawn from Nagaz et al., 2012).

The highest water-use efficiencies were observed for I30 treatment in both irrigation methods (Figure 15). WUEs values decreased significantly from I30 to I100 with increase in water supply.

The deficit irrigation treatment I30 gave a higher WUE because yield reduction (32% for SDI and 30% for SSDI in spring and 48% for SDI and 43% for SSDI in autumn) was less than that for irrigation (70%) and total water supply (62% in spring and 56% in autumn). In this type of studies, generally, the lower the amount of water received, the higher the water use efficiency (Fabeiro et al., 2001). Data show also that for all irrigation treatments WUE was highest with SSDI (Figure 15), due to higher yields compared to SDI. Higher IWUE and TWUE of SSDI demonstrate that SSDI method provides some advantage on yield and WUE in potato production although differed not significantly from SDI ($P<0.05$) for each irrigation treatment.

6. Deficit Irrigation of the Carrot Crop

Carrot (*Daucus carota* L.) is an important short duration root vegetable grown for fresh market food, but it is considered as relatively susceptible to salinity (Bernstein et al., 1974). This crop is grown in arid regions of Tunisia during autumn and winter in individual plantings usually not exceeding 1 ha and irrigated with water from shallow wells.

Because carrot crop has high economic value, the irrigation management strategy seeks maximum yield by supplying all requirement of the crop.

The impact of irrigation scheduling and deficit irrigation with saline water on carrot yield and quality has been studied with the objective to promote appropriate irrigation methods among farmers' communities. All field work was conducted within private farms and with farmers' participation (Nagaz et al. 2012). The experiments took place during the fall-winter seasons over three consecutive years (2007-2010). Carrot was grown on a sandy soil and surface-irrigated with water from a well having an ECi of 3.6 dS/m. Starting with a soil at field capacity, simulation starts with a soil water content at field capacity at planting. Four irrigation methods based on the use of soil water balance (SWB) to estimate irrigation amounts and timing were compared to traditional farmers' practices. SWB methods consisted in replacement of cumulated ETc when readily available water is depleted with levels of 100% (I100), 80% (I80) and 60% (I60). I100 was considered as full irrigation while I80 and I60 as deficit irrigation regimes. Regulated deficit irrigation regime where 40% reduction is applied only during ripening stage (I60-ripening) was also used. Farmer method (Farmer) consisted in giving fixed amounts of water (25 mm) every 7 days from planting till harvest.

Results on carrot production and soil salinization are globally consistent between the three years and show significant differences between irrigation regimes. During the three experiments, changes in ECe values at harvest were observed under all irrigation treatments and compared to initial soil salinity (Figure 16).

Figure 16 shows a decrease in ECe values between planting and harvest for full irrigation treatment (I100). I60-maturity irrigation treatment resulted also in low ECe values at harvest without significant difference with I100 during the three years. However, higher soil salinity levels were observed for deficit irrigation regimes. ECe values were, in a decreasing order, I60 > I80 > I100. The highest ECe values were observed for farmer's irrigation method where more water is applied without adequate scheduling. The high frequency of application during the first stage seem to concentrate salts in the root zone.

The capacity of fall-winter rainfall to leach salts in the region is variable and depends on the total amount and distribution of rainfall events (Figure 17).

This is illustrated by the lowest ECe values observed in the first year which corresponds to the highest amount of total rainfall during carrot growing season that seemed to be effective in removing salts accumulated in the root zone. The decrease of ECe values observed during the first year is attributed to the leaching of soluble salts by fall and winter rains (57 mm).

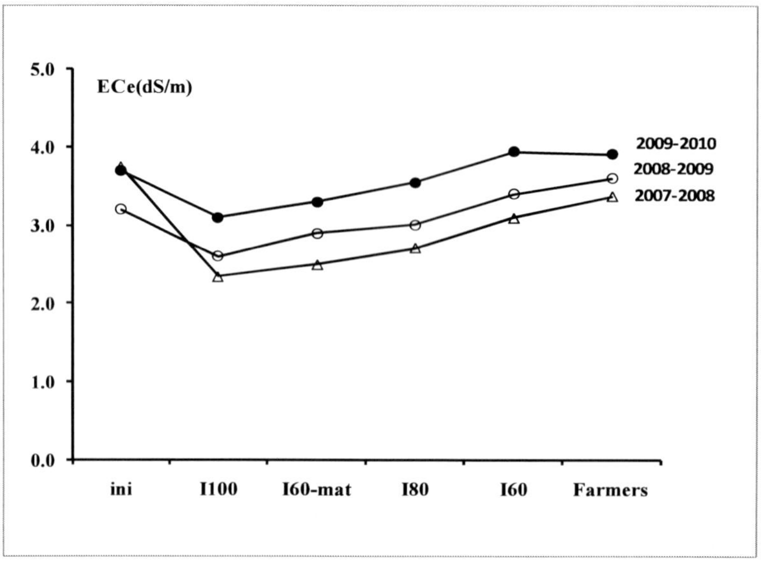

Figure 16. Soil salinity (ECe, dS/m) under different irrigation treatments of carrot (after Nagaz et al., 2012).

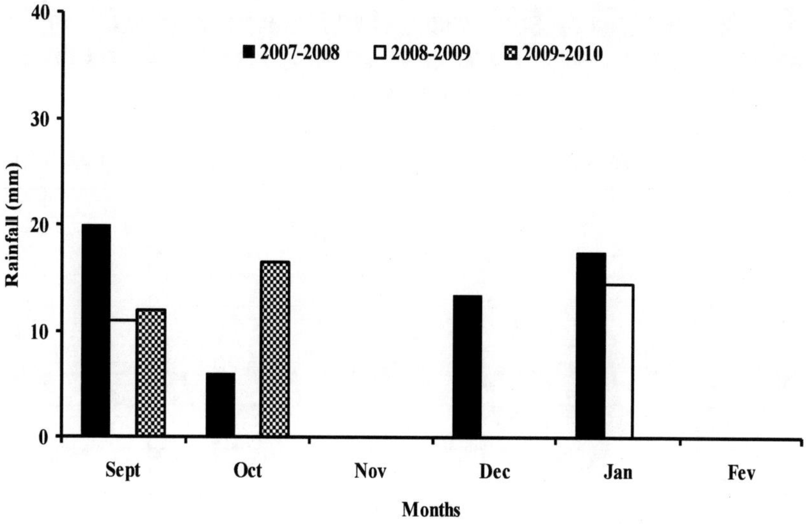

Figure 17. The rainfall received during the carrot cropping periods (2007-2010) (after Nagaz et al., 2012).

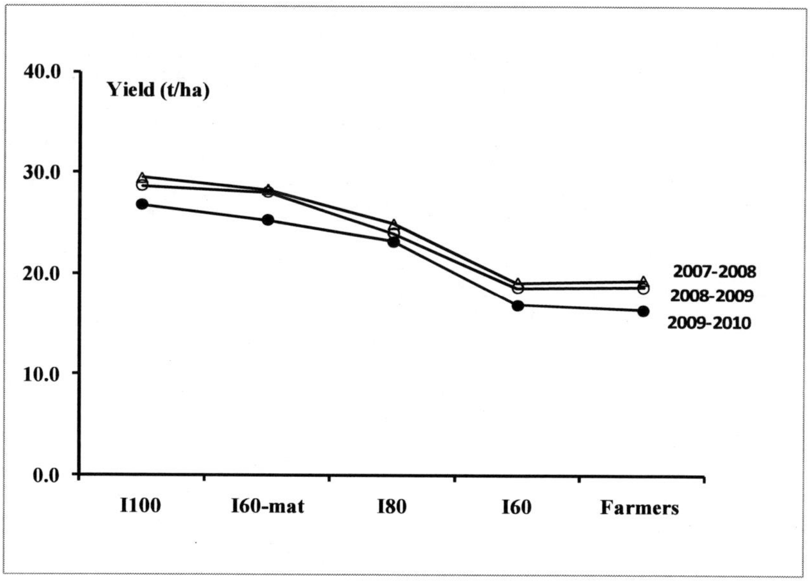

Figure 18. Yields of carrot in relation to the irrigation treatment and the year of production (redrawn from Nagaz et al., 2012).

In 2009 and 2010, ECe decreased only for treatments I100, I80 and I60-maturity while its value was relatively higher at harvest than the initial ECe for I60 and farmer's method because of lower fall and winter rains.

Under actual farming conditions, the impact of using saline waters for irrigation of short-cycle crops on soil salinization is mitigated by rainfall, as salts added by irrigation are removed from the root zone by natural leaching during the rainy season. In this experiment ECe values under the different irrigation treatments were lower than the ECi of irrigation water during the first year, but increased steadily the following years.

Table 3. Irrigation water supply, irrigation and total water use efficiency under different irrigation treatments during the growing period of carrot for 3 years (after Nagaz et al., 2012)

Treatment	Irrigation water (mm)	IWUE (kg/m^3)	TWUE (kg/m^3)
	2007-2008		
I100	328	8.9	7.6
I80	262	9.5	7.8
I60	197	9.7	7.5
I60-maturity	297	9.5	7.9
Farmers	509	3.8	3.4
LSD (5%)	-	0.78	0.54
	2008-2009		
I100	330	8.6	8.0
I80	264	9.1	8.3
I60	198	9.4	8.3
I60-maturity	315	8.9	8.2
Farmers	518	3.6	3.4
LSD (5%)	-	0.62	0.39
	2009-2010		
I100	328	8.1	7.5
I80	263	8.8	7.9
I60	197	8.6	7.5
I60-maturity	299	8.4	7.7
Farmers	471	3.4	3.2
LSD (5%)	-	0.50	0.41

Highest carrot yields for the three years were obtained with I100, (29.5, 28.7 and 26.8 t/ha) although there are not significant differences with the regulated deficit irrigation regime (I60-maturity) (Figure 18). Carrot yields of I80 treatments were significantly reduced and lowest yields were observed with the farmer's method and I60. The two latter treatments have similar root yields, but 40% less irrigation water than full irrigation regime is used in the first, while the farmer used 40% more water than I100. Also, with 20% less water I80 treatment resulted in 22% to 29% more yield than farmer' method. Reduction in root yields is associated with lower root number/m^2 and root weight. Relatively good yields obtained under I100 and I60-maturity treatments are attributed to better growth and yield components due to appropriate water supply during sensitive stages, as reported by Hartmann et al. (1986), Parabhakar et al. (1991), Imtiyaz et al. (2000) and Paradiso et al. (2002) who obtained higher total marketable, root size, and number with 100% ETc, full irrigation treatment.

Yields were highest in the first year because of the low soil salinity and the higher amount of rainfall (57 mm). During the three experimental seasons, differences in yield under

I100 and I60-maturity treatments were not significant. Reduction of water supply after mid-season stage by 40% (I60-maturity) seems to have low impact on soil salinity and yield of the carrot crop, compared to full irrigation. The use of SWB strategy for managing irrigation water resulted in better yields than the method using fixed frequency and amounts. Lower yields obtained under empirical method may be attributed to the fact that farmers apply water to the crop regardless of the effective plant needs. They seem to relate irrigation occurrences to days after planting rather than to progress in crop growth processes. The corresponding irrigation applications are often characterized by periods of over- and under-irrigation. Raes et al, (2002) reported that excess watering in saline conditions may cause loss of valuable nutrients out of the root zone and soil salinization, especially during crop sensitive periods, which results in limited growth and reduction in crop yield. Severe deficit irrigation as well as farmers' strategy resulted in higher salinity in the rooting zone compared to I100 and I60-maturity, inducing substantial yield reduction of carrot. However, the carrot crop seems to give acceptable yield with saline water, if irrigation management practices maintain a deficit not exceeding 20% of ETc (I80). Obtained WUE values were similar to those reported for carrot by many authors (Parabhakar et al., 1991; Imtiyaz et al., 2000) and were affected by irrigation treatments (Table 3). There is also a variation in WUE values between years. For all irrigation treatments, yield was higher in the first year compared to the two following years. Values of water productivity of irrigation (IWUE) reflect this difference, they varied typically around 3.8-9.7, 3.6-9.4 and 3.4-8.8 kg/m^3, respectively, in the first, second and third year. For all experiments, the WUE values obtained with I100 treatment were not significantly different from those obtained with I80 and I60-maturity treatments but were statistically different from those obtained with I60 and farmers' treatments. IWUE of 3.4-3.8 kg/m^3 obtained with farmers' treatments were statistically different from those obtained with SWB treatments. The low IWUE corresponding to farmers method during the experiments can be attributed to both reduced yields and also higher irrigation water use. Results obtained over 3 years of field experiment showed that full-irrigated carrot grown in the arid condition of southern Tunisia used about 330 mm of in-season irrigation water. Water amounts can be reduced by adopting regulated and moderate deficit irrigation (I60-maturity and I80). Deficit irrigation treatments I60 resulted in much lower yields and in higher salinity in the root zone. The "fixed amount approach" gave the lowest root yields i.e. 34-38% less than I100 with 43 to 57 % more water and resulted in higher salinity in the rooting zone. High WUE observed for the most restricted regime (I60) is counterbalanced by reduced yield and quality. The relatively good yields and WUE values obtained under I80 and I60-maturity indicate the potential of the carrot crop to valorize irrigation waters of limited quality. Full irrigation scheduling technique I100 could be recommended for irrigation of carrot crop under the arid climate of southern Tunisia with the possibility to reduce supply up to 20% (I80) and to 40% only during ripening stage (I60-maturity) in case of limited water availability.

7. Deficit Irrigation of Millet

Two years experiment (2005-2006) were conducted to analyze soil salinization, yield and water efficiency of millet (*Pennisetum glaucum* (L.) R. Br.) grown in a commercial farm during summer season on a sandy soil and drip-irrigated with water having an ECi of 7.6

dS/m (Nagaz et al., 2009). Irrigation regimes consisted in replacement of cumulated ETc with levels of 100% (I100), 80% (I80), 60% (I60) and 40% (I40). I100 was considered as full irrigation, while the others as deficit irrigation regimes. Irrigation was applied whenever readily available soil moisture for I100 treatment reached 60% of total available water in the root zone.

Comparison between measurements of soil salinity data shows that there were increases in ECe from initial value under (I100). A further increase in ECe occurred under I80 and I60 regimes, with the highest value observed under I40 treatment. The higher soil salinity obtained under the least irrigated treatments is attributed to limited leaching of the top soil layer (Figure 19). Millet growth, expressed by leaf area index (LAI), was affected by irrigation treatments (Figure 20). LAI was directly related to the amount of water applied and values under deficit-irrigated treatments (I80, I60 and I40) were proportionally reduced. In both experiments, the difference between treatments in LAI became significant 30 days after planting.

These results suggest that the amount of water applied during the critical period of growth which corresponds to panicle differentiation and initiation, flowering and grain filling, is particularly important for millet crop. Mahalakshmi and Bidinger (1985) and Mahalakshmi et al. (1988) reported that millet growth was most sensitive to water supply during these periods, although insufficient water also slowed growth at earlier stages. Millet plants receiving only 40 and 60% of their ETc requirements are not only subject to water deficit but also to a saline stress (Figure 19). Plant growth is reduced when the root zone is subjected to the combination of water deficit and salt accumulation (Devitt et al., 1993).

Yield data show that for both experiments the maximum final dry matter production and grain yield occurred in the I100 treatment. Most drastic reduction in yields occurred with the I60 and I40 treatments (Figure 21).

Millet yields obtained under the I100 treatment were statistically different from those obtained with I80. For both years we observed the same trend in relative yield reduction as related to relative water supply.

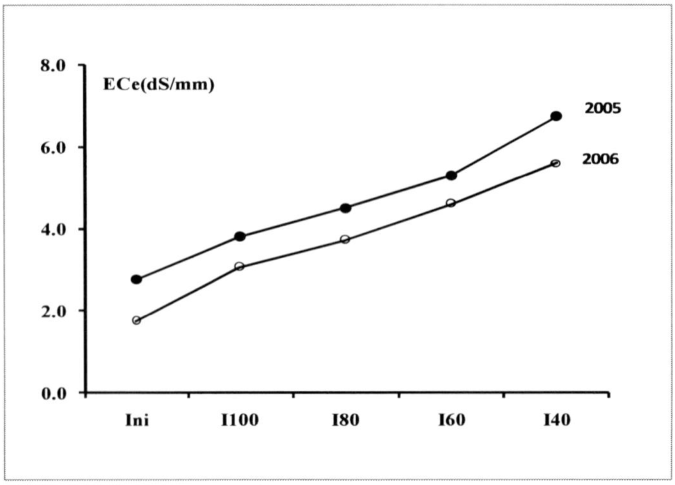

Figure 19. Mean soil salinity for decreasing irrigation supplies compared to initial values (redrawn from Nagaz et al., 2009).

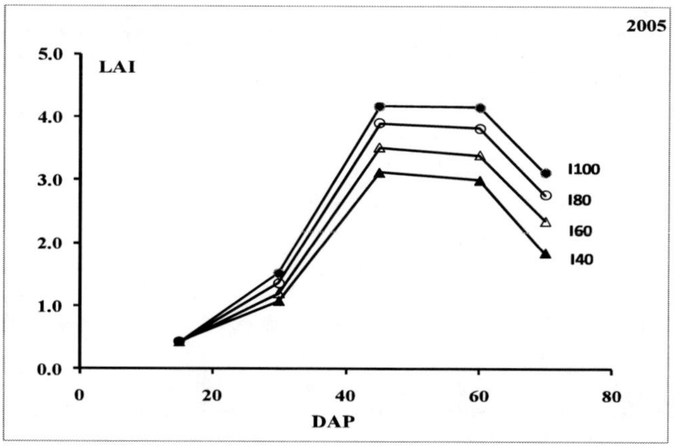

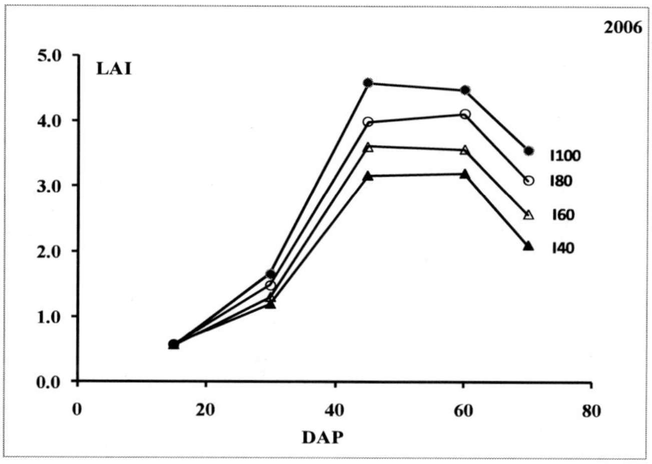

Figure 20. Millet leaf area index (LAI) under different irrigation treatments in first and second year experiments (after Nagaz et al., 2009).

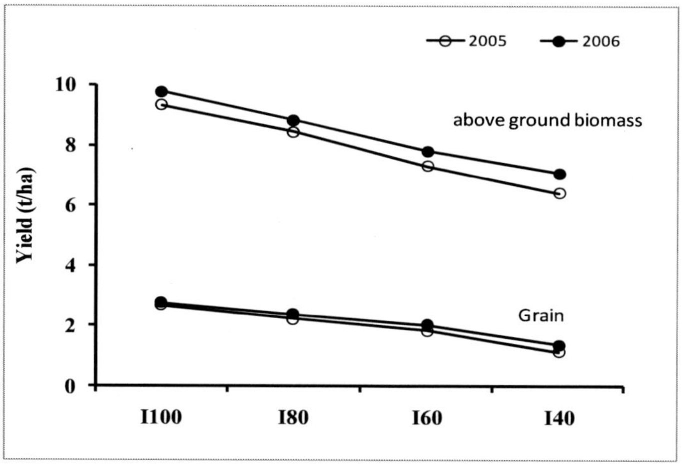

Figure 21. Yield of pearl millet obtained over two cropping seasons under different irrigation treatments (after Nagaz et al., 2009).

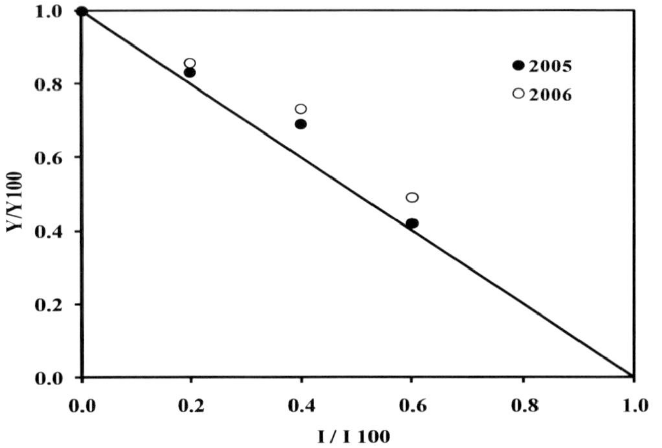

Figure 22. Relative grain yield (Y/Y100) decrease in relation to relative water supply (I/I100) (redrawn from Nagaz et al., 2009).

Millet yields observed under I100 are 17, 31, 58%, and 14, 27 and 51% greater than those obtained with I80, I60 and I40, respectively, for the first and second experiments (Figure 22). The reduction in grain yield was mainly attributed to reduction in panicle number/m^2, kernel number/panicle and kernel weight as a consequence of water shortage during panicle differentiation and initiation, flowering and grain filling.

Previous studies have shown that adequate irrigation water supply during these periods increases the panicle number and kernel number and weight (van Oosterom et al., 2002; Mahalakshmi and Bidinger, 1985, 1986). There were differences between results obtained for the two experimental years. Yields were higher in the second year because of lower initial soil salinity. To assess their respective sensitivity to salt, grain yields data for each experiment were statistically analyzed with a piecewise linear response model (van Genuchten, 1983).

The first analysis outcome indicated that thresholds were nearly the same for both experiments. Therefore, both yield sets were combined for analysis. Based on these data it can be stated that under our conditions Yr = 100- 17 (ECe-3.46) for ECe greater than 3.4 dS/m and Yr= 100 for ECe less than 3.4 dS/m. The threshold is close to the value calculated from published salt tolerance data (3.4 vs. 3.6) but the slope is considerably steeper (17 vs. 6.7) (Nagaz et al., 2006), most likely because of the combined effect of salinity and water deficit in an arid climatic context characterized by a high evaporative demand.

For both experiments, the WUE for grain yield was the lowest under I100 and the highest under I40 and I60. These latter irrigation treatments gave a higher WUE because grain yield reduction (31 and 58% for 2005 and 27 and 51% for 2006) was less reduced than the irrigation water supply (40 and 60%). The relatively high yields and water use efficiencies noted under I80 treatment in both experiments indicate an acceptable response of millet crop to moderate water deficit. These results, obtained under actual farming conditions, sustain the practicality of using the I100 methodology to implement an efficient use of saline water for millet production. Under severe shortage of water, irrigation could be reduced voluntarily. Irrigation water reduction during the whole season of 20, 40 and 60% has resulted in grain yield losses of lower proportions respectively, for the first and second year experiments.

8. Deficit Irrigation of Barley

Barley (*Hordeum vulgare* L.) is classified as tolerant to water stress caused by deficit irrigation or salinity (Maas and Hoffman, 1977; Hajor et al., 1996). It is a major cereal crop in the arid regions of Tunisia and occupies a large cultivated area during the winter period. However, productivity is usually low and irrigation with saline waters is often applied, for the purpose of improving yield, according to local empirical practices. The experiment presented here was carried out on the site of Medenine, during the season November 2008 to June 2009, to study the impacts of deficit irrigation regimes on soil salinization and yield of barley cultivated on a sandy soil and drip-irrigated with water having an ECi of 7.6 dS/m. Five irrigation regimes were evaluated. Full irrigation treatment (I100) irrigated with a quantity equal to 100 % of cumulated ETc. The second treatment irrigated at the same frequency as treatment I100, but with quantity equal to 50% of cumulated ETc (I50) during the whole growing period. The three other treatments consisted in irrigation water restriction during the development stage (I50-Dev), maturity ((I50-Mat) and both development and maturity stages (I50-Dev+Mat). Deficit irrigation at a particular phase was equal to 50 % of cumulated ETc. The SWB method was used to estimate the day when the target soil water depletion would be reached and the amount of irrigation water needed to replenish the soil profile to field capacity (I100). Soil salinity was measured before planting, during mid-season and after harvest to evaluate the impact of deficit irrigation on salt accumulation in the root zone. Obtained results show a reduction in soil salinity in the root zone during vegetative growth, the ECe for the most irrigated treatments FI-100 and I50-Mat decreased from 5.7 at plantation in November to approximately 5.2 dS/m at the mid-season stage. For I50-Dev and I50-Dev+Mat treatments, the soil salinity (5.5-5.6 dS/m) was not significantly different with that at plantation.

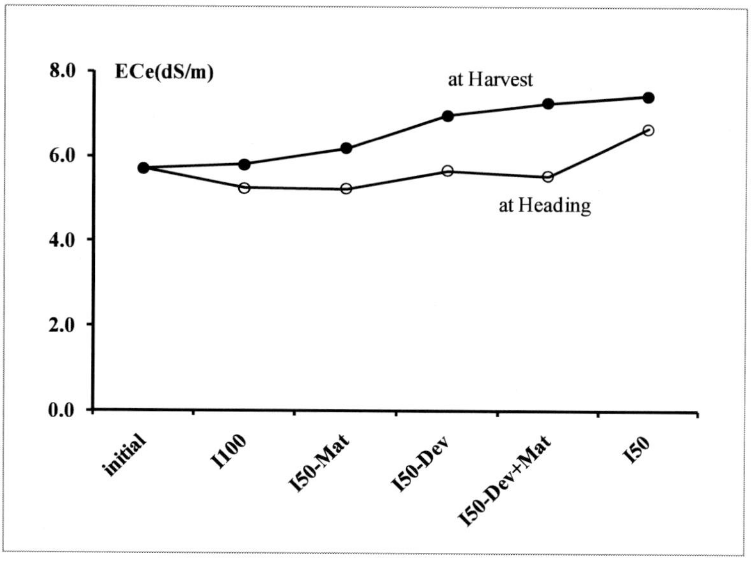

Figure 23. Soil salinity at mid-season stage and after harvest under different irrigation treatments (redrawn from Nagaz et al., 2008).

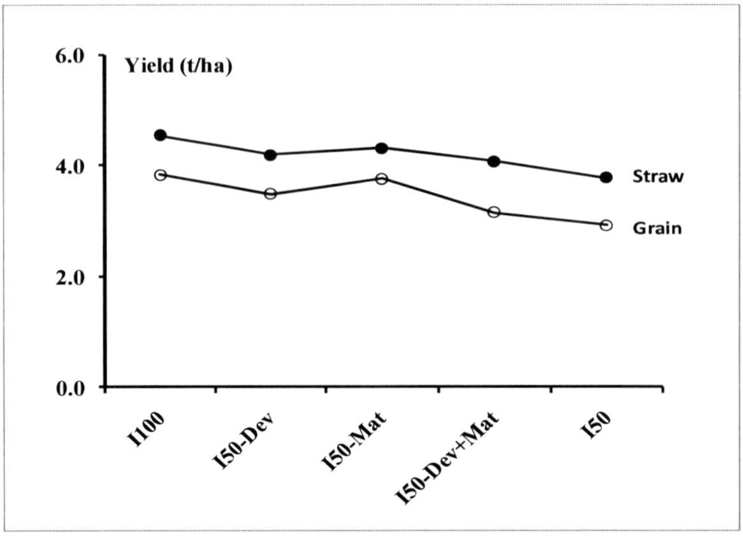

Figure 24. Grain yield and final dry matter of straw (stems+ leaves) under full (I100) and deficit irrigation (I50, I50-Dev, I50-Dev+Mat and I50-Mat) (redrawn from Nagaz et al., 2008).

However, the ECe obtained with continuous deficit I50 increased from 5.7 dS/m to 6.6 dS/m (Figure 23). All irrigation treatments, except I50 seem to have benefited from the leaching effects of 89 mm rainfall, received during the initial and development phases.

At harvest, soil salinity values are comparable to the initial ECe for I100 and I50-Mat. Whereas ECe values obtained with the rest of treatments I50, I50-Dev and I50-Dev+Mat are higher than the initial ECe in spite of important rainfall received during the maturity phase (26 mm), which was supposed to cause significant leaching of salts accumulated in the root zone. The increase can be explained by the sampling date which corresponds to a period of high evaporation demand.

The practice of continuous deficit irrigation (I50) increased ECe in the root zone. Water supply reduction by 50% reduces certainly salt amounts brought to the soil, but does not allow for a deep enough leaching. Salinity remains lower in the top layer under J100 treatment, due to a situation of drainage which ensures the evacuation of salts beyond the sampled depth of 80 cm.

Similarly the increase in soil salinity for I50-Mat was less important than that observed with I50-Dev and this latter lower than ECe corresponding to I50-Dev+Mat treatment.

Combined effects of salinity and water deficit stresses were observed on biomass and grain productions. Highest total dry matter and grain yields were obtained with I100 followed by I50-Mat and I50-Dev irrigation treatments (Figure 24).

Previous studies have shown that moderate water deficit during the vegetative and maturity stages don't have significant effects on dry matter accumulation of barley and wheat (Jamieson et al., 1995).

This means that the water restriction applied under I50-Mat and I50-Dev irrigation treatments were not too severe to affect considerably the crop. However barley dry matter production was affected significantly under continuous deficit irrigation (I50) compared to I100 and I50-Mat.

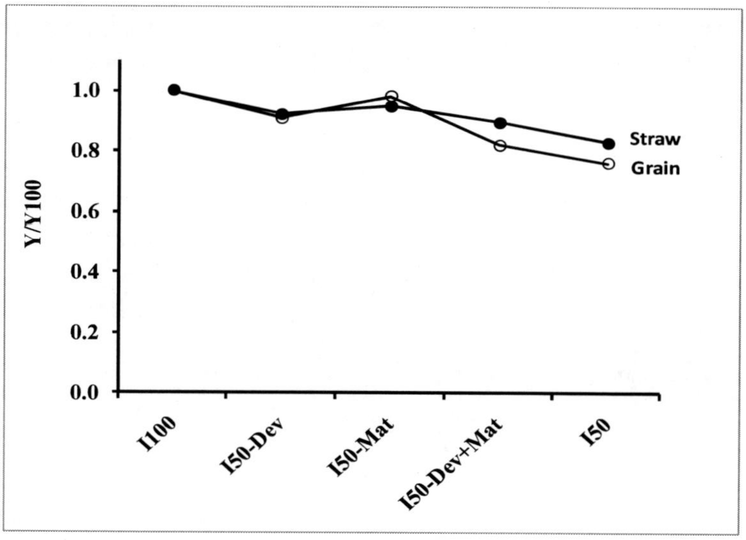

Figure 25. Relative yield of grain and straw under different irrigation strategies of barley (after Nagaz et al., 2008).

Also, yields obtained with regulated deficit irrigation treatments (I50-Dev and I50-Dev+Mat) were higher than those obtained with I50.

The reduction in grain yield associated with the treatment I50 is mainly attributed to a significant reduction in yield components, as a consequence of water shortage during flowering and grain formation. Indeed, continuous deficit irrigation (I50) caused a reduction in tillers number and subsequently a reduction in the number of spikes per m^2. Also, reduction in seeds number/spike and 1000-seeds weight was noticed and could be attributed to grain filling failure as a consequence of stress caused by both water supply shortage as well as increased salt concentration in the root zone.

Water deficit conditions can aggravate the stress placed on plants growing under saline conditions. Yields losses associated to the different regulated deficit irrigation strategies are presented in (Figure 25). The highest yield losses were observed with the continuous deficit irrigation treatment and was about 24 and 17% compared to full irrigation treatment, respectively, for grain yield and final dry matter.

The water restriction during the development and maturity (I50-Dev+Mat) caused a reduction about 18 and 9 %, respectively, for grain and dry matter yields.

The I50-Mat and I50-Dev treatments resulted in barley yields comparable to those obtained under I100 treatment, most likely due to their effect on reducing the build-up of salinity. Note that the I50 and I50-Dev+Mat deficit irrigation strategies resulted in higher salinity in the rooting zone than the other irrigation strategies (Figure 23). Barley productivity is most sensitive to water stress during flowering and grain filling (Sepaskhah, 1978; Weltzien and Srivastava, 1981; Ceccarelli, 1987; Baheri et al., 2005).

The higher salinity associated with severe deficit irrigation strategies (I50 and I50-Dev+Mat) were sufficient to cause important reduction in grain yield (Figure 25). Under conditions of water shortage, irrigation of barley can be reduced during maturity stage.

Water supply consisted in 78 mm pre-irrigation for all treatments and in season irrigation water supply ranged from 169mm (I50) to 338 mm (I100), that complemented a total rainfall of 115 mm. In-season irrigation water reduction obtained under deficit treatments was about 50, 10, 12 and 22%, respectively, for the treatments I50, I50-Dev, I50-Mat and I50-Dev+Mat.

Table 4. Irrigation and total water use efficiency under different treatments (after Nagaz et al., 2008)

Treatments	Irrigation* (mm)	IWUEg (kg/ha/mm)	TWUEg (kg/ha/mm)	IWUEb (kg/ha/mm)	TWUEb (kg/ha/mm)
I100	338	11.3	8.4	13.4	10.0
I50	169	17.3	10.3	22.3	13.2
I50-Dev	304	11.4	8.3	13.8	10.0
I50-Mat	296	12.7	9.1	14.5	10.5
I50-Dev+Mat	262	12.0	8.3	15.5	10.7
LSD (5%)		1.84	1.31	1.83	1.28

* An irrigation of 78 mm supplied just before planting is not included in these totals. In season rainfall is 115mm.

Deficit irrigation applied to barley is intended to provide means of reducing irrigation water supply without affecting significantly the yield. Water use efficiency is a measure of attainment of this goal. As expected, the highest water use efficiencies (IWUE and TWUE) for grain yield and dry matter production were observed for DI-50 treatment and the lowest WUEs were obtained under full irrigation (I100) (Table 4). The high WUE for I50 is attributed to the fact that yield reduction under this treatment was less than reduction of the seasonal water applied. However, the difference between the treatments I100, I50-Dev, I50-Mat and I50-Dev+Mat was not significant ($p<0.05$) although the treatments I50-Dev, I50-Mat and DI50-Dev+Mat have higher values of WUE. Values of WUE, based on dry matter of straw, were significantly different between I100 and I50-Dev+Mat treatments ($p<0.05$).

The barley yields obtained with full irrigation (I100) are comparable with those obtained with I50-Dev and I50-Mat treatments. The deficit irrigation treatment with less irrigation water of about 22% (I50-Dev+Mat) gives WUE values comparable with those obtained under I100, I50-Dev and I50-Mat treatments. These results show how irrigation water supply for the barley crop can be reduced without a significant yield reduction by adopting deficit irrigation strategies during maturity and development stages (I50-Mat, I50-Dev and I50-Dev+Mat). They also provide an assessment of the consequences of reducing by half the water supply on yield and salt build up in the soil.

9. Alternative Water Management Strategies

In many arid regions, there could be important amounts of saline ground waters, whereas the supply of better quality waters is limited. In such situations alternation of waters with different salinity in a cyclic mode may be necessary to meet the irrigation requirements of the entire area. This study investigates the usefulness of alternating and blending waters of different origins for crop production. Two water resources available in the area of interest were considered: canal water (CW, 3.2dS/m) and saline ground water (SW, 5.2 dS/m).

The water management strategy based on blending and alternation was tested for two years (1994 to 1996) under actual commercial-farming conditions on a well-drained sandy soil (Nagaz and Ben Mechlia, 1998). Four irrigation treatments were applied: SW, irrigation with saline water (ECiw=11.7 dS/m); CW, irrigation with canal water as a control (ECiw=3.4 dS/m); SW/CW, irrigation with saline and canal waters alternately; and SW-CW, blending canal water and saline water to an intermediate salinity (ECiw= 6 to 7 dS/m).

The annual distribution of rainfall and evaporation during 1994 to 1996 is given in Figure 26. Most of the rainfall occurred during the cool season from the beginning of October to the end of April. Maximum evaporation occurred during June-August, when the field remained fallow.

For irrigation scheduling the class A-pan evaporation (ETpan), measured at an adjacent site, was used to estimate evapotranspiration. ETpan recorded daily was modified by Kc and Kp, the crop and pan coefficients, respectively, to determine the crop evapotranspiration (ETc). It was assumed that ETc between two successive irrigation is given by the formula ETc=I+P, where I was the amount of irrigations water applied, P is the amount of precipitation. Soil depth is limited to 70 cm because of a cracked rocky layer in the experimental site. At each irrigation, 4 cm depth of water was applied by surface method in order to limit the water losses through deep percolation, and thus to keep drainage at the depth of 70 cm close to negligible. Irrigations were applied when the cumulative crop evapotranspiration after accounting for rainfall reached 40 mm. In order to differentiate between the 4 strategies, the amount of irrigation water was kept the same in all treatments.

Soil salinity data obtained after two years of experimentation (Figure 27) show that the use of saline water increased the ECe in the entire soil profile as compared with the control (CW). The average ECe values even after 2 years of continuous use of saline water (SW) were lower than the EC of the irrigation water.

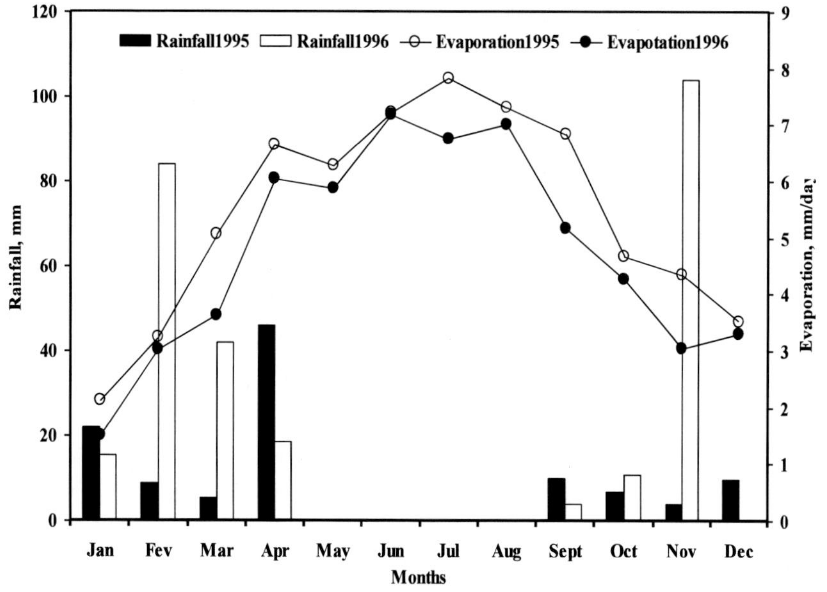

Figure 26. Annual distribution of rainfall and evaporation (1994-1996) (after Nagaz and Ben Mechlia, 1998).

Table 5. Yield of barley under different irrigation treatments (after Nagaz and Ben Mechlia, 1998)

	Grain yield (q/ha)			Final dry matter (q/ha)			1000 seed weight (g)		
Treatment	1995	1996	Mean	1995	1996	Mean	1995	1996	Mean
CW	25.2	28.9	27.0	41.5	49.0	45.2	40.2	44.1	43.0
SW	19.8	24.6	22.2	35.7	45.2	40.4	36.0	38.6	37.3
SW/CW	23.0	27.2	25.1	37.9	47.9	42.9	41.6	42.2	40.9
SW-CW	21.4	26.5	23.9	36.7	47.0	41.9	38.3	40.8	39.5
LSD (.05)	2.54	2.94		6.53	5.88		1.8	3.0	

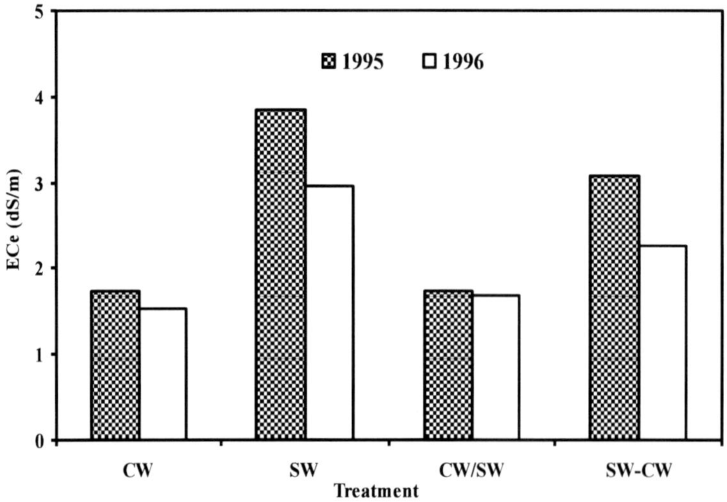

Figure 27. Mean soil salinity values under different irrigation treatments (after Nagaz and Ben Mechlia, 1998).

A lower value of ECe under the current climatic conditions was due to the leaching of soluble salts with the relatively high amount of rainfall received and irrigation. The leaching by rain water occurred during November and January-April. Leaching seemed to be sufficient to control the build-up of soluble salts in the profile of the well drained soil.

Dry matter production was not affected significantly by irrigation treatments. However, grain yield was lower in the SW and SW-CW treatments compared to CW treatment (Table 5). The decreased grain yields are associated with lower 1000-seed weight caused by saline stress. A further reduction in production parameters occurred with the SW treatment.

The mean grain yield and dry matter production for SW in relation to CW were 18 % and 11 %, respectively, indicating that salinity of irrigation water had a larger effect on grain yield than on total dry matter. Yields were highest the second year because of the low initial soil salinity and the higher amount of rainfall received (270 mm). During the period 1994-1995 and 1995-1996 the differences in yield and its components under CW and SW/CW treatments were not significant. Due to its effect on reducing the build-up of salinity the SW/CW treatment resulted in barley yields comparable with those obtained under CW treatment. It was rxpected that the SW/CW strategy would increase yields compared to the SW-CW strategy. Note that blending results in higher salinity in the rooting zone than the cyclic mode (Figure 27). The higher salinity associated with blending was sufficient to cause important reductions in yield. These results support the adoption of cyclic water strategy to facilitate the use of highly saline water for irrigation as recommended by numerous reports (Grattan et al., 1987; Rhoades, 1985, 1987, 1989; Rhoades et al., 1988a; Bradford and Letey, 1992).

In the SW/CW treatment, saline water accounted for 29 and 23 % of the total water applied in 1995 and 1996, respectively. Rainfall was significant. These data show that less canal water was used for irrigation by the cyclic strategy (SW/CW) in 1995 than 1996. This cyclic use of 'low' and 'high' salinity waters prevents the soil from becoming excessively saline while permitting, over a long period, substitution of saline water for a low-salinity water for a large fraction of the crop requirements (Rhoades, 1983, 1984, 1985).

The highest water-use efficiencies (WUEg) were observed for CW treatment uniformly in all the years (Table 6). Decreased WUE with SW treatment is the consequence of significant yield reduction. Irrigation treatments (SW/CW and SW-CW) had little effect on water-use efficiencies, except for 1995 when WUEg was less for SW-CW than SW/CW treatment. WUE can be increased to a greater degree by cyclic strategy than by blending.

This two-year field investigation showed that more substantial amounts of saline water were substituted for canal water in the irrigation of barley without yield loss compared to those obtained from continuous irrigation with canal water (CW). No significant differences in yield or crop quality occurred when saline water substituted for canal water for up to 20 to 30% of the irrigation requirements of the crop. Note that the blending strategy (SW-CW) has higher salinity in the rooting zone than the cyclic strategy (SW/CW). The higher salinity associated with the blending was sufficient to cause reduction in barley yield. This conclusion is consistent with the experimental results of Meiri (1984) and Shalhevet (1984) and confirms the credibility of the recommended cyclic water strategy to facilitate the use of saline water for irrigation.

Table 6. Water use (mm) and water use efficiency (kg/ha/mm) under different treatments (after Nagaz and Ben Mechlia, 1998)

	Irrigation water (mm)		Rainfall (mm)	Total water applied (mm)	% of saline water	WUEb (kg/ha/mm)	WUEg (kg/ha/mm)
Treatment	Saline	Canal					
				(year 1995)			
CW	-	247	96	343	-	12.0	7.3
SW	207	40	96	343	60	10.3	5.7
SW/CW	100	147	96	343	29	11.0	6.8
SW-CW	103	143	96	343	30	10.6	6.2
				(year 1996)			
CW	-	132	270	402	-	12.1	7.2
SW	92	40	270	402	23	11.2	6.1
SW/CW	92	40	270	402	23	11.9	6.7
SW-CW	46	86	270	402	11	11.6	6.6

The findings are oriented to managing supplemental irrigation (SI) with saline water in order to optimize barley production in arid lands where most of the water available for irrigation is saline. The conjunctive use of different water resources i.e. saline irrigation water to supplement rainfall, can increase considerably rainfall efficiency. The use of saline water in SI is particularly attractive due to the limited water amounts needed for irrigation, and therefore the reduced risk of soil salinization, compared to full irrigation.

CONCLUSION

Scarcity of water resources and increasing needs for the intensification of agriculture is constraining many farmers to use waters of high salinity. Irrigation of sensitive and tolerant crops to water deficit and salinity is expanding rapidly, carrying the danger of rapid soil degradation and heavy losses for the existing farms. Good management practices are required to save water, improve production and reduce risks of soil salinization. The reinforcement of farmers' capacities to face the problem of increasing water scarcity starts with helping them improve their knowledge about the integration of information on soil, evapotranspiration, and salinity in water management.

The participatory experiments carried out on two vegetables and two cereals, show that deficit irrigation offers a potential way to improve water productivity and to control soil salinization. Crops should be grown preferably during the winter season when they can benefit from the rains along with leaching the salts. Simplified irrigation scheduling methods based on soil water balance have proved to be highly effective under real conditions of farming. Future investigation should focus on this issue and evaluate the efficiency of small amounts of rain that occur in fall-winter for improving productivity and natural leaching of added salts.

REFERENCES

Alexiou, I., Kalfountzos, D., Kotsopoulos, S., Vyrlas, P. and Kambeli, S. (2003). Comparison of subsurface and surface drip irrigation of cotton crop. In: Proceedings of the 9th National Congress of the Hellenic Hydrotechnical Association, vol. A., April 2-5, Thessaloniki, Greece, pp. 199-206.

Allen, R.G., Perreira, L.S., Raes, D., and Smith, M. (1998). Crop evapotranspiration: Guidelines for computing crop water requirements. Irrigation and Drainage Paper N° 56, FAO, Rome, Italy, 300p.

Al-Tahir, O.A., Al-Nabulsi, Y.A. and Helalia, A.M. (1997). Effects of water quality and frequency of irrigation on growth and yield of barley (Hordeum vulgare L.). Agric. Water Manage., 34, 17-24.

Ayars, J.E., Schoneman, R.A., Dale, F., Meso, B. and Shouse, P. (2001). Managing subsurface drip irrigation in the presence of shallow ground water. *Agric. Water. Manage.,* 47, 243-264.

Ayers, J.E., Hutmacher, R.B., Schoneman, R.A., Vail, S.S. and Felleke, D. (1986). Drip irrigation of cotton with saline drainage water. *Trans. Am. Soc. Agric. Eng.*, 29, 1668-1673.

Baheri, S.F., Javanshir, A., Kazemi, H. and Aharizad, S. (2005). Effects of irrigation at different phenological stages on some traits in spring barley genotypes. *Iranian Journal of Agricultural Sciences,* 36 (1), 169-176.

Bastiaanssen, W.G.M., Singh, R., Kumar, S., Schakel, J.K. and Jhorar, R.K. (1996). Analysis and recommendations for integrated on-farm water management in Haryana, India: a model approach. Report 118, DLO Winand Staring Center, Wageningen, 152 pp.

Bernstein, L. and Francois, L.E. (1973). Comparisons of drip, furrow, and sprinkler irrigation. *Soil Sci.,* 115, 73-86.

Bernstein, L., Francois, L.E. and Clark, R.A. (1974). Interactive effects of salinity and fertility on yields of grains and vegetables. *Agron. J.,* 66, 412-421.

Bradford, S. and Letey, J. (1992). Cyclic and blending strategies for using non-saline and saline waters for irrigation. *Irrig. Sci.,* 13, 123-128.

Camp, C.R. (1998). Subsurface drip irrigation, a review. *Trans. ASAE,* 41, 1353-1367.

Cappaert, M.R., Powelson, M.L., Christensen, N.W., Stevenson, W.E. and Crowe, F.J. (1992). Influence of irrigation severity on potato early dying tuber yield. *Phytopathplogy,* 82, 1448-1453.

Ceccarelli, S. (1987). Yield potential and drought tolerance of segregating population of barley in contrasting environments. Euphetica, 36, 265-273.

Cetin, O. and Bilgel, L. (2002). Effects of different irrigation methods on shedding and yield of cotton. *Agric. Water Manage.,* 54, 1-15.

DeMalach, Y. and Pasternak, D. (1993). Agriculture in a desert saline environment–a case study. In: G.B. Marani-Bettolo (Ed.), Agriculture and the Quality of Life–New Global Trends. Proceeding of an International Conference held at The Ponitificia Academia Scientiarum, Ex Aedibvs acadimicis in Civitate Vaticana, Rome, Italy, pp. 97-126.

Devitt, D., Bowman, D. and Schulte, P. (1993). Response to Cynodon dactylon to prolonged water deficits under saline conditions. *Plant Soil,* 148, 239-251.

English, M., and Nakamura, B. (1989). Effects of deficit irrigation and irrigation frequency on wheat yields. *J. Irrig. Drain. Eng.,* 115 (2), 172-184.

English, M. and Raja, S.N. (1996). Perspectives on deficit irrigation. *Agric. Water Manage.,* 32, 1-14.

Fabeiro, C., Martin De Santa Ollaza, F. and De Juan, J.A. (2001). Yield and size of deficit irrigated potatoes. *Agric. Water Manage.,* 48, 255-266.

Feibert, E.B.G., Shock, C.C. and Saunders, L.D. (1998). Nitrogen fertilizer requirements of potatoes using carefully scheduled sprinkler irrigation. *Hort. Sci.,* 33(2), 262-265.

Feigin, A. (1985). Fertilization management of crops irrigated with saline water. *Plant and Soil,* 89, 285-299.

Fereres, E., Cuevas, R. and Orgaz, F. (1985). Drip irrigation of cotton in southern Spain. Proceedings Third Drip/Trickle lrrigation Congress, ASAE Publ. 10-85, ASAE, St. Joseph, Mi., pp. 185-192.

Ferreira, T.C., Malheiro, A.N.C., Freixo, F.A.M.F.P., Bernardo, A.A.S. and Carr, M.K.V. (1999). Variation in the response to water and nitrogen of potatoes (Solanum tuberosum L.) grown in the highlands of Northeast Portugal. In Proceedings of 14th Triennial

Conference of the European Association for Potato Research, 2-7 May 1999, Sorrento, Italy, pp. 401-411.

Geerts, S. and Raes, D. (2009). Deficit irrigation as an on-farm strategy to maximize crop water productivity in dry areas. Agric. *Water Manage.*, 96, 1275-1284.

Ghinassi, G. and Trucchi, P. (2001). Deficit irrigation trials on maize in a Mediterranean semi-arid environment. *Int. Water Irrig.*, 21 (1), 12-17.

Goffart, J.P., Destain, O. and Frankinet, M. (2002). Stratégie de gestion de la fertilisation azotée de la pomme de terre de consommation. Centre de Recherches Agronomiques de Gembloux, Septembre 2002, 118 p.

Goldberg, D. and Shamueli, M. (1970). Drip irrigation - a method used under arid and desert conditions of high water and soil salinity. *Trans. ASAE,* 33, 38-41.

Grattan, S.R., Shennan, C., May, D., Roberts, B., Mitchell, J.P. and Burau, R.J. (1987). Use of drainage water for irrigation of melons and tomatoes. *Cal. Agric.*, 41, 27-28.

Hajor, A.S., Al-Hatalani, L.S. and Khafagi, O.A. (1996). A comparative study on salt tolerance of millet (Pennisetum glaucum L.) and sorghum (Sorghum bicolor L.). *Alexandria Journal of Agriculture Res.*, 41, 23-39.

Halitligil, M.B., Akın, A. and Ilbeyi, A. (2002). Nitrogen balance of nitrogen-15 applied as ammonium sulphate to irrigated potatoes in sandy textured soils. *Biol. Fertil. Soils,* 35 (5), 369-378.

Hamdy, A. and Samti, H. (1995). Interaction salinité-fertilisation et son effet sur la production de la pomme de terre. Proceedings of Workshop on Farm Sustainable Use of Saline Water in Irrigation: Mediterranean Experiences, Hammamet, Tunisia, a9.1- a9.24.

Hanson, B.R. and May, D.E. (2004). Effect of subsurface drip irrigation on processing tomato yield, water table depth, soil salinity, and profitability. Agric. *Water Manage.*, 68, 1-17.

Hartmann, H.D., Werum, H., Bruckuer, U. and Zengerle, K.H. (1986). Effect of irrigation on carrot yield. *Gemuse,* 22, 62-65.

Hassanli, A.M., Ebrahimizadeh, M.A. and Beecham, S. (2009). The effects of irrigation methods with effluent and irrigation scheduling on water use efficiency and corn yields in an arid Region. Agric. *Water Manage.*, 96, 93-99.

Imtiyaz, M., Mgadla, N.P., Manase, S.K., Chendo, K. and Mothobi, E.O. (2000). Yield and economic return of vegetable crops under variable irrigation. *Irrigation Sci. J.*, 19 (2), 87-93.

Jamieson, P.D., Francies, G.S., Wilson, D.R. and Martin, R.J. (1995). A drought influences on grain yield of barley, wheat and maize. N. Z. J. *Crop Hortic. Sci.*, 23, 55-66.

Kaman, H., Kirda, C., Cetin, M. and Topcu, S. (2006). Salt accumulation in the root zones of tomato and cotton irrigated with partial root-drying technique. *Irrig. Drain.*, 55, 533-544.

Kirda, C.R. (2002). Deficit irrigation scheduling based on plant growth stages showing water stress tolerance. In: Deficit Irrigation Practices, *FAO Water Reports,* 22, pp. 3-10.

Maas, E.V. and Hoffman, G.J. (1977). Crop salt tolerance: current assessment. *J. Irrig. Div. Am. Soc. Civ. Eng.*, 103, 115-134.

Mahalakshmi, V. and Bidinger, F.R. (1985). Water stress and time of floral initiation in pearl millet. *Journal of Agricultural Sciences,* 105, 437-445.

Mahalakshmi, V. and Bidinger, F.R. (1986). Water deficit during panicle development in pearl millet: Yield compensation by tillers. *J. Agric. Sci.*, 6,113,119.

Mahalakshmi, V., Bidinger, F.R. and Rao, G.D.P. (1988). Timing and intensity of water deficient during flowering and grain filling in pearl millet. *Agronomy J.*, 80,130-135.

Mao, X., Liu, M., Wang, X., Liu, C., Hou, Z. and Shi, J. (2003). Effects of deficit irrigation on yield and water use of greenhouse grown cucumber in the North China Plain. *Agric. Water Manage.*, 61, 219-228.

Meiri, A. (1984). Plant response to salinity: Experimental methodology and application to the field. In: J. Shainberg, and J. Shalhevet (Eds.), Soil Salinity and Irrigation. Ecological Studies 51, *Springer,* Heidelberg, pp. 284-297.

Mugabe, F.T. and Nyakatawa, E.Z. (2000). Effect of deficit irrigation on wheat and opportunities of growing wheat on residual soil moisture in southeast Zimbabwe. *Agric. Water Manage.,* 46, 111-119.

Nagaz, K., Toumi, I., Mahjoub, I., Masmoudi, M.M. and Ben Mechlia, N. (2009). Yield and water-use efficiency of pearl millet (Pennisetum glaucum (L.) R. Br.) under deficit irrigation with saline water in arid conditions of Southern Tunisia. *Research Journal of Agronomy,* 3(1), 9-17.

Nagaz, K., Toumi, I., Masmoudi, M.M. and Ben Mechlia, N. (2008). Soil salinity and barley production under full and deficit irrigation with saline water in arid conditions of southern Tunisia. *Research Journal of Agronomy,* 2(3), 90-95.

Nagaz, K., Toumi, I., Masmoudi, M.M. and Ben Mechlia, N. (2008). Comparative effects of drip and furrow irrigation with saline water on the yield and water use efficiency of potato (Solanum tuberosum L.) in arid conditions of Tunisia. *Agricultural Journal,* 3(4), 272-277.

Nagaz, K., Masmoudi, M.M. and Ben Mechlia, N. (2007). Soil salinity and yield of drip-irrigated potato under different irrigation regimes with saline water in arid conditions of Southern Tunisia. *Journal of agronomy,* 6(2), 324-330.

Nagaz, K. and Ben Mechlia, N. (1998). Barley production under different irrigation regimes with two water qualities. *Agricoltura Mediterranea,* 128(1), 28-33.

Nagaz, K., Masmoudi, M.M. and Ben Mechlia, N. (2006). Stratégies de gestion et d'utilisation des eaux salées en irrigation: Cas des céréales dans les régions arides de sud tunisien. *Revue des Régions Arides,* 21 (3), 1100-1106.

Nagaz, K., Masmoudi, M.M. and Ben Mechlia, N. (2004). Etude de la réponse de la pomme de terre de saison, de primeur et d'arrière-saison aux régimes d'irrigation à l'eau salée dans le Sud Tunisien. *Revue des Régions Arides,* 294-304.

Nagaz, K., Masmoudi, M.M. and Ben Mechlia, N. (2009). Formes d'adaptation à des conditions du déficit hydrique en milieu aride du sud Tunisien. Actes du Colloque International "Sociétés en transition et développement local en zones difficiles" DELZOD, Jerba, 22-24 Avril, 2009, pp. 619-629.

Nagaz, K., Khlouj, K., Toumi, I., El-Mokh, F., Masmoudi, M.M. and Ben Mechlia, N. (2011). Irrigation water quality and nitrogen for yield and water-use efficiency of potato in the arid conditions of Tunisia. *Journal of Applied Horticulture,* 13(2), 127-133.

Nagaz, K., Masmoudi, M.M. and Ben Mechlia N. (2012). Effects of surface and subsurface drip irrigation and irrigation regimes with saline water on yield and water use efficiency of potato in arid conditions of Tunisia. *Accepted for publication in Journal of Applied Horticulture.*

Nagaz, K., Masmoudi, M.M. and Ben Mechlia, N. (2012). Impacts of irrigation regimes with saline water on carrot productivity and soil salinity. *Journal of the Saudi Society of Agricultural Sciences,* 11, 19-27.

Oron, G., DeMalach, Y., Gillerman, L. and David, I. (1995). Pear response to saline water application under subsurface drip irrigation. Proceedings of the 5th International Microirrigation Congress, Orlando, FL, 2-6 April, pp. 97-103.

Oron, G., DeMalach, Y., Hoffman, Z., Keren, Y., Hartmann, H. and Plazner, N. (1990). Wastewater disposal by subsurface trickle irrigation. Proceedings, 15th Biennial Conference, IAWPRC, Kyoto, Japan, 29 July-3 August, pp. 2149-2158.

Oron, G., DeMalach, Y., Hoffman, Z. and Cibotaru, R. (1991). Subsurface microirrigation with effluent. Journal Irrigation and Drainage Engineering Division, *ASCE,* 117, 25-36.

Oron, G., DeMalach, Y., Hoffman, Z. and Manor, Y. (1992). Effect of effluent quality and application method on agricultural productivity and environmental control. *Water Sci. Technol.,* 26, 1593-1601.

Oron, G., DeMalach, Y., Gillerman, L., David, I. and Rao, V.P. (1999)). Improved saline-water use under subsurface drip irrigation. *Agric. Water Manage.,* 39, 19-33.

Oron, G., DeMalach, Y., Gillerman, L., David, I. and Lurie, S. (2002). Effect of water salinity and irrigation technology on yield and quality of pears. *Biosyst. Eng.,* 81, 237-247.

Panda, R.K., Behera, S.K. and Kashyap, P.S. (2003). Effective management of irrigation water for wheat under stressed conditions. *Agric. Water Manage.,* 63 (1), 37-56.

Parabhakar, M., Srinivas, K. and Hegde, D.M. (1991). Effect of irrigation regimes and nitrogen fertilization on growth, yield, N uptake and water use of carrot (Daucus carota L.). Gartenbauwissenschaft, 56(5), 206-209.

Paradiso R., De Pascale, S. and Barbieri, G. (2002). Effects of water regime and nitrogen supply on yield and mineral composition of carrot. *Itlus Hortus,* 8(4), 8-13.

Patel, N. and Rajput, T.B.S. (2008). Effect of drip tape placement depth and irrigation level. *Agric. Water Manage.,* 88, 209-223.

Patel, A.D. and Pandey, A.N. (2008). Effect of soil salinity on growth, water status and nutrient accumulation in seedling of Cassia Montana (Fabaceae). *J. Arid Environ.,* 70, 174-182.

Phene, C.J., Davis, K.R., Hutmacher, R.B., Bar-Yosef, B., Meek, D.W. and Misaki, J. (1991). Effect of high frequency surface and subsurface drip irrigation on root distribution of sweet corn. *Irrig. Sci.,* 12, 135-140.

Phene, C.J. (1993). Subsurface drip irrigation on row crops. Proceedings Microirrigation Workshop and Trade Show, Santa Maria, CA, 29 October, pp. 14-32.

Phene, C.J. (1995). Sustainability and potential of subsurface drip irrigation. In: Proceedings of the Fifth International Micro Irrigation Congress, Orlando, Florida, pp. 359-367.

Phene, C.J., Blume, M.F., Hile, M.M.S., Meek, D.W. and Re, J.V. (1983). Management of subsurface trickle irrigation systems. *ASAE Paper* No. 83-2598.

Raes, D., Smith, M., De Nys, E., Holvoet, K. and Makarau, A. (2002). Charts with indicative irrigation intervals for various weather conditions. Proceedings of workshop organized by FAO-ICID on Irrigation Advisory Services and Participatory Extension in Irrigation Management, July 2002, 23-24, Montreal, Canada, pp. 65-72.

Rains, D.W, Goyal, S., Weyranch, R. and Lauchli, A. (1987). Saline drainage water reuse in a cotton rotation system. *Calif. Agric.,* 41, 24-26.

Rhoades, J.D., Kandiah, A. and Mashali, A.M. (1992). The use of saline waters for crop production. FAO Irrigation and Drainage Paper No. 48, Rome, Italy.

Rhoades, J.D. (1983). Reusing saline drainage waters for irrigation: A strategy to reduce salt loading of rivers. Proc. Int. Symp. State of the Art Control of Salinity, Salt Lake City, Utah, 12-15 July, Chap. 3, pp. 455-464.

Rhoades, J.D. (1984). Use of saline water for irrigation. *Calif. Agric.,* 38, 42-43.

Rhoades, J.D. (1985). New strategy for using saline waters for irrigation. In: J.A. Replogle, and K.G. Renard (Eds.), Proc. ASCE Irrigation and Drainage Spec. Conf.: Water Today and Tomorrow, July 2-26, 1984, Flagstaff, AZ. ASCE, New York.

Rhoades, J.D. (1987). Use of saline water for irrigation. *Water Quality Bull.,* 12, 14-20.

Rhoades, J.D. (1989). Intercepting, isolating and reusing drainage waters for irrigation to conserve water and protect water quality. *Agric. Water Manage.,* 16, 37-52.

Rhoades, J.D., Bingham, F.T., Letey, J., Dedrick, A.R., Bean, M., Hoffman, G.J., Alves, W.J., Swain, R.V., Pacheco, P.G. and LeMert, R.D. (1988a). Reuse of drainage water for irrigation: Results of Imperial Valley Study. I. Hypothesis, experimental procedures and cropping results. *Hilgardia,* 56(5): 1-16.

Roltson, D.E., Rains, D.W., Biggar, J.W. and Lauchli, A. (1988). Reuse of saline drain water for irrigation. In: K.K. Tanji (Ed.), Agricultural Salinity Assessment and Management. *Am. Soc. Civil Engineers,* 70, 432-439.

Saggu, S.S. and Kaushal, M.P. (1991). Fresh and saline water irrigation through drip and furrow method. *Int. J. Trop. Agric.,* 9, 194-202.

Schoups, G., Hopmans, J.W., Young, C.A., Vrugt, J.A., Wallender, W.W., Tanji, K.K. and Panday, S. (2005). Sustainability of irrigated agriculture in the San Joaquin Valley, California. In: Proc. National Academy of Sciences USA, 102, pp. 15352-15356.

Sepaskhah, A.R. (1978). Time and frequency of irrigation in relation to growth stages of barley. *Agron. J.,* 70, 731-734.

Shalhevet, J. (1984). Management of irrigation with brackish water. In: I. Shainberg and J. Shalhevet (Eds.), Salt Salinity under Irrigation, Processes and Management (pp. 298-318). New York: Springer Verlag.

Shock, C.C., Feibert, E.B.G. and Saunderds, L.D. (1998). Potato yield and quality response to deficit irrigation. *HortScience,* 33, 655-659.

Singh, B. and Bhumbla, D.R. (1968). Effect of quality of irrigation water on soil properties. *J. Res. Punjab Agric Univ.,* 5, 166-172.

Singh, S.D., Gupta, J.P. and Singh, P. (1977). Water economy and saline water use by drip irrigation. *Agronomy J.,* 70, 948-951.

Ünlü, M., Kanber, R., Şenyigit, U., Onaran, H. and Diker, K., (2006). Trickle and sprinkler irrigation of potato (Solanum tuberosum L.) in the Middle Anatolian Region in Turkey. *Agric. Water Manage.,* 79, 43-71.

van Genuchten, M.Th. (1983). Analyzing crop salt tolerance data: model description and user's manual. Research Report N° 120, p.50, USDA-ARS, U.S.S. Laboratory, Riverside, CA.

van Oosterom, E.J., O'Leary, G.J., Carberry, P.S. and Craufurd, P.Q. (2002). Simulating growth, development, and yield of tillering pearl millet. III. Biomass accumulation and partitioning. *Field Crops Res.,* 79, 85-106.

Waddel, J.T., Gupta, S.C., Moncrief, J.F., Rosen, C.J. and Steel, D.D. (1999). Irrigation and nitrogen management effects on potato yield, tuber quality, and nitrogen uptake. *Agron. J.,* 91, 991-997.

Weatherhead, E.K. and Knox, J.W. (1998). Irrigation potatoes three trickle irrigation for potatoes. *Irrigation News,* 27, 19-28.
Weltzien, H.C. and Srivastava, J.P. (1981). Applications in breeding strategies, ICARDA, Aleppo (Syria). In: barley genetics IV, Fourth. International Barley Generics Symposium, Edinburgh, Scotland, pp. 351-369.
Yohannes, S. and Tadesse, T. (1998). Effect of drip and furrow irrigation and plant spacing on yield of tomato at Dire Dawa. *Ethiopia. Agric. Water Manage.,* 35, 201-207.
Zhang, Y., Kendy, E., Qiang, Y., Changming, L., Yanjun, S. and Hongyong, S. (2004). Effect of soil water deficit on evapotranspiration, crop yield, and water use efficiency in North China Plain. *Agric. Water Manage.,* 64, 107-122.

In: Irrigation Management, Technologies …
Editor: M. Hossain Ali

ISBN: 978-1-62417-862-7

Chapter 5

WATER CONSERVATION INTERVENTIONS IN PAKISTAN AGRICULTURE FOR ACHIEVING FOOD SECURITY: SYNTHESIS AND NEW EVIDENCE

***Rai Niaz Ahmad*[1,2], *M. Arshad*[2] *and Munir A. Hanjra*[*,3,4]**

[1]Water Management Research Centre,
University of Agriculture, Faisalabad, Pakistan
[2]Department of Irrigation and Drainage,
University of Agriculture, Faisalabad, Pakistan
[3]Charles Sturt University, Wagga Wagga Campus, Australia
[4]Future Directions International,
Perth, WA, Australia

ABSTRACT

Conserving water, preserving land can make a sterling contribution towards achieving food security in Pakistan. Water competition among agriculture and other sectors is intensifying the need to conserve irrigation water through productivity enhancing interventions for boosting food production and supporting rural livelihoods. Water conservation interventions are everywhere, yet social and economic incentives are lacking for their adoption and up scaling for revitalizing irrigation, and to support socioeconomic transformation and uplift the standard of living across South Asian countries – where bulk of the world's irrigation is concentrated and holds huge importance for food security and peace in the region. There is also a need to understand the impacts of these interventions and economic incentives in terms of enhancement in productivity, crop output, input use, economic returns, gender empowerment and social equity effects for various crop-mix, farm-size and social classes which condition and mediate the technology adoption decisions. Such comprehensive analysis of water conservation investments is yet to be done. This chapter presents a synthesis of water conservation interventions with emphasis on the Indus Basin where bulk of Pakistan agriculture is concentrated. The interventions include: watercourse improvement program; laser land leveling; bed and furrow irrigation/bed planting; raised bed

[*] Corresponding author: E-mail: mahanjra@hotmail.com.

technology; permanent raised beds; reduced or zero tillage; drip/irrigation; and sprinkler system. Synthesis and new evidence is drawn from a number of resource conservation projects and interventions implemented at farmer fields in a number of settings across the Basin. Data shows that these interventions conserve water and other inputs, boost crop yield and output, and improve food security and livelihoods of the farmers. In particular, new evidence shows that bed furrow planting system improves crop yield for wheat, rice, sugarcane and cotton and offers practical opportunities for producing "more crop per drop" and contributing towards achieving food security in Pakistan.

1. Background

Conserving water, preserving land is the main pathway to ensuring global food security (Molden et al., 2010). Over the past 50 years, gains in agricultural production have provided a platform for rural and urban growth worldwide including South Asia (Pretty et al., 2011). The challenge to produce "more crop per drop" is more urgent than ever before due to growing population and lesser land and water resources per capita. There is a need for moving towards a Blue-Green Revolution through sustainable agricultural modernization process (Horlings and Marsden, 2011) involving social, economic, cultural, and spatial and political aspects that affect technology adoption decisions and productivity at farmer fields. Agriculture should not be seen as a servant of economic growth (McMichael and Schneider, 2011) rather should be developed as foundational source of food security (Pretty et al., 2003; 2006), social and economic development, and environmental sustainability.

To meet the food demand of ever increasing population in Pakistan, conservation of available water resources has become a prime concern. This chapter highlights the present and future status of "Water Conservation Interventions" for sustainable agriculture in the country. Out of 21mha of land under agriculture, about 17mha is irrigated for which the demand of irrigation water is being met from three western rivers i.e. Indus, Jhelum and Chenab, and groundwater. Land resources are available to expand irrigated agriculture and significant potential exists for improving crop productivity but water is becoming a significant constraint to the way forward for further expansion in agriculture and increase in production. However, there is a consistent increase in cropping intensity with little increase in agricultural area. It is important to mention that irrigation system in the country was designed for 75% cropping intensity and is unable to meet the irrigation demand of present cropping intensity of about more than 150%. Due to increase in cropping intensity, irrigation water is becoming a permanent root cause of conflict among provinces, farmers and clans. Most of rainfall occurs during the monsoon season (July to August) in the Indus Basin catchment. Most of this rainfall cannot be used for crop production and there is a need for building a number of water reservoirs to capture and use the rainwater. Due to ongoing debate among provinces for irrigation water distribution, there is less scope for building new reservoirs in the near future. Under the present situation, sustainability of agriculture in the country is facing a big threat due to water scarcity. Presently, flood irrigation is a traditional method of irrigation and there is lot of scope to introduce improved irrigation methods i.e. bed furrow irrigation, sprinkler irrigation and drip irrigation. Thus, on short term basis, there is enough scope for improved water saving interventions to address food security issue of increasing populace. In addition, there is a need to employ each and every option of water conservation technique like laser land leveling, farm layout, cropping pattern, residue management and zero tillage etc. Another option of increasing on farm water for agriculture is the safe use of wastewater and skimming of fresh water overlying the saline water. There is a list of different technologies which can enhance water availability and water productivity on the farm. Some of those technologies already being tested in the country are discussed in this chapter.

1.1. Past, Present and Future Water Scenarios

Agriculture in the Indus Basin plains is historic and was mostly rainfed or irrigated by river's outrage and groundwater in its early stages. From late eighteen century onward, irrigation network was constructed in the area without drainage which resulted in an intensive agriculture but with a sever problem of twin menace: waterlogging and salinity. As a result vertical drainage was introduced, later on strengthened with horizontal drainage as well. In its earlier stage after introduction of canal system in the area, there was low intensity agriculture with an abundant supply of surface water, groundwater and sufficient rainfalls. Under this scenario, flood irrigation was adopted preferably as a low cost irrigation practice under gentle slope suitable for gravity flow. The flood irrigation practice got its place as a common irrigation practice in Indus basin. Along with extensive irrigation network, routine floods and flood irrigation were the major causes of water-logging and salinity. Enough efforts were made to control waterlogging and salinity in Indus part of country but each one was limited to macro scale, larger area than the micro or field scale. As an outcome of this drainage practice, the vertical drainage practice, tubewells, was adopted by the farmers not for the drainage purpose but for irrigation purpose to meet irrigation demand of their expanding agriculture to help produce more for the increasing population. However, limited efforts were made to improve traditional flood irrigation for any improved irrigation practice or technology. Over time, both agriculture land and water resources per capita squeezed and resulted in crop diversification and higher cropping. The water shortage was aggravated under drought like situation in the country from 1998 and did not come back so far. The canal or surface supply irrigation based agriculture went to groundwater based agriculture as tubewell number increased from 0.3 million to 1.2 million during last decade. There is over-exploitation of groundwater as pumping rate is too higher than the recharge resulting in fast decline in groundwater table, making expensive to pump groundwater, enhancing intrusion of saline water into fresh water and above everything creating threat of secondary salinity a serious threat to the agriculture of Pakistan. If this situation is not addressed properly under the climate change scenario of drying rivers, there could be a serious threat to sustainable agriculture.

1.2. Traditional Irrigation Practices

Surface irrigation is a traditional irrigation practice in the country. Basin size depends on the farm size and source of irrigation. Mostly small farmers (<2 ha), and tube well irrigated farms have smaller basin to irrigate than the medium farmers (> 2 to 5 ha) of tube well and/or canal irrigated farms. Although Indus Basin Irrigation System (IBIS) itself was designed to apply less amount of water on larger area; thus system represents a case of managing water scarcity and spreading limited water supplies amongst the largest number of farmers across the Basin. Negligible efforts have been made to introduce modern flood irrigation practices or high efficiency irrigation systems. Under population pressure, farmers adopted some measure of saving water like ridge-furrow irrigation practice but the method of irrigation remained limited to some specific crops mostly for vegetable and maize or cotton. Border irrigation could be seen with the medium and large farms (> 5 ha) who have farm power like tractor or some other alternates.

Policy makers realized that irrigation water shortage in early 1970s were posing a threat to agriculture across the Indus Basin. Watercourse lining programs were initiated in all the country which is still continuing under "National Watercourse Lining Program".

During last few years, awareness about increasing water shortage has increased, both at the level of government and common farmer. Various efforts are being made to save water from main canal to tertiary systems and at farm level. Work on rehabilitation of main canal systems is in progress with other reforms like involvement of farmers to operate the system mainly to address water inequity issue. The vital part of recent water saving activities at government level is the introduction of improved irrigation practices/high efficiency irrigation systems (HEIS)". Pakistan government has launched a "HEIS" where farmers have the facility of subsidy for installing drip and sprinkler system at their farms. At the same time, Government is helping farmers for acquiring other water saving technologies like laser land leveling and bed-furrow planting. Regarding these activities, smaller projects are being funded in each province to help farmers for providing laser levelers and bed-furrow planting machines. The main milestones of all these programs are to replace traditional floods irrigation with improved irrigation practices.

1.3. Current Crop and Water Productivities

The increase in cropped area and cropping intensity is not only resulting in a wider gap between surface water demand and supply but also increasing dependence of agriculture on groundwater (Hussain et al., 2004). Groundwater quality in the Indus basin varies from good to poor as system leads from head to middle and middle to tail. The situation is not different at all levels like the main system to tertiary, head, middle and tail of the individual main canal, distributary, minor or even a watercourse. The other groundwater issue exists in the Indus Basin is that water quality deteriorates as one moves away from the river's bed to middle of Doab (area between to rivers). It is important to report that agriculture of Pakistan is no more surface supply water based but groundwater based as share of groundwater is about 60-70 percent. The area of crop irrigated with poor quality groundwater has not only less crop and water productivity but also is under severe threat of secondary salinity. Their cost of production is higher and economic returns are lower such that potential livelihood opportunities are lost.

Under increasing gap between demand and supply of irrigation water, yields of major crops are decreasing and finally reducing crop and water productivity. Relationship holds generally true for all crops with some exceptions where farmers are using agricultural chemicals for better crop's yield and soil health. The relationships between irrigation application, yield, and cropped area especially for land and water productivity may worsen under changing climate of low water supplies in rivers and more dependence on groundwater.

Table 1. Area, production and yield per hectare of agricultural crops in irrigated areas of Punjab

	Wheat			Cotton			Rice			Sugarcane		
Year	Area (000 ha)	Production (000 tonnes)	Yield (kg/ha)	Area (000 ha)	Production (000 tonnes)	Yield (kg/ha)	Area (000 ha)	Production (000 tonnes)	Yield (kg/ha)	Area (000 ha)	Production (000 tonnes)	Yield (t/ha)
2000-01	6255.5	15419	2465	2386.4	8540	609	1627.2	2577	1584	615.5	26740	48
2001-02	6101.8	14594	2518	2526.4	8846	542	1475.9	2266	1535	656.8	31803	45
2002-03	6097.3	15355	2500	2208.3	7664	590	1512.3	2580	1706	735.3	33169	45
2003-04	6255.5	15639	2724	2386.8	7702	549	1687.9	2871	1701	709	34023	48
2004-05	6378.9	17375	2724	2518.3	11149	753	1754.2	2980	1699	644.6	33048	51
2005-06	6483.4	16776	2588	2426	10268	720	1762.4	3180	1804	625.2	28969	46
2006-07	6432.8	17853	2775	2462.9	10350	715	1728.4	3075	1779	711.8	37542	53
Average	6286.46	16144.43	2613.43	2416.44	9217.00	639.71	1649.76	2789.86	1686.86	671.17	32184.86	48.00
Increase (%)	2.83	15.79	12.58	3.21	21.19	17.41	6.22	19.32	12.31	15.65	40.40	10.42

Data source: Punjab Agricultural Statistics (various years).

Table 1 presents areas of four major crops of Punjab i.e. wheat, cotton, rice and sugarcane, their total production and land productivity in terms of kg per hectare and tones per hectare in case of sugarcane. A trend of increase in area of each crop can be seen in the last seven years. Table further shows that sugarcane (15.6 %) increase was greater than the other crops which are 2.4, 3.2 and 6.2 for wheat, cotton and rice respectively. It is notable that area of high delta crops (sugarcane and rice) had increasing trends during last decade even with significant reduction of water supplies.

To enhance water productivity and discourage high delta crops under increasing irrigation water demand scenario, there is a dire need to declare irrigation water "as a valuable input commodity. This is only possible by introducing irrigation water price for both surface and groundwater and providing economic incentives to promote more efficient use of available water resources. Action may reduce the irrigation losses on the farm, enhance the uptake of water conservation technology, improve crop and water productivity and discourage increasing trends of cultivation of high delta crops.

2. Water Conservation Technologies

2.1. Scope and Constraints

Conservation agriculture is an approach for design and management of resource conservation agriculture system. Main objectives of conservation agriculture is to conserve, improve and make more efficient use of natural resources through integrated management of soil, water, crop and other biological resources in combination with selected external inputs. Such a technological package represents a resource saving and efficient/ effective agriculture that contributes to environmental conservation and at the same time enhances production on sustainable basis. Elements of conservation agriculture include on farm water management, minimum tillage and direct seeding through crop residue (Govaerts et al., 2009).

Water conservation technologies (WCTs) are a part of conservation agriculture and play significant role in achieving goals of conservation agriculture. The commonly used WCTs include watercourse improvement which is a basin level activity in the country but it is important to mention that only a portion of main watercourses are being partially lined and there is no lining at the level of farmers field. LASER land leveling, bed and furrow irrigation system/ bed planting (Raised Bed Technology), zero tillage and high efficiency pressurized irrigation system etc are the "On Farm Water Saving Interventions" for saving irrigation water. The surface irrigation water reaches on the farm is the most precious one and needs to be used on "crop per drop basis" as loss of such water on the farm owes a much higher value due to its cost paid to carry it on the farm and suboptimal returns on farming inputs and investments.

On the other hand, farmers especially small farmers can not afford to adopt such technologies as they need high initial/capital cost, operational power, specific training for proper operation, repair and maintenance of the system etc. It is common that new technologies have limited supply, higher cost due to less commercial values and as a result industry involved are reluctant for the manufacturing of new machinery etc and finally the new introduced technologies cost higher than routine business.

2.2. Water Saving Technologies

Any technology or package of technologies, for its wide spread acceptability among farming community, should have clear cut advantages over the traditional or its competitive technologies. Use of WCTs has been evaluated widely throughout the world as well in Pakistan. Although many of these studies have indicated the superiority of these technologies yet there are apprehensions and doubts regarding the adaptability of these technologies into the existing cropping practices of majority of the farmers. Furthermore, there are few research studies that have indicated that the use of WCTs in their current form might not be useful for water saving. So there exists a need for further evaluation of these technologies on a wider scale to assess their feasibility in socio-economic perspective. Different water saving technologies are under way to save irrigation water especially to replace traditional flood irrigation, improve irrigation application efficiency, reduce on farm water losses especially conveyance losses in Pakistan. Some of those are explained below briefly.

2.2.1. Watercourse Improvement

Watercourse is a designed channel to take allocated irrigation surface water from minor or distrbutary to the farmer fields. The tertiary irrigation conveyance network in Pakistan mainly includes minors and watercourses, whereas watercourses are generally community channels, off-taking from government controlled irrigation system. The watercourses are operated and maintained by the farmers receiving water through these channels. A watercourse commands an area of about 150-250 ha and is shared by 40-50 farm families. Gill (2004) reported that studies have indicated that about 40% of water is lost during its conveyance through nation's 135,000 watercourses because of their aging and deteriorated condition.

Improvement of watercourse consists of complete demolishing of community channel and its rebuilding/ realigning according to the design with clean compacted soil. Portions of channels are lined and necessary control structures are installed to save the losses of irrigation water. In first instance, about 1330 watercourses under the pilot project over 5 year's period (1976-1981) were improved. It was further reported by Gill (2004) that annual water saving in an improved watercourse is about 12.5 hectare meter besides other socioeconomic benefits (Table 2).

Table 2. Major benefits of watercourse improvement

S. No.	Particulars	Extent (%)
1	Time saving in irrigation time	28
2	Labor saving	50
3	Increase in cropping intensity	23
4	Increase in cropped area	17
5	Increase in yield	16-37
6	Increase in net farm income	20

Considering the future water requirements and to ensure the food security of the country a project entitled "National Program for Improvement of Watercourses (NPIWC)" costing Rs.

66000 million for improving and partial lining of watercourses was launched. The detail of province wise watercourse improvement achieved under NPIWC is presented in Table 3.

Table 3. Watercourse improvement achieved under NPWIC till 2007-08

Description	Physical targets (Nos.)	Physical progress (Nos.)					Total
		2003-04	2004-05	2005-06	2006-07	2007-08	
Punjab	30000	3	2606	5105	5004	5026	17744
Sindh	29000	1	300	5071	5740	5473	16585
NWFP	10000	75	1190	3177	3708	2235	10385
Balochistan	13466	-	744	3755	4142	2252	10893
Watercourses	7100	-	278	2452	2071	804	5605
Water storage tanks	6366	-	466	1303	2071	1448	5288
AJK	1000	-	76	17	69	76	238
FATA	1600	-	131	95	196	199	621
FANA	600	-	35	-	167	150	352
ICT	337	-	-	24	50	75	149
Total	86003	79	5082	17244	19076	15486	56967

2.2.2. LASER Land Leveling

Land leveling has always been an important component of agricultural operations as it helps in uniform irrigation application; improve irrigation application efficiency thereby avoiding over- or under-irrigation thus increasing the overall water and crop productivity. Farmers use local made planner to level their field up to some extent which is not sufficient to get the goal of uniform application of irrigation water and lose both the water crop yield. Moreover, with the increasing pressure on land and land preparation implements due to increase in cropping intensity, it would not be possible to level the land with traditional implements.

Precision land leveling (PLL) is a topographic modification, to attain grading and smoothing of land to an even level with an elevation difference of 2 cm from the average elevation. Precision land leveling improves irrigation application efficiency. Many local studies reported an increase of land use intensity from 8-63 percent and cropping intensity from 6-70 percent after the use of laser land leveling. Therefore, leveled field helps reduce the amount of irrigation water required and helps reduce the labor requirements. Jat *et al.*, (2006; 2009) provided few more interesting facts about laser land leveling. According to their findings, it increased cropping intensity by 40 percent, water application efficiency up to 50 percent, wheat yield by 15 percent, sugarcane yield by 42 percent and rice yield by 61 percent. Laser land leveling also helped in weed control thus reducing the expenditures on chemical weed control.

In Pakistan, Punjab On Farm Water Management Department has recently procured laser-guided PLL technology for attaining higher accuracy in leveling work and land leveling work in almost all parts of Punjab is in progress. Table 4 shows the updated laser leveling progress of the On Farm Water Management Department up to the year 2008.

Table 4. Laser land leveling progress in the Punjab

Sr. #	Year	Provision of LASER Units (Nos.)		Area Leveled (Acres)
		Target	Achievements	
1.	2005-06	500	500	40,000
2.	2006-07	1,000	1,000	155,000
3.	2007-08	1,000	1000	5,54500
Total		2,500	2,500	749,500

Source: OFWM, 2009.

Table 5 presents results of impact assessment studies conducted to reveal the benefits of precision land leveling.

Table 5. Major benefits of LASER land leveling

Particulars	Extent (%)
Curtailment in irrigation application	25
Reduction in labor requirements	35
Enhancement in irrigated area	2
Increase in crop yield	20

Source: OFWM, 2009.

2.2.3. Zero Tillage Technology

Minimum/zero tillage is not a relatively new technology in the country, but is still making its place in the rice wheat and cotton wheat zones of Pakistan. Late sowing of wheat, especially after rice and cotton, results in low production of wheat. Keeping the area of rice and wheat in view, there is a big potential to introduce the zero/minimum technology. Moreover, it is an innovation that not only conserves water and energy sources a prime need of the country but also helps sowing of wheat in short span of time after rice and cotton; a significant desire of the farmers. This technology has been practiced in many parts of the world. In America 17.3 Mha are grown with zero tillage whereas in Brazil 5 Mha are sown with this technique. This technique was first developed in Brazil in early 1970s to check massive soil erosion. Moreover, 3 Mha of land is sown with zero tillage in Argentina. In Africa, zero tillage is applied mainly in South Africa and Zimbabwe and to a lesser extent in Kenya and Tanzania (Steiner *et al.,* 1998) and Lesotho (Silici et al. 2011). To date, the most widely adopted resource conservation technology in the Indo-Gangetic Plains of South Asia has been zero tillage (Erenstein and Laxmi, 2008) Currently there is significant progress in this direction in Australia with 35% farmers in New South Wales and 90% in Western Australia adopting such systems (Maraseni and Cockfield, 2011).

Zero tillage was introduced in Pakistan during 1980's. Initial trials were confined to progressive farmers. Although yield results were impressive but uptake of technology remained limited. There is a dire need to conduct long term experiment for fact finding for enhancing the uptake of this important technology in the country.

Advantages of Zero Tillage

Following key advantages have been identified in various studies:

i. Enables timely sowing of wheat thus can add to the yield by about 20 percent
ii. Water saving of about 36 percent in rice-wheat system has been reported through zero tillage technology (Sharma *et al.*, 2009)
iii. On an average, there is a net benefit of US$150 per ha, through higher yields and less land preparation cost (Gupta, 2002; Gray, 1996).

But farmers and experts have concerns about the efficacy of this technology in the existing production technologies due to the incidence of heavy weed infestation of fields and also the greater incidence of soil borne diseases.

2.2.4. Bed & Furrow Irrigation System / Bed Planting

Application of surface irrigation through flood method is the most ancient way that is still in practice for applying water to field crops. It was estimated that about 90 percent of the irrigated area in the world is being irrigated through flood irrigation. In the face of increasing pressure on water resources, due to increasing population and cropping intensity, flood irrigation methods are losing their efficacy thus resulting in low application efficacy of irrigation due to changing soil characteristics (Berkhout, 1997).

Researchers suggest modified surface irrigation techniques, by developing crop specific field layouts, such as bed and furrow irrigation system helps to save water. In bed planting systems, wheat or other crops are planted on the raised beds in ridge- furrow system. This system is often considered more appropriate for growing high value crops that are more sensitive to temporary water logging stress. Farmers often raise crops such as cotton, maize-soybean and wheat on the raised beds. However, the practice of growing rice, the major water-using crop in rice-wheat systems, on narrow raised beds was introduced only very recently in the Indo-Gangetic Plains (IGP) to reduce water use, conserve rainwater and improve system productivity. Recent work shows that system of raised bed planting of crops may be particularly advantageous in areas where groundwater levels are falling and herbicide-resistant weeds are becoming a problem. This tillage and crop establishment option also facilitates crop diversification and intercropping of wheat, chickpea and Indian-mustard with sugarcane, maize with potato, mint with wheat, rice with soybean, and pigeon pea with sorghum or green gram. Results of farmer participatory trials indicate that significant water savings can be affected by planting rice (major consumer of irrigation water in rice-wheat systems) on raised bed besides improvement in crop yields.

Different studies have indicated the advantages of raised bed planting in many aspects. Few of these advantages are given below:

i. In Pakistan, wheat yield losses are considerable and one of the major causes of wheat lodging is the timing and method of irrigation. So raised bed planting of wheat could result in the increased yield of wheat by 11-16 percent (Kahloon *et al.*, 2002; Ahmad and Mehmood, 2005).
ii. Management of irrigation water is improved, simpler and more efficient. On an average it requires 30% less water than flat bed methods and improves crop yields by more than 20% (Department of Soil and Water Conservation, Punjab, India).
iii. Furrow Irrigated Raised Bed (FIRB) planting saves 30% to 50% wheat seed compared to flat bed planting.

iv. Better upland crop production is possible in the wet monsoon because of better drainage.
v. Fertilizer efficiency can be increased because of better placement including top dress applications (OFWM, 2009).

Situation in Punjab and Pakistan

In Pakistan, several experiments were initiated by University of Agriculture, Faisalabad (UAF), Water Resource Resreach Institute (WRRI), NARC/PARC in demonstrative research plots at farmers' fields in N.W.F.P (Mardan District) on waterlogged, salt-affected soils and Department of On Farm Water Management (OFWM) Punjab in cotton-wheat and maize-wheat cropping systems. Results of the experiences have shown that on an average the Permanent Raised Beds (PRB) saves around 25 mm of water per irrigation for maize crop and 30 mm of irrigation water for wheat crop. Similar amount of irrigation water saving is anticipated for cotton crop as well. In their fresh-initiatives, PARC and UAF in collaboration with OFWM department, Punjab has started projects for FIRB in rice-wheat cropping system. OFWM has planned to distribute about 400 furrow bed planters on 50 percent subsidy to the farmers in the cotton-wheat cropping zone by the year 2009-10.

2.3. Pilot Areas Studies

Realization about food security of increasing population has made essential to introduce WRTs at all levels of farming for saving irrigation water to bring more area under cultivation and also to improve crop water productivity of the current cropped areas. A number of pilot areas studies are under way where different institutions are involved conducting research studies for data base and before launching these technologies at macro-level or farmer's scale in different cropping zones in the country.

2.3.1. Rice-Wheat Consortium Project

Rehman (2007) reported about package of resource conservation technologies introduced under Rice Wheat Consortium (RWC) project in collaboration with PARC, UAF and OFWM in six villages of Tehsil Depalpur, District Okara-Pakistan from 2004 to 2007. The primary objective of the project was to enhance the livelihoods through integrated crop and resource management in rice-wheat cropping system. In this regard, various farmers participatory research trials were conducted on resource conservation technologies in the selected project villages. Precision land leveling was done on 1050 hectares through laser land leveling technology. To strengthen the machinery pool, two zero tillage drills, one bed planter, and one power sprayer were purchased from the project budget. RWC provided two turbo multiple crop seeders whereas OFWM department provided four laser land levelers, one bed planter and one zero tillage drill to project area. The available pooled machinery was provided to the farmers for mechanized farm operations on nominal charges.

Promotion of resource conservation technologies were made in the project area through individual contacts, field seminars, organizing television program publishing and distributing brochures with full packages of improved production technologies to the farmers and stakeholders. Dissemination of Resource Conservation Technologies included demonstration of Laser land leveling, zero tillage technology, bed planting for wheat, Leaf Color Chart Technology (LCC). He also reported experiments on farmers fields related to RCTs and crop residue management.

2.3.2. Permanent Raised Bed Project

Zahid (2009) reported in the brochure of the on-going project in the country that Permanent Raised Bed (PRB) farming systems aims at creating and maintaining a seed bed with near ideal physical and fertility conditions. The objectives of the PRB are to develop root zone for crops that absorbs rain and irrigation water more quickly, makes water easily available to plants, and allows plants roots to freely take the soil water and nutrients. Minimum soil operation retains the root of the previous crops not exposing soil organic matter or biota to the atmosphere. PRB implies that the beds stay in place for season, in comparison with being plowed down and re-constructed every season as with more intensive tillage system. The PRB does not however, imply that all bed disturbances are totally excluded. Some pre-sowing minimum cultivation may be necessary to renovate or reshape the beds. The project activities are being undertaken in all four provinces including FATA and AJK of Pakistan. The PRB adaptation and research activities are mainly for maize, wheat and cotton zones of Pakistan (Zahid, 2010).

Project Objectives

The specific objectives of the project are to:

i) Increase agricultural productivity of land and water resources by stimulating the adaptation of PRB farming practices for maize, wheat and cotton crops and explore its potential for other crops
ii) Optimize bed and furrow configurations for maize, wheat, cotton and other crops under various soil types in relation to the designed PRB farm machinery
iii) Study the fate/behavior of water, nutrients and solute movement within the soil and advise mechanisms/management practices to ensure the favorable balances of these parameters
iv) Commercialize locally low-cost, multipurpose and efficient PRB machinery suitable to the local conditions
v) Study the long term potential for adaptation of PRB farming systems in the severally/partially water logged areas
vi) General and broadly disseminate knowledge and skills of PRB farming systems, including quantitative data on the water and crop productivity benefits and socio-economic aspects.

Project Targets

Following are the major targets of the project:

- Mobilization of the farmers/private sector to actively participate in strengthening of PRB technology services in the target areas;
- Provision of 1000 sets of PRB machinery to the Farmers/Service Providers in the target area;

- Development of manufacturing facilities and after sale services by the PRB machinery manufacturers for effective utilization of PRB machinery along.
- Impart training to about 1500 farmers/services providers in operation and maintenance of PRB machinery and farm layout planning.

2.3.3. Cereal Systems Initiative in South Asia (CSISA) Project

Gupta (2009) reported that the RWC efforts to disseminate the Conservation Agriculture (CA) based technologies have been recognized on large scale and successful, proliferated in the field, far beyond its own control area. The established base of RWC has encouraged for the development of the Cereal Systems Initiative for South Asia (CSISA) project being funded by Bill & Melinda Gates Foundation to the tune of US$19.6 millions with USAID co-financing US$10 m. The CSISA is initially for a 3 years and will cover the South Asia region including Bangladesh, India, Nepal and Pakistan. Project manual indicates that the RWC has been successful in covering nearly 2 Mha of land under zero-till for wheat sowing in the rice-wheat system. Technologies and equipment introduced by RWC to achieve these goals included, zero till drill, laser land leveler, bed former and planter, happy seeder, turbo-seeder among the more prominent. In order to extend the reach of the CA based RCTs to large numbers of farmers it requires swift changes in the way research and extension systems are conducted by the National Agricultural Research Stations (NARS) in South Asia.

2.3.4. On Farm Research and Development Component LCC Rehabilitation Project

Ahmad et al. (2009) reported about the work being carried out under "Research and Development Component" of JICA funded project for Rehabilitation of Lower Chenab Canal System (East), one of the big and intensive canal system in Pakistan and facing lot of irrigation water shortage and salinity problems. The farmers in the lower part of the system, located at the tail of the tail system are suffering a lot due its poor quality groundwater being used for agriculture. The research and development component mainly address the following objectives:

- Improve existing farming/cultural practices and on-farm water management.
- Introduction of new technologies compatible with local socio-economic conditions,
- Confidence building of farmer community through involvement and skill development, there by strengthening institutional reform process.

The over all objective is social uplifting and in turn poverty reduction of rural communities

2.3.5. Future Thrust and Realization

Technologies reported in the above sections were compared for water saving and adoption level at the farmer's level. The most liked technology/practice at the farmers' level is the one which not only addresses the water saving but it also increases crop yield. Both the issues are of national importance under increasing demand of irrigation water. Looking in the face of water saving there is ample scope to improve irrigation application efficiency by replacing traditional flood irrigation with improved irrigation methods like bed-furrow planting (Raised Bed technology), HEIS i.e. drip & sprinkler irrigation. As reported in various studies; both the raised bed and HEIS not only save irrigation but also increase crop yield. Thus, the future thrust for agriculture in the country is to indigenization of bed-planting machine suitable for the available farm power and local development and design of low cost, low energy drip irrigation system suitable for small farms as 83% farmers hold small farms.

3. Raised Bed Technology

3.1. Definition

Raised bed technology is an advanced form of ridge furrow irrigation in which irrigation water is applied in furrows and crop is planted on the top or side of ridge. A lot of work has been done on the design of ridge furrow irrigation especially for the size and length of furrow. The most important design parameter in ridge furrow system is the infiltration rate of soil which affects advance movement of water in the furrow and finally irrigation application. Bed-furrow irrigation system is similar to ridge furrow in many aspects except the ridge size, as bed is bigger than the ridge i.e. 2-6 times of ridge depending upon the soil and crop to be planted on the beds. Thus, raised bed system is a combination of bed and furrow where furrow is used to apply water and bed is used to plant crop on the both sides and/or top of beds keeping distances in view for plant to plant and row to row of plants.

3.2. Size of Bed-Furrow Systems

Size of bed-furrow system is major concern for the indigenization prospects of raised bed technology. Both the bed and furrow sizes are very much depended to the type of soil and crop as well as the stream size. Keeping stream size constant, the first concern to be considered is the type of crop and its sowing practice whether the crop is sown at the edges of furrow or on the top of bed and how many lines. If crop is to be sown on both edges of the bed, the furrow size must addresses the row to row distance among plants. In this situation, furrows size can be controlled by planting the crop on top side of bed. Thus, major concern in this case is the number of plant population rather than lateral movement of irrigation water in the bed and dry stress issue. The lateral movement of water in the bed and dry stress to plants are significant design parameters of bed-furrow system when the crop needs to be grown on the bed in more than two lines. In case crops sown on the top of beds in more than two lines, sufficient soil moisture must reach to the central rows otherwise crop may suffer from the dry stress resulting in less yield. Moreover, if there is one or two irrigations with saline water (poor groundwater), as is common case in Pakistan, central part of the bed must not be planted as it acts as salt concentration area.

3.3. Design of Bed-Furrow Planting Machine

Pakistan, once having abundant supply of surface and groundwater, is now suffering with water shortage which may jolt the sustainability of its agriculture. Tackling of the issue at micro level has been found reasonably inexpensive, viable, and feasible under prevailing water use efficiency level. Forty to fifty percent water supplies on farm are lost due to inefficiency of traditional basin flood irrigation method. Loss of such big quantum of water on the farm requires urgent replacement of this traditional irrigation method. Adoption of raised bed technology could be a feasible option to improve water use efficiency and save enough water to delay acute water shortage scenario. Bed Planting Machine designed and developed at Water Management Research Center–University Agriculture, Faisalabad is being promoted for sowing different crops on bed-furrow system. Proper bed-furrow system and bed planting geometry were the main concerns for improving water use and meeting plant populations. A substantial saving in irrigation water (about 40–50%) with increase in yield (10–25%) encouraged the adoption of the 90 cm bed-furrow planting (60 cm x 30 cm) for all crops at larger scale. Water saving from transplanted rice on raised beds as compared to flat was 29% whereas yield increase was more than 25%. Similar results were obtained for maize, cotton and sugarcane, when inter-planted in wheat. The brief on potential of bed planting and its impact on water productivity of various grain crops in Pakistan reported by Ahmad et al.(2009) has been reproduced in this chapter.

3.3.1. Salient Features of the Bed Planting Machine

The machine has been equipped with three adjustable furrow openers (Figure 1). These openers have the provision to change both the depth and top width of each furrow, separately. The provision of this adjustment allows the user to obtain the required size of furrow considering the type of crop, soil and its seed bed preparation. The machine has also a provision of adjustable seed planting mechanism as well, which not only allows adjusting the planting depth but also allows adjusting the spacing within the line to line distance of the crop if required. The seed planting system consists of discs rather than the traditional tines. The disc system in comparison with the tines does not disturb the shape of bed and provides suitable small opening like fine channels to place the seed at proper depth. All of the four discs are bolted on a rectangular iron bar through a spring covered iron strips, which have been fixed on the main frame from where the sowing depth can be adjusted as a whole and/or on an individual disc bases. The spring provided on the iron strip between the disc and iron bar helps keep the proper depth of opening in the bed. There is an adjustable planer fixed on the main frame prior to the seed planting mechanism, which not only presses and levels the bed top but also provides flat surface on the bed for precise planting. As the planer covers all three furrows and two beds at a time, it results showing an in-built furrow-bed system with compacted edges of the furrows. Later on, the level beds are helpful for the harvesting of crop.

The planting machine has the provision to apply fertilizer underneath the beds. However, fertilizer is applied in the center of two adjacent rows on each side of the bed at deeper depth than the seed. Moreover, fertilizer applicators are fixed on the front side of the main frame while the seed sowing mechanism is fixed on the side of the main frame.

3.4. Planting of Crops on Bed-Furrow System

The bed planting machine developed for wheat has been used for all major crops and the machine is being commercialized by the industry. The results obtained under JICA funded project from experimental and farmer's areas, presented in ICID conference held at Dheli, India by Ahmad et al (2009) are presented here.

Figure 1. Four-row wheat bed planting machine and irrigation application.

Wheat

Plant Population

Plant population was one of the concerns in bed planting system especially in case of wheat. Efforts were made to address the concern by adjusting geometry of planting crop lines. Different, but commonly used bed-furrows systems and wheat lines per 0.4 ha for each bed-furrow system are presented in Table 6. Traditionally, there are 264 lines of wheat in 0.4 ha (one acre). Following two and three lines bed furrow system not only results in less number of wheat lines/population in comparison with traditional sowing of wheat but also results in less water saving. It was observed in experiments conducted for the evaluation of different bed-furrow systems that water saving increases with the increase of bed size; however, lateral intake of irrigation water into the bed limits the bed size. Calculation indicates that 90 cm bed furrow system matches the traditionally required number of the wheat lines/population. There are four lines on 60 cm bed, two lines on each side of bed. First line is at 7.5 cm and second line is at 15 cm from the furrow, leaving 21 cm buffer area in the center of the bed. Thus, each furrow has to irrigate only 21 cm of the adjacent beds while there is buffer zone (21 cm) in the center of the bed to take care of salts if there is irrigation with low quality water. The proposed wheat sowing geometry not only addresses the dry stress problem but also issue of salt stress to the crop.

Experiments on farmer's fields revealed that the suitable size of bed-furrow for wheat is 83 cm (60 cm x 23 cm) as it saves more than 50% water with irrigation application efficiency above 75%.

Table 6. Different bed-furrows systems for wheat

Traditional/bed furrow system	Size of bed (cm)	Size of furrow - depth x width (cm)	Nos. of lines/ acre (0.40 ha)	Avg. irrigation application efficiency (%)
Rabi (winter) drill			264	50
Two lines bed furrow	30	30 x 30	198	60
Three lines bed furrow	45	30 x 30	158	66
Four lines bed-furrow	60	30 x 30	264	72
Four lines with small furrow	60	20 x 20	280	+75

Source: determined in experiments conducted at WMRC-UAF.

Irrigation Water Saving in Wheat

To apply adequate irrigation depth in experimental/ demonstration areas, mostly on farmers' fields, soil samples were taken from furrows and beds for soil moisture analysis. To determine irrigation water saving, measured amount of irrigation depths were applied using cut-throat flume. As the wetting/contact area for irrigation in bed-furrow planting is about one third than that of the basin flood irrigation, less depth of irrigation was required. Moreover, furrow irrigation excelled the water advance phase and resulted in shorter irrigation time and increased irrigation application efficiency.

As the soil of beds is more porous, water intake and storage in the beds was relatively higher which increased the water use efficiency of crop. Normal numbers of irrigations (three/ four for wheat) were applied following conventional practice.

Wheat experiment conducted in the experimental area of Water Management Research Centre (WMRC), University of Agriculture, Faisalabad, showed that there was 51.3% irrigation water saving in comparison with the flat sowing.

Table 7 presents the water saving results obtained from a number of field trials conducted during 2007-08 and 2008-09 under bed planting in comparison to flat sowing at farmers' fields in the command area of three distributaries, one in district Faisalabad (Khurrianwala) and two in district T. T. Singh (Killianwala and Mungi). During first year, the water savings under bed planting were found in the range of 40-50%. The results show that more water saving was obtained at all the sites during the second year, which could be attributed to the farmer's experience with bed planting in the second year.

Wheat Grain Yield

Table 8 shows the wheat grain yields under conventional and bed planting for the trials conducted on farmers' fields in Rabi 2007-08 and 2008-09 in the command area of three distributaries. Table indicates that at all the locations; there was an increase in grain yield under bed planting as compared to flat planting. The difference in grain yields within the bed planted fields/locations could be attributed to the difference in management practices at farmer level i.e. agricultural inputs, land fertility, time of sowing, irrigation, irrigation water quality, etc.

Table 7. Water saving in wheat bed planting at farmers' fields

Irrigation #	Water saving in bed planting (%)						
	Khurrianwala		Killianwala		Mungi		Average
	2007-08	2008-09	2007-08	2008-09	2007-08	2008-09	
1st	36	49	34	51	36	52	43
2nd	39	54	38	49	38	49	45
3rd	44	51	43	47	44	48	46
4th	48	53	47	49	48	51	49
Average	42	52	41	49	42	50	46

Table 8. Wheat grain yield results at farmers' fields

Distributary (area in ha)	Bed planting (kg/ha)		Conventional planting (kg/ha)		Increase in grain yield (%)		
	2007-08	2008-09	2007-08	2008-09	2007-08	2008-09	Average
Khurrianwala (70)	4518	4987	3752	4143	17	20	18.5
Killianwala (60)	4079	4919	3359	4141	18	19	18.5
Mungi (115)	4688	5110	4169	4305	13	19	16.0
Average	4428.33	5005	3760	4196	16	19.	17.6

The overall yield increase in bed planting in comparison with flat planting was found 16% and 19% in 2007-08 and 2008-09, respectively. These results reveal the advantages of bed planting over flat planting. The greater production on raised beds resulted due to better and strong crop stand, which could be due to high fertilizer use efficiency under low intensity of weeds. Besides, there is less chance of nitrate leaching due to non-flooding irrigation under bed-furrow system. Similarly, lodging is reduced as a result of drainage of rainwater from the furrows and tough anchorage of plants/roots in the soil. Similar advantages of planting wheat on beds have been reported by Sayre and Moreno Ramos (1997) and Hobbs *et al.* (1998).

Intercropping of Sugarcane in Bed Planted Wheat

Planting wheat on raised beds provides an opportunity of intercropping sugarcane in the furrows between the beds during February (while wheat crop is growing in the field). Wheat crop is harvested in Apr-May, while sugarcane crop keeps growing in the furrows. Farmers have not to wait for sowing sugarcane till the wheat crop is harvested and the season becomes too hot for the establishment of sugarcane. Thus, farmers can obtain both crops from the same area without losing time to plant sugarcane. Furthermore, irrigation and nutrients unused for wheat are also utilized for sugarcane which is an additional benefit for the intercropping of sugarcane. Keeping these facts in view, intercropping of sugarcane was practiced at some farmers' fields. Results are shown in Table 9. The best part of this practice was that the area under wheat was efficiently utilized for timely sowing of sugarcane without affecting grain yield of wheat. In addition, sugarcane yield was also satisfactory. Table shows that the intercropped sugarcane gave 20% higher yield than the sugarcane planted during late April or early May. Results concluded that intercropping of sugarcane in wheat avoids and competition among wheat and sugar, while it results in increasing yield of sugarcane without reduction in wheat grain yield. So this practice was found to increase farm return and area under sugarcane. The practice may also address problem of our increasing demand of sugar in the country.

Table 9. Intercropping of sugarcane in bed planted wheat (2007-08)

Location #	Area sown (acres)	Yield		
		Wheat (kg/ha)	Sugarcane (1000 kg/ha)	Sugarcane planting in May (1000 kg/ha)
1	7	4150	79.04	65.50
2	1	4150	83.98	68.45
3	2	4446	69.16	59.10
4	1	3557	79.04	66.60
5	2.5	4347	79.04	65.50
6	0.5	4347	83.98	68.50
Average		4166	79.04	65.61

Table 10. Grain yield of bed planted and ridge planted maize under farmers' fields condition (2008)

Sr. #	Grain yield					
	Spring sowing			Autumn sowing		
	Bed planting (kg/ha)	Farmer practice (kg/ha)	Increase over farmer practice (%)	Bed planting (kg/ha)	Farmer practice (kg/ha)	Increase over farmer practice (%)
1.	12449	7904	57.50	9386	7904	18.75
2.	10374	9386	10.53	7904	6916	14.29
3.	10572	9386	12.63	9386	8398	11.76
4.	11856	9386	26.32	7904	6916	14.29
Average	11313	9016	27	8645	7534	15

Maize

Farmers usually sow maize on ridges, therefore, both methods, maize on beds and on ridges were compared at farmers' field. The same bed planter was used to prepare beds (90 cm x 30 cm) for planting maize. The results indicated 10 to 26% increase in maize grain yield on beds as compared to farmers' practice for spring sown maize, while the increase ranged from 11 to 19% for autumn maize.

Irrigation depth was measured on irrigation time basis similar to the case of wheat under farmers' conditions. Results revealed that there was 30% water saving in furrow-bed sowing in comparison with ridge sowing. It is important to mention that furrow-ridge method of irrigation saves more water than flat sowing. The water saving is less as compared to the case of wheat because furrow-ridge is also an improved method of irrigation. However, increase in grain yield encouraged the adoption of bed-furrow system of maize; and as mentioned above, furrow-bed sowing not only resulted in an increase in yield but also saved more water than the ridge-furrow irrigation method (Table 10).

Cotton

Cotton trials were conducted at farmers' fields in the command area of Killianwala and Mungi distributaries. Table 11 shows the detail of sowing of cotton at both sites alongwith percent water saving achieved under bed planting. There were 54 and 31 trails of cotton at Mungi and Killianwala sites, respectively. Average water saving in cotton bed planting in comparison to conventional method was observed to be 43.5%.

Table 11. Water saving in cotton bed planting at farmers' fields

Site	Number of farmers	Total area (ha)	Water saving
Mungi	54	253	41%
Killianwala	31	150	46%
Average			43%

Table 12. Cotton yield under bed and flat planting at farmers' fields

Sr. No.	Sowing months	Grain yield (kg/ha)		Increase in yield (%)
		Bed planting	Flat planting	
1	March	5072	4347	17
2	May	3728	3300	13
3	June	3404	3054	11

Table 13. Rice yield under bed and flat planting at farmers' fields (2008)

Sr. #	Farmer's name	Village No.	Rice yield (kg/ha)		Increase in rice yield (%)
			Bed planting	Flat planting	
1	A Din	53/RB	6422	5039	27
2	M Hussain	53/RB	6718	5434	24
3	M Ali	54/RB	5533	4446	24
4	M Jamil	51/RB	7509	5434	38
5	M Naeem	41/RB	6224	4644	34
6	M Saleem	51/RB	5533	4940	12
7	M Sarwar	54/RB	5730	4940	16
8	N Hussain	54/RB	6126	5236	17
9	Z Iqbal	54/RB	5928	4940	20
10	Z Iqbal	53/RB	5434	4940	10
11	Z Randhawa	48/RB	5928	3853	54
Average			6183	4940	26

Yield results of cotton bed planting in comparison to flat planting have been shown in Table 12. Three traditional sowing timings were practiced for cotton under both methods i.e. March, May and June sowing. The yields under both methods were found highest in March sowing. Moreover, the impact of bed planting in enhancing the yield was also found more prominent in March sowing, presenting 17.% increase in yield under bed planting in comparison to flat sowing.

Rice

Results regarding paddy yield under bed and flat planting at farmers' fields have been presented in Table 13. Average paddy yield on beds was found to be 6183kg/ha in comparison to 4940kg/ha under conventional flat planting with percent increase of 26%. Water saving was also recorded at all the sites and average value was found to be 29%. The results represent data from various farmers' fields of about 30 ha (71 acres). Water measurements were made by the farmers by filling the water application record Performa with the help of field team. Similarly, grain yield data was obtained by weighing total grain per unit area basis.

4. Up-Scaling Water Conservation Interventions

4.1. Water Shortage Scenario and Water Management Options

Pakistan is facing the most severe irrigation water crisis in the history of the country, as the gap between demand and supply has increased to the levels creating unrest among provinces. Irrigated agriculture was, still is, and will remain in future the backbone of Pakistan's economy. In the country, the average annual rainfall is 250 -350 mm against a potential demand of about 1750 mm of water for agriculture. To meet the demand and supply gap we need to address each and every option of conserving water and improving water use efficiency including augmenting alternate resources than surface and groundwater.

The historical data on surface water shows that the River Indus and its tributaries are providing about 103 MAF (127 BCM) water annually. Out of this, 38.11 MAF (47 BCM) comes to Punjab according to its 37% share (Water Apportionment Accord 1991) but only 11.81 MAF (14.5 BCM) is available at farm for crop production, as only about 30% of flows reach to farm. Crop water requirements (CWR) for major crops are mentioned in Table 14. Rest of irrigation water is being contributed by the groundwater resource (tubewells). Thus, major contribution of irrigation water for crops in Punjab is being met by groundwater and rainfall (mostly monsoon season rainfall and sporadic rain in winter). Groundwater table is consequently falling in many areas and farmers are using saline water to irrigate crops. This situation clearly indicates that groundwater resource is being over-exploited and need urgent attention of its judicial use, which otherwise, can put Pakistan agriculture at no return point – a disaster to agriculture. Therefore, there is a dire need of legislation for groundwater use. For instance, it is possible to add tubewells for community use, "one tubewell at individual canal outlet of watercourse" in the area of good to marginal groundwater quality which should be operated as per need of suitable cropping pattern proposed by the Agriculture Department.

4.2. Water Management Options

There are different management options to cope with the existing water shortage scenario, which can be explained one by one as follows:

4.2.1. Bed Furrow Planting (Raised Bed Technology)

Raised Bed Technology has immense potential to save water and needs urgent adoption. The technology has been promoted in different areas of Punjab. After testing at the experimental area of Water Management Research Center (WMRC), University of Agriculture, Faisalabad, the results regarding water saving (Ahmad et al., 2009) are presented in Table 15.

Table 14. Water Requirements for major crops in Punjab

Crop	CWR in root zone (ft)	CWR at farm-gate (71% Efficiency) (ft)	Area (Ma)	Requirement at farm (MAF)
Wheat	1.07	1.51	15.82	16.93
Cotton	2.1	2.96	5.99	12.58
Maize	1.4	1.97	1.32	1.85
Rice	3.54	4.99	4.26	15.08
Sugarcane	3.68	5.18	2.04	7.51
Total				53.94

Source: Water Resources Development and Management, Report of National Committee, 2003.

Table 15. Water saving for different crops under raised bed technology sites

Item		Wheat	Cotton	Maize	Rice
Area under trial (acres)	WMRC Experimental Area	50	50	40	25
	Faisalabad	350	150	50	150
	T.T. Singh	520	775	280	
	Okara	80			
	Total	1000	975	370	175
Water saving (%)	WMRC Experimental Area	51.30	47	45.4	31.1
	Faisalabad	46.93	46.5	43.2	29
	T.T. Singh	45.30	35.5	38.5	
	Okara	48.83			
	Average	48.1%	43.0%	42.4%	30.0%

Table 16. Water saving potential in bed-furrow planting on Punjab basis

Crop	Wheat	Cotton	Maize	Rice
Total area under crop (million ha)	6.40	2.43	0.53	1.72
Water saving under Raised Bed Technology (%)	48.10	43.00	42.40	30.00
Potential of increasing area under crop (mha) (million ha)	3.08	1.04	0.23	0.52
Estimated water saving potential (BCM)	6.56	5.71	0.76	4.71
Total water saving potential:	17.74 BCM			

It was further reported that if the raised bed technology is adopted only for major crops, total water saving at Punjab level would be17.7 BCM (14.41 MAF). The detail for water saving potential in bed-burrow planting on Punjab basis is presented in Table 16.

4.2.2. Drip Irrigation

Drip irrigation is another option for reducing irrigation water loss. However, it is more expensive than bed furrow planting, but it could be one of our future needs to meet the worsening water shortage. Drip irrigation is important High Efficiency Irrigation System (HEIS) in which 90% irrigation application efficiency can be achieved. The technology can be beneficially used for cotton, maize and sugarcane, as well as for garden and vegetables. Keeping in mind 90% application efficiency of this system, its water saving potential can be assessed as given in Table 17. However, there is a need to refine these estimates on potential water savings through hydrological modeling.

Table 17. Water saving potential in drip irrigation on Punjab basis

Crop	Water requirement at farm (BCM)	Water requirement under conventional irrigation (55% efficiency) (BCM)	Water requirement under drip irrigation (90% efficiency) (BCM)	Water saving potential under drip irrigation (BCM)
Cotton	15.47	28.13	17.20	10.93
Maize	2.28	4.13	2.53	1.61
Sugarcane	9.24	16.79	10.26	6.53
Total	26.99	49.06	29.99	19.08

Table 18. Wastewater flows from urban areas of Punjab

Sr. No.	Name of city	Estimated quantity of wastewater per annum (BCM)
1	Lahore	0.74
2	Faisalabad	0.18
3	Gujranwala	0.14
4	Multan	0.11
5	Rawalpindi	0.15
6	Sargodha	0.05
7	Sialkot	0.05
8	Gujrat	0.04
9	Sheikhupura	0.05
10	Jhang	0.04
11	Other Punjab	0.41
	Total	1.94

It can be seen from the Table 17 that the adoption of drip irrigation for only cotton, maize and sugarcane crops will result in 19BCM (15.51MAF) water saving, while the use of this technology for other crops is also open. It is proposed that adoption of drip irrigation be made compulsory for vegetables and garden and optional to other crops.

4.2.3. Use of Treated Wastewater for Agriculture

It is important to mention that a considerable quantity of wastewater is being generated in major cities of Punjab, which can be used for agriculture after proper treatment. Annual wastewater flows being generated in different cities of Punjab are given in Table 18.

Table 18 shows that 1.94 BCM (1.58 MAF) wastewater is being generated in Punjab which can be utilized for agriculture after proper treatment. Preferably peri-urban agriculture, after suitable treatment of wastewater, can spare surface and groundwater by using 1.94 BCM (1.58 MAF) of this wastewater flow. Nutrient capture and energy savings will be the additional benefits of wastewater reuse for irrigation (Hanjra et al., 2011).

4.2.4. Change in Cropping Pattern

Under the prevailing water shortage situation, another important option may be to partially replace high delta crops like sugarcane and rice with low delta as well as with high value crops like wheat and vegetables. In Table 19, water requirements have been calculated on the basis of 25% and 50% reduction in the sugarcane and rice cropped areas. The reduced cropped area is replaced with vegetables for rice and with wheat and vegetables for sugarcane. Table shows that 25% replacement will result in 2.55 MAF water saving and 50% replacement will result in 5.07 MAF water saving. However, the reduction in sugarcane and rice cropped areas must be decided keeping in view their consumption requirements in the province as well as economic returns to the farmers across various agro-climatic zones.

4.3. Water Saving Potential under RCTs

Crop water requirements, demand, supplies and deficit reported were calculated for the supplies/flows of all canals during 2009, thus deficit showed in Table 19 represents recent irrigation water shortage in comparison with canal supplies. After the contribution of rainfall, the deficit is being met from the groundwater and resulting in groundwater decline and secondary salinity in case of using low quality groundwater. Although sugarcane can also be planted in bed-furrow system, however, Table 20 presents potential of water saving from RCTs for all minors and majors crops of Punjab other than sugarcane. In addition, the following the results reported by Gill (2004) and Mujeeb (2007), water saving potential from laser land leveling and rice residue management were also included in the calculation.

Table 20 shows that 52.39 BCM of water can be saved by adopting a simple package of RCTs consisting of bed-furrow planting alongwith laser land leveling and rice residue management. Drip irrigation can be another but the last option to include in RCTs, as the option is expensive and needs capacity building at all levels, i.e. training of trainers and training of service providers as well as the beneficiaries. As drip irrigation can be more beneficial for some crops, conjunctive use of both the bed-furrow planting and drip irrigation in the country could be a long term solution to address the water shortage.

Table 19. Water saving under different cropping scenarios

Crop	CWR in RZ (ft)	Existing area (Ma)	Requirement at farm level (MAF)	Proposed area (25% replacement)	Requirement at farm in MAF (BCM)	Proposed Area (50% replacement)	Requirement at farm level in MAF (BCM)
Rice	3.54	4.26	15.08	3.19	11.29 (13.89)	2.13	7.54 (9.27)
Sugarcane	3.68	2.04	7.51	1.53	5.63 (6.93)	1.02	3.75 (4.62)
Wheat	1.07	15.82	16.93	16.33	17.47 (21.49)	16.84	18.02 (22.16)
Vegetables	1.63	---	---	1.58	2.58 (3.17)	3.15	5.13 (6.32)
Total			39.52		36.97 (45.47)		34.45 (42.37)

Table 20. Water saving potential under RCTs for all major and minor crops of Punjab

Crop	Actual CWR (BCM)	CWR under conventional Irrigation method (50% efficiency)	Water saving potential under RCTs (BCM)			
			Bed-furrow irrigation	Laser land leveling	Rice residue management	Drip irrigation
Wheat	13.30	26.60	8.87	2.42	1.51	
Cotton	16.20	32.40	10.80	2.95		14.40
Maize	2.30	4.60	1.53	0.42		2.04
Rice	5.40	10.80	3.60	0.98		
Sugarcane	8.70	17.40		1.58		7.73
Sorghum	4.00	8.00	2.67	0.73		
Oil seed	0.9	1.8	0.60	0.16		
Rabi Minor	5.8	11.6	3.87	1.05		
Kharif Minor	10.2	20.4	6.80	1.85		
Total	66.8	133.6	38.73	12.15	1.51	24.18

4.4. Summary and Recommendation for Institutional and Policy Reforms

1. Studies conducted on bed-furrow planting at farmers fields showed a substantial saving in irrigation water (about 40–50%) with increase in yield (10–25%) and higher water productivity (2.35 kg/m^3). Results are convincing to adopt the technology at larger scale as only wheat on bed-furrow planting can save 8.87 BCM water addressing seasonal water crisis in the country.
2. Results regarding intercropping of sugarcane in bed planted wheat are promising to adopt the same practice at the level of country addressing the increasing demand of sugar.
3. Water savings in raised beds for rice as compared to transplanting on flat (29%) are marvelous as its adoption at national level can save 3.60 BCM of water to address water shortage in the country. Results revealed that transplanting of rice on raised beds increases yield by 26%, thus improving water productivity of rice.
4. The increase in yield of maize and other major and minor crops encourage adopting bed-furrow planting for increasing land and water productivity in the country.
5. Laser land leveling not only saves irrigation water, but also improves yields, thus its adoption is very much needed to save irrigation water as well as to meet future food requirements of increasing population.
6. Studies revealed that rice residue has potential to save water (1.51 BCM) in addition to adding organic matter to soil. Thus, burning of rice residues should be discouraged to avoid environmental pollution problems and increase organic matter for soils, a desired activity for improving land productivity.
7. Drip irrigation is a promising technology for saving irrigation water and is already giving the same results of saving irrigation water (90% application efficiency) and 40-60% increase in yield under the pilot projects in the country.

The widening gap between irrigation demand and supply is resulting in over exploitation of groundwater resource, a special buffer reserve which already proved instrumental in helping out the country from the drought during 1998-2001. Presently, groundwater table is declining on one hand, and causing secondary salinity problem on the other, thus, judicious use of groundwater must be a priority of the country for saving agriculture. This is only possible if available surface water supplies are used with minimum unavoidable losses, which is possible by adopting a package of WCTs. This effort can make the system sustainable as bed-furrow planting, a component of RCTs, results in substantial saving of irrigation water (from 40–50%) with increase in yield (10–25%) for instance. The same irrigation method offers intercropping of sugarcane in bed planted wheat which can help address increasing demand of sugar in the country. Water saving in raised beds as compared to transplanting rice on flat soil is another significant advantage of saving irrigation water upto 29%, as well as yield increase of 26%. Moreover, increase in maize grain yield on beds (10-26%) as compared to farmers' practice with substantial amount of water saving proposes the wider adoption of raised bed technology. For sustainable agriculture in Pakistan, adoption of RCTs is strongly recommended for phasing out inefficient agricultural practices for improving land and crop water productivities.

It is further recommended that resource conservation technologies like watercourse improvement, precision land leveling, higher efficiency irrigation techniques and zero tillage may also be promoted as contribution of each can help mitigate adverse effects of shrinking water on our agriculture and food security. The most desired activities related to the RCTs are continued research at farmer fields, its education and extension at the stakeholder level. Capacity building of staff involved in education and extension of RCTs is highly desired for which educational and training centers should play their role to address increasing water shortage in the country. More emphasis may be on mass awareness and participatory approach in planning and decision making process for clear rather than grey message of RCTs benefits in the country. Socioeconomic and environmental issues related studies for the RCTs should be routine for further improvements till its full adoption at the beneficiary level. Comprehensive economic analysis is needed to estimate the costs and returns of RCTs including their benefits in terms of livelihoods and food security.

ACKNOWLEDGMENTS

This Chapter was written at the Leadership Award under the project "Knowledge Sharing for Food and Water Security in Asia" funded by AusAID, Australian Government, and ACIAR project "Optimising Canal and Groundwater Management to Assist Water User Associations in Maximizing Crop Production and Managing Salinisation in Australia and Pakistan". Financial support from JICA, Punjab Irrigation and Drainage Authority (PIDA) and others for "On Farm Research and Development Project" Part-B Lower Chenab Canal System reported in this Chapter is also acknowledged. We also appreciate funding from AAB for putting some of the related work together. We are grateful for useful comments and valuable suggestions from the Leaders who participated into the AusAID funded capacity development project at Charles Sturt University, Australia.

REFERENCES

Ahmad N., Arshad M. and Shahid A (2009). Bed-Furrow System to Replace Conventional Flood Irrigation in Pakistan. *Proceedings of 5th Int. ICID Conference,* New Delhi, India.

Ahmad, R.N and Mehmood, N. 2005. Impact of raised bed technology on water productivity and lodging of wheat. *Pak. J. Water Res.*, 9(2): 7-15.

Berkhout, M. 1997. Farmer's Use of Basin, Bed and Furrow-and-furrow Irrigation Systems and the Possibilities for Traditional Farmers to Adopt the Bed and Furrow Irrigation Method.

Erenstein, O., Laxmi, V., 2008. Zero tillage impacts in India's rice-wheat systems: A review. *Soil and Tillage Research* 100, 1-14.

Gill, M. A., 2004. Water and Disasters: Sustainable Irrigated Agriculture through Resource Conservation Technology. Proceeding of one day Seminar on World Water Day held on April 8,2004 at University of Agriculture, Faisalabad, Pakistan. pp 3-11.

Govaerts, B., Verhulst, N., Castellanos-Navarrete, A., Sayre, K.D., Dixon, J., Dendooven, L., 2009. Conservation agriculture and soil carbon sequestration: Between myth and farmer reality. *Critical Reviews in Plant Sciences* 28, 97-122.

Gray, R. S., J. S. Taylor and W. J. Brown, 1996. Economic factors contributing to the adoption of reduced tillage technologies in central Saskatchewan. *Canadian Journal of Plant Sciences*. 76: 661-668.

Gupta, R. (2009). Operative Guidelines and Work Plans of Delivery and Adaptive Research in Cereal System Initiative for South Asia (CSISA), CIMMYT, India.

Gupta, R.K.2002. The Rice Wheat Consortium for Indo-Gangetic Plains. Highlights 2001-02. CIMMYT. *International Irrigation Management Institute,* Report No. R-33.

Hanjra MA, Blackwell J, Carr G, Zhang F, Jackson TM. 2011. Wastewater irrigation and environmental health: Implications for water governance and public policy *International Journal of Hygiene and Environmental Health* doi:10.1016/j.ijheh.2011.10.003

Hobbs, P.R., G.S. Giri and P. Grace. (1998). Reduced and Zero tillage Options for the Establishment of Wheat after Rice in South Asia. RWCIGP Paper No. 2. Mexico, D.F.: RWCIGP and CIMMYT.

Horlings, L.G., Marsden, T.K., 2011. Towards the real green revolution? Exploring the conceptual dimensions of a new ecological modernisation of agriculture that could 'feed the world'. *Global Environmental Change* 21, 441-452.

Hussain I, Mudasser M, Hanjra MA, Amrasinghe U, Molden D (2004). Improving wheat productivity in Pakistan: econometric analysis using panel data from Chaj in the upper Indus Basin. *Water International* 29: 189-200.

Jat, M.L., Chandna, P., Gupta, R., Sharma, S. K. and Gill, M.A. 2006. Laser Land Leveling: A Precursor Technology for Resource Conservation. Rice-Wheat Consortium Technical Bulletin Series 7. New Delhi: Rice-Wheat Consortium for the Indo-Gangetic Plains.

Jat, M.L., Gathala, M.K., Ladha, J.K., Saharawat, Y.S., Jat, A.S., Kumar, V., Sharma, S.K., Gupta, R., 2009. Evaluation of precision land leveling and double zero-till systems in the rice-wheat rotation: Water use, productivity, profitability and soil physical properties. *Soil and Tillage Research* 105, 112-121.

Kahloon, M.A., Gill, M.A and Ashraf, M. 2002. Evaluation of resource conservation technologies in rice-wheat system of Pakistan. *Research Report* 2-2002, PCRWR, Islamabad, Pakistan.

Maraseni, T.N., Cockfield, G., 2011. Does the adoption of zero tillage reduce greenhouse gas emissions? An assessment for the grains industry in Australia. *Agricultural Systems* 104, 451-458.

McMichael, P., Schneider, M., 2011. Food security politics and the millennium development goals. *Third World Quarterly* 32, 119-139. Pretty, J.N., Morison, J.I.L., Hine, R.E., 2003. Reducing food poverty by increasing agricultural sustainability in developing countries. *Agriculture, Ecosystems and Environment* 95, 217-234.

Molden D, Oweis T, Steduto P, Bindraban P, Hanjra MA, Kijne J. 2010. Improving agricultural water productivity: Between optimism and caution. Agricultural Water Management, Comprehensive Assessment of Water Management in Agriculture (Special Issue) 97: 528-535

OFWM. 2009. Presentation on ongoing projects of RCTs in Punjab, Department of Agriculture (Water Management Wing), Govt. of Punjab, Lahore.

Pretty, J., Toulmin, C., Williams, S., 2011. Sustainable intensification in African agriculture. *International Journal of Agricultural Sustainability* 9, 5-24.

Pretty, J.N., Noble, A.D., Bossio, D., Dixon, J., Hine, R.E., De Vries, F.W.T.P., Morison, J.I.L., 2006. Resource-conserving agriculture increases yields in developing countries. *Environmental Science and Technology* 40, 1114-1119.

Rehman, M. H. 2007. Increasing productivity through permanent raised beds in rice-wheat cropping system in Pakistan. Proceedings of the INPIM's Ninth International Seminar on Participatory Irrigation Management "*Institutional and Technological Interventions for Better Irrigation Management in the New Millennium*", Lahore, Pakistan.

Sayre, K.D. and O.H. Moreno Ramos. 1997. Application of raised bed planting system to wheat. Wheat Special Report No. 31. Mexico, DF: CIMMYT.

Sharma, B. R. and Xueliang, C. 2009. Water saving technologies in Indo-Gangetic Basin: New developments. Invited paper at 15th Regional Technical Coordination Committee meeting, Rice-wheat consortium, new Delhi, India.

Silici, L., Ndabe, P., Friedrich, T., Kassam, A., 2011. Harnessing sustainability, resilience and productivity through conservation agriculture: The case of Likoti in Lesotho. *International Journal of Agricultural Sustainability* 9, 137-144.

Steiner K. G., Derpsch, R. and Koller, K. H. 1998. Sustainable Management of Soil Resources Through Zero Tillage. *Agriculture + Rural Development*, 5: 64-67.

Zahid, H. (2009). Brochure on " Permanent Raised Bed Technology in Pakistan".

Zahid, H, Khan MA, Irfan M. 2010. Water energy and economic analysis of wheat production under raised bed and conventional irrigation systems: A case study from a semi-arid area of Pakistan. *Soil and Tillage Research* 109: 61-67.

In: Irrigation Management, Technologies …
Editor: M. Hossain Ali
ISBN: 978-1-62417-862-7

Chapter 6

ASSESSING SUITABILITY OF GROUNDWATER FOR IRRIGATION PURPOSE IN COASTAL JORF AQUIFER, SOUTHEASTERN TUNISIA: GEOSTATISTICAL APPROACH

Belgacem Agoubi[1], Adel Kharroubi[1] and Habib Abida[2]
[1]University of Gabes, Higher Institute of Water Sciences and Techniques, Tunisia
[2]University of Sfax, Faculty of Sciences,
Department of Earth Sciences, Tunisia

ABSTRACT

In Jorf peninsula, southeastern Tunisia on Mediterranean Sea, groundwater represents the main water resource to satisfy the needs of various sectors. The Jorf coastal aquifer is subjected to seawater intrusion and groundwater quality degradation. The object of this study is to assess the suitability of groundwater of Jorf aquifer for irrigation purposes through geostatistical approach. With this aim, water samples were collected from hand-dug wells, and chemical parameters such as electrical conductivity, pH, total dissolved solids (TDS), Na, K, Ca, Mg, Cl, HCO_3 and SO_4 were analyzed. Sodium adsorption ratio (SAR), magnesium adsorption ratio (MAR), sodium percentage (%Na) and permeability index (PI) were computed. Results showed that groundwater salinity was controlled by sodium and chloride. Groundwater was dominated by Na–Cl–Ca–SO_4 hydrochemical facieses, mainly as a result of mineral dissolution (halite and gypsum) and seawater intrusion. The water of coastal zone is characterized by fair to poor quality, and is therefore just suitable for chemical amendments in open soils and with excessive leaching. A refined geostatistical assessment map of the groundwater quality indicated that only the water of central and northwest of the study area is suitable for irrigation purposes.

[1] Belgacem.Agoubi@isstegb.rnu.tn; Adel.Kharroubi@isstegb.rnu.tn.
[2] Habib.Abida@fss.rnu.tn.

1. INTRODUCTION

1.1. Background

Water is the most important input required for plant growth, and hence in agricultural production. Under rainfed conditions, soil water storage is continuously replenished with natural rainfall (Kirda, 1997); however, irrigation is essential in arid and semi-arid climates to maintain soil water storage at an optimum level to get higher yield. In arid and semi-arid zones of Tunisia (i.e. Jorf area, southeastern Tunisia), fresh groundwater has long been regarded as the main source of water supply. In this area, surface water is very scarce or absent, and groundwater represents the main source of water used to meet the ever-growing needs of agricultural sectors. The shallow aquifer is exposed to various anthropogenic and natural constraints such as urbanization, agricultural development, and very high evaporation rates. The shallow aquifer is not deep enough to protect against salinity and pollution effects, which is further accentuating by continuous evaporation.

Highly soluble salts in the soil make soil moisture more difficult for plants to extract, and crops become water stressed even when the soil is moist. Assessment of irrigation water quality and its salinity hazard is a very important step to formulate appropriate management strategies. Irrigation water quality generally varies depending upon the types and quantities of dissolved salts. Numerous parameters are used to define irrigation water quality, and assess salinity hazards. With the FAO assessment method (Phocaides, 2007), complete water quality analysis includes the determination of the total concentration of soluble salts, the relative proportion of sodium to the other cations - expressed by sodium adsorption ratio, the bicarbonate concentration as related to the concentration of calcium and magnesium, and the concentrations of specific elements and compounds.

Geostatistical assessment of hydrochemical variables is an important tool in the analysis of groundwater chemistry and salinity hazards. Geostatistics designate the statistical study of natural phenomena which can often be characterized by a distribution in space of one or more variables (EC, pH, TDS and majors ions). These variables are called regionalized variables (Journel and Huijbregts 1978; Olea, 1999). In the recent works, few studies dealt with the geostatistical assessment of groundwater quality databased on different irrigation indices (Kooveei et al., 2005; Shi et al., 2005; Assaf and Saadeh, 2009; Monego et al., 2010; Saoras 2010; Agoubi et al., 2011).

A case study of coastal Jorf aquifer is presented in which groundwater quality for irrigation use is assessed. Groundwater quality parameters such us Sodium Adsorption Ratio (SAR), Sodium Percentage (%Na), Magnesium Adsorption Ratio (MAR), Permeability Index (PI), and Kelly's Index (KI) were computed. Finally, geostatistical tools were used to map the salinity hazard of groundwater.

1.2. Approaches of groundwater suitability assessment

Geostatistics, a spatial statistical technique used in environmental monitoring, is widely applied to analyze and map distributions of concentrations and variations in space and time. Geostatistics uses variograms to characterize and quantify spatial variability, perform rational interpolation, and estimate the variance in the interpolated values (Yeh et al., 2006). A variogram quantifies the commonly observed relationship between the values of data, pertaining to the samples, and the samples' proximity. Prior to any geostatistical estimation, the semivariogram must be determined from the regionalized variable data. Semi-variogram is the basic geostatistic tools for visualizing the spatial autocorrelation of a regionalized variable (Matheron, 1962). The experimental variogram (Figure 1) is the key of any geostatistical analysis. It helps to highlight the structural correlation between variables. The semi-variogram $\gamma(h)$ is given by eqn. [1].

$$\gamma(h) = \frac{1}{2N(h)} \sum_{i=1}^{N(h)} [(Z(x_i) - Z(x_i + h))]^2 \quad (1)$$

Where N(h) is the number of pairs of data points separated by h, $Z(x_i)$ is the sampled value at x_i location and $Z(x_i + h)$ value at point distant h.

A discontinuity showed at the origin of semivariogram is called the nugget effect. The nugget effect can be further divided into measurement error and microscale variation. The nugget effect is simply the sum of measurement error and microscale variation and (Johnson et al., 2001). The distance at which the semivariogram levels off to the sill is called the range.

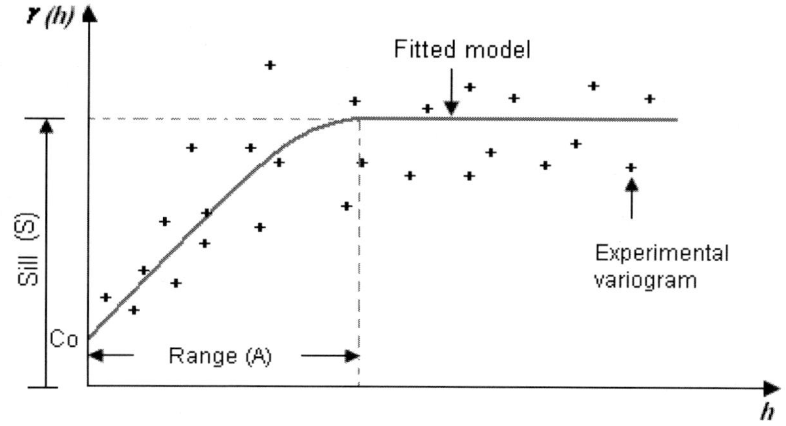

Figure 1. The anatomy of a typical semivariogram.

The interest in modeling of experimental variogram is essentially practical to move from a defined function points to a continuous function in space and with a mathematical expression (Rentier, 2002). The anisotropy of the random process was examined through the variogram for different distances and for different directions. The directional variogram is defined as [eqn.2]:

$$\gamma(h,\theta) = \frac{1}{2N(h,\theta)} \sum_{i=1}^{N(h,\theta)} [Z(x_i) - Z(x_i + h)]^2 \quad (2)$$

Where h is a distance vector between two locations, N(h,θ): number of pairs in direction θ.

Kriging was a useful tool to estimate the groundwater parameters quality (Agoubi et al., 2011). Kriging compared to other contouring techniques is its ability of quantifying the estimation variance, which will lead to define the precision of the resulting estimates (Issaks and Srivastava, 1989). The standard deviation map shows the confidence envelope that surrounds the estimated values.

The kriging estimate is of the form [eqn.3]

$$Z^*(x_0)= \sum_{i=1}^{n} \lambda_i \; Z(x_i) \tag{3}$$

In which $Z^*(x_0)$ is an estimator of the unknown true value of the functional and λ_i are weighting coefficients to be determined.

Assuming a second-order stationarity the minimum-variance condition becomes [eqn.4]

$$Var[Z^*(x_0)-Z(x_0)]=\sum_{i=1}^{n}\sum_{j=1}^{n}\lambda_i\lambda_j C(x_i-x_j)-2\sum_{i=1}^{n}\lambda_i C(x_i-x_0)+C(0) \tag{4}$$

in which x_i - x_j = h.

The kriging variance is given by [eqn.5]

$$\sigma_k^2 = \mathrm{var}\left[Z^*(x_0)-Z(x_0)\right]=C(0)-\sum_{i=1}^{n}\lambda_i C(x_i-x_0)+\sum_{l=1}^{k}\mu_l f^l(x_0) \tag{5}$$

The kriging system can be expressed in terms of the semivariogram function $\gamma_R(h)$, particularly when the random function $R(x)$ is intrinsic only and the covariance function $C(x_i - x_j)$ is not defined.

$$\begin{cases} \sum_{j=1}^{n}\lambda_j\gamma_R(x_i-y_j)-\sum_{l=1}^{k}\mu_l f^l(x_i)=\gamma(x_i-x_o) & i=1,2,\ldots,n \\ \sum_{i=1}^{n}\lambda_i f^l(x_i)=f^l(x_0) \quad l=1,2,\ldots,k & \end{cases} \tag{6}$$

The details of this theory can be found in Matheron 1962, 1969, 1970; Kitanidis, 1997 and Isaaks and Srivastava, 1989.

For the geostatistical assessment of water quality, data was first subjected to elementary statistical analysis. It consists of drawing the histogram for every variable which provides the first information about the distribution of the probability distribution of every variable and secondly producing a 2D dots map, for every variable which gives an overview about its real dispersion.

Using cross-validation is the optimal selection of model performance. A cross-validation technique was used to test the model performance. For each model Root Mean Squared Error (RMSE) was computed. This parameter is often used to compare the performance of interpolation methods (Rakotomalala, 2008). RMSE is computed according to [eqn.7]:

$$RMSE = \sqrt{\frac{1}{n}\sum_{i=1}^{n}\left[z(x_i) - z^*(x_i)\right]} \quad (7)$$

In which z(xi) is the observed value and z*(xi) corresponds to the predicted value of observation i in location x_i. RMSE is frequently used to evaluate errors in mapping aquifer parameters (Gundogdu and Guney, 2007). It was tested whether the differences between the lowest RMSE value and the others are important. The method that yields the smallest value of RMSE is considered as the best fitted one.

2. Chemical Water Composition

The problem of salinity manifests itself in the environment (saline soil, saline groundwater, rock-water interactions, and seawater intrusion in coastal aquifers, saline irrigation and drainage water). Geochemical properties and principles that govern the behavior of dissolved chemical constituents in groundwater are referred to as hydrogeochemistry (Obiefuna and Sheriff, 2011).

The major solutes comprising dissolved mineral salts are the cations Na, Ca, Mg and K and the anions Cl, SO_4, HCO_3 and NO_3.

The chemical composition of groundwater is related to the solid product of rock weathering and changes with respect to time and space. Therefore, the variation on the concentration levels of the different hydrogeochemical constituents dissolved in water determines its usefulness for domestic, industrial and agricultural purposes.

In this way, Pie diagram and major ions relationship such us Na/Ca versus Cl/SO_4, Ca versus HCO_3+SO_4, Cl versus Na and Ca/(HCO_3+SO_4) versus Na/Cl were established in order to identify groundwater chemical compositions.

3. Water Quality Assessment Parameters

Geochemical assessment of groundwater quality for irrigation use is based on several parameters such us Sodium Adsorption Ratio (SAR), Sodium Percentage (%Na), Magnesium Adsorption Ratio (MAR) and Permeability Index (PI).

3.1. Sodium Absorption Ratio (SAR)

High sodium concentrations become a problem when the infiltration rate is reduced to such a rate that the crop does not have enough water available or when the hydraulic conductivity of the soil profile is too low to provide adequate drainage. The sodium or alkali hazard potential in irrigation water may be expressed in terms of the sodium adsorption ratio (SAR), which can be estimated by equation-8 given by Richards (1954), where all the ions are expressed in meq/L:

$$SAR = \frac{Na}{\sqrt{\frac{(Ca + Mg)}{2}}} \quad (8)$$

SAR expresses the relative activity of sodium ions in the exchange reactions with the soil. This ration measures the relative concentration of sodium to calcium and Magnesium.

There is a significant relationship between SAR values of irrigation water and the extent to which sodium is adsorbed by the soils.

If water used for irrigation is high in sodium and low in calcium, the cation-exchange complex may become saturated with sodium (Sarkar and Hassan, 2006). This can destroy the soil structure owing to dispersion of the clay particles.

Continued use of water with a high SAR value leads to a breakdown in the physical structure of the soil caused by excessive amounts of colloidally absorbed sodium (Khodapanah et al., 2009).

This breakdown results in the dispersion of soil clay that causes the soil to become hard and compact when dry and increasingly impervious to water penetration due to dispersion and swelling when wet. Fine-textured soils, those high in clay, are especially subject to this action.

3.2. Magnesium Adsorption Ratio (MAR)

Raghunath (1987) proposed the MAR ratio to characterize irrigation waters. The Magnesium Adsorption Ratio (MAR) is given by equation [9], in which all the ions are expressed in meq.L^{-1}:

$$MAR = \frac{Mg}{Ca + Mg} x100 \quad (9)$$

Waters with MAR >50 are considered to be harmful and unsuitable for irrigation use.

3.3. Sodium Percentage (%Na)

EC and sodium concentration are very important in classifying irrigation water. Besides affecting the growth of the plants directly, salts also affect soil structure, permeability and aeration, which indirectly affect plant growth (Singh et al. 2008). Wilcox (1955) Diagram uses sodium percentage (%Na) and EC values to classify irrigation water quality. Na % is calculated using equation [10], in which all the ions are expressed in meq.L^{-1}:

$$\%Na = \frac{Na + K}{Ca + Mg + Na + K} x100 \quad (10)$$

3.4. Permeability Index (PI)

Permeability index is calculated according to Doneen (1964) and Rahgunath (1987), as given by equation [11]:

$$PI = \frac{Na + \sqrt{HCO_3}}{Ca + Mg + Na} x100 \quad (11)$$

All chemical elements here are expressed in equivalent per liter.

Where ion concentrations is expressed in epm. Knowing the TDS and PI, the soil can be classified as: i) highly permeable soils (sands and sandy loam) ii) medium permeable soils (silts) and iii) low permeability soils.

3.5. Kelly's index (KI)

Kellys index (KI) is used for the classification of water irrigation purposes. Sodium measured against calcium and magnesium is considered for calculation of this parameter. A KI >1 indicates an excess level of sodium in water. Therefore, water with a KI <1 is suitable for irrigation, while those with greater ratio are unsuitable (Sundary et al., 2009). The KI is computed using the following formula [12] in which Na, Ca and Mg are expressed in meq/l.

$$KI = \frac{Na^{2+}}{Ca^{2+} + Mg^{2+}} \quad (12)$$

4. A Case Study: Coastal Jorf Aquifer, Southeastern Tunisia

4.1. Study Area and Geological Setting

The study area consists of Jorf Peninsula which is part of the vast coastal plain of coastal Jorf (South-Eastern Tunisia) (Figure 2). The overall structure is characterized by faults with small discharges, which are organized in parallel bands of NW-SE orientation (Jedoui, 2000). These faults are steep of the crust and reveal Villafranchian place by the Mio-Pliocene bedrock.

The Jorf peninsula is characterized by an arid to semi-arid climate influenced by hot dry air mass coming from the desert. Rainfall is variable and irregular from year to year. The mean rainfall varies from 225 to less than 100 mm/year. Relative humidity ranges from 43 to 84%, temperature varies from 10°C in winter to 45°C in summer (Ouessar et al., 2004; OSS, 2005). The evapotranspiration rate exceeds 1,700 mm/year. The surface drainage has an intermittent flow regime because of long drought periods. Therefore, groundwater constitutes the main water resource.

The Quaternary cover consists largely of red silt attributed to the Holocene-Pleistocene or fine silt and conglomerate of late Pleistocene gypsum crust (Jedoui, 2000; Bouaziz et al., 2003). The general stratigraphy of the region is marked by the crust Villafranchian that overcomes the thick Mio-Pliocene series.

4.2. Data Sampling and Analysis

Thirty-six (36) groundwater samples were collected from hand-dug well used for domestic, agricultural, and domestic/agricultural purposes, and were found uniformly distributed over the study area (Figure 2). Groundwater sampling was performed in March 2009. Electrical conductivity (EC), pH and total dissolved solids (TDS) were measured *in situ* using a multi-parameter (C933 Multi-Parameter) Analyzer. The major ions were analyzed using Ionic Chromatography (Methrohm 850 Professional IC).

Basic statistics of major ions concentrations, EC, pH, TDS are summarized in Table 1. The pH of water is an indicator of its quality and geochemical equilibrium for solubility calculation (Hem, 1991; Umapathy, 2011).

The pH indicates the state of equilibrium reaction in which the water precipitates. In study area, pH of water is slightly acidic to alkaline in nature. It varies from 6.7 to 7.8 with a mean of 7.6. Electrical Conductivity (EC) is an indirect measurement of ionic strength and mineralization of natural water. EC values vary from 3.6 S/cm to 16.9 S/cm. TDS values vary between 3.40 and 13.5 g/L, with a mean of 6.3 g/L.

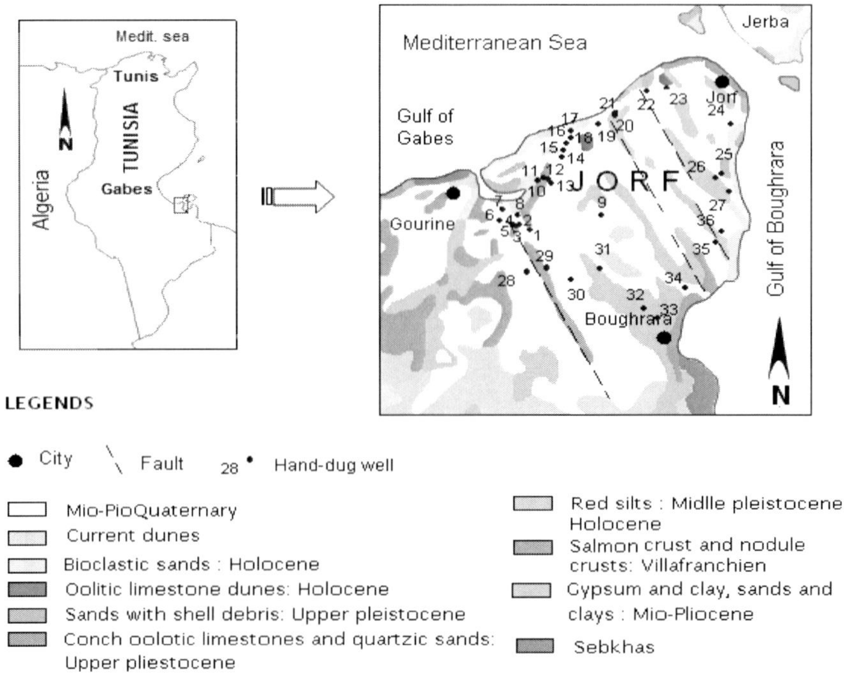

Figure 2. Location and geological map of study area and wells location (After, Jedoui, 2000).

Table 1. Basic statistics of the groundwater chemical composition

	Min	Max	Mean	SD
EC (ms/cm)	3.6	16.9	7.3	2.6
pH	6.7	7.8	7.6	0.2
TDS (g/L)	3.4	13.5	6.3	2.1
Ca	64	1032	561.3	224.1
Mg	124	672	263.6	117.7
Na	225	2392	790.4	446.8
K	14	92	25.3	14.1
SO_4	1248	2640	1801.2	350.7
Cl	335	5041	1584	915.7
HCO_3	51	442	223	101

All major ions are expressed in mg/L.

The large variations in the EC, TDS and ionic concentrations may be attributed to geochemical processes and hydrogeological conditions controlling groundwater mineralization.

4.3. Groundwater Chemical Composition and Origin of Salinity

The abundance of anions and cations in groundwater from Jorf aquifer is shown in Pie diagram (Figure 3). It clearly indicates that Na^+, Cl^-, $SO_4^{2-}$and Ca^{2+} are the dominant ions of groundwater composition.

The diagram also shows that chlorine, sodium and sulfate ions are more abundant. Cl and SO4 represent 44% and 50%, respectively, of the cations while Na represents 48% of the anions. A correlation between Na and Cl concentrations suggest that these ions have probably a common origin.

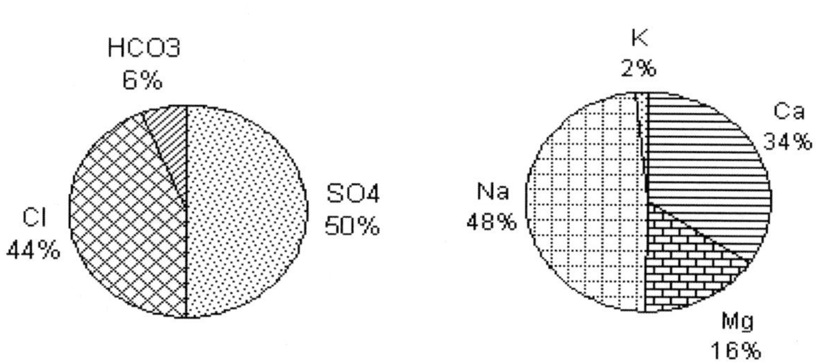

Figure 3. Pie diagram of groundwater (in 2009).

The presence of Na^+ and Cl^- in natural water is attributed to the dissolution of halite. However, sodium chloride may have other origins (natural or anthropogenic). The evolution of the Na^+ concentration with that of Cl^- (Figure 4) shows an excess of Cl^- level which is explained by the existence of another origin for chlorine other than halite. The chlorine content is mainly attributed to seawater intrusion. Na/Cl versus Cl/SO4 plotted on bilog diagram (Figure 5) shows three dominant groundwater facieses, namely Na-Cl, Na-SO_4 and Ca-SO_4. The abundant major ions (Na, Cl, Ca, and SO4) in groundwater from the coastal Jorf Aquifer resulted from the dissolution of Mio-Pliocene evaporite deposits and saltwater intrusion.

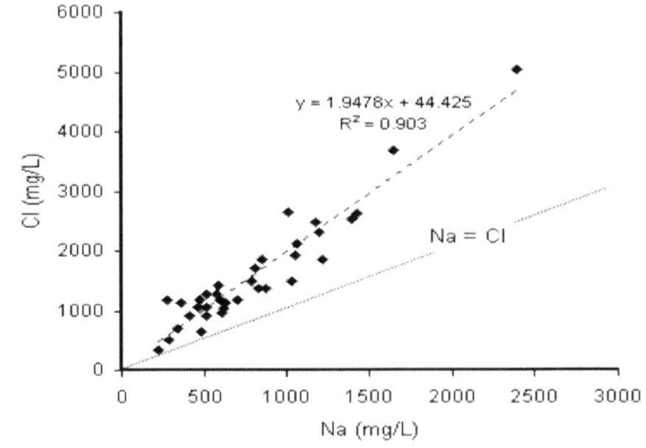

Figure 4. Relationship between Na and Cl.

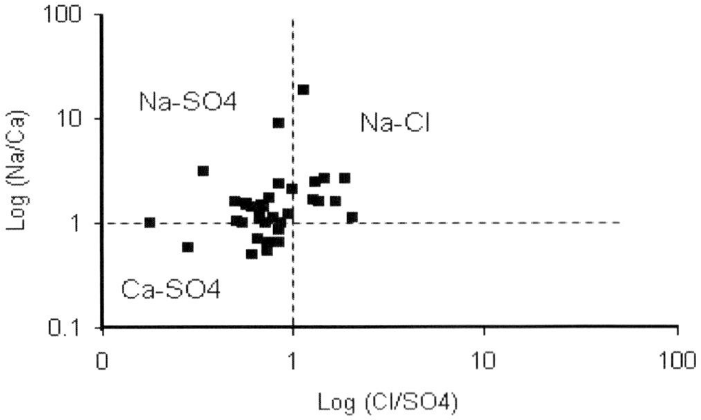

Figure 5. Chemical facies distribution of groundwater from Jorf coastal aquifer.

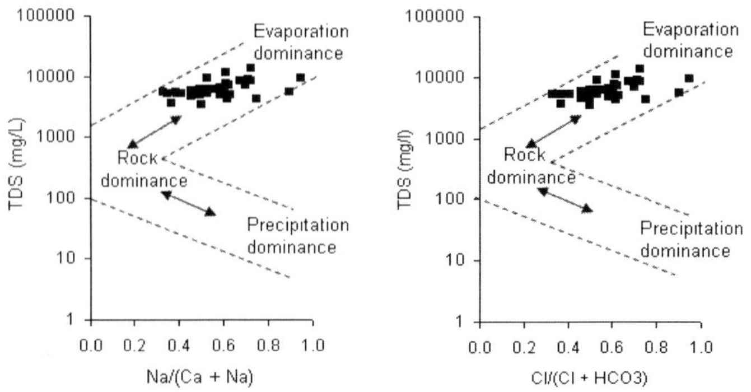

Figure 6. Gibbs plot indicating the mechanism controlling the groundwater chemistry.

The groundwater chemical data were also plotted on Gibbs diagram (Figure 6) to determine the different processes controlling the groundwater chemistry (Gibbs, 1970).

Data points on Gibbs diagram suggest that groundwater chemistry is controlled by rock-water interaction and evaporation which are the dominant factors, leading to the poor quality of groundwater.

This is expected, as evaporation greatly increases the concentrations of ions formed by chemical weathering of the rock, leading to higher salinity. High concentrations of calcium and sulfate in groundwater may probably be related to dissolution of gypsum ($CaSO_4,2H_2O$) and anhydrite ($CaSO_4$) while high concentrations of Na and Cl could be related to dissolution of halite (NaCl).

4.4. Assessment of Water Quality for Irrigation Purpose

4.4.1. Geostatistical Analysis

Geostatistical analysis was performed using GS+ software version 7.0 (GDS, 2007) to understand the spatial variability of irrigation parameters (TDS, SAR, MAR, %Na, PI and KI). Geostatistical tools allow spatial interpolation and therefore mapping spatial variability of quality parameters. The theoretical fundamentals of the geostatistical methods are described elsewhere, from a hydrogeochemical point of view, they are widely-used techniques (Ceron et al, 2000; Sanchez-Martos et al., 2001; Wackernagel, 2003; Agoubi et al., 2010 and 2011).

Selecting a semivariogram model is an iterative process that involves calculating experimental semivariograms, fitting an alternative semivariogram, calculating an alternative ordinary kriging prediction and carrying out a cross-validation statistical analysis to assess the performance of the prediction phenomena (Assaf and Saadeh, 2009). The semivariogram model was selected using two main evaluation criteria: Nugget/Sill ratio (Hani and Karimineja 2010) and the mean standardized prediction error (MSPE). Variographic parameters are summarized in Table 2. The experimental variograms and fitted model are shown in Figure 7. TDS, SAR and KI are fitted exponential, %Na and PI are modeled spherical and MAR is fitted gaussian. Table 2 shows that the MSPE values for the prediction models range from -10^{-4} to 14^{-4} and the Nugget/Sill ratio varies between 9% to 18%, implying thereby a relatively significant unbiased estimations and a strongly model prediction. Fitted semivariograms (Figure 7) show a stationary phenomenon with an unknown mean. This implies that the ordinary kriging can be used to map spatial development of all studied parameters.

Table 2. Variographic parameters

	Model	Sill	Range	Nugget	N/S	SD	MSPE
TDS	Exponential	3.93	2.07	0.71	18%	1.312	0.0014
%Na	l acirehpS	73.8	1.19	7.7	13%	0.212	0.0005
SAR	Exponentiel	7.98	1.26	0.22	3%	1.62	0.0012
MAR	Gaussian	916.9	16.63	153	17%	3.22	-0.0001
PI	Spherical	63.5	1.016	6.3	9%	0.412	0.0007
KI	la at ni noepS	39 3	2.07	0.71	18%	49 1	0.00003

N: Nugget, S: Sill, MSPE: Mean Standardized Prediction Error.

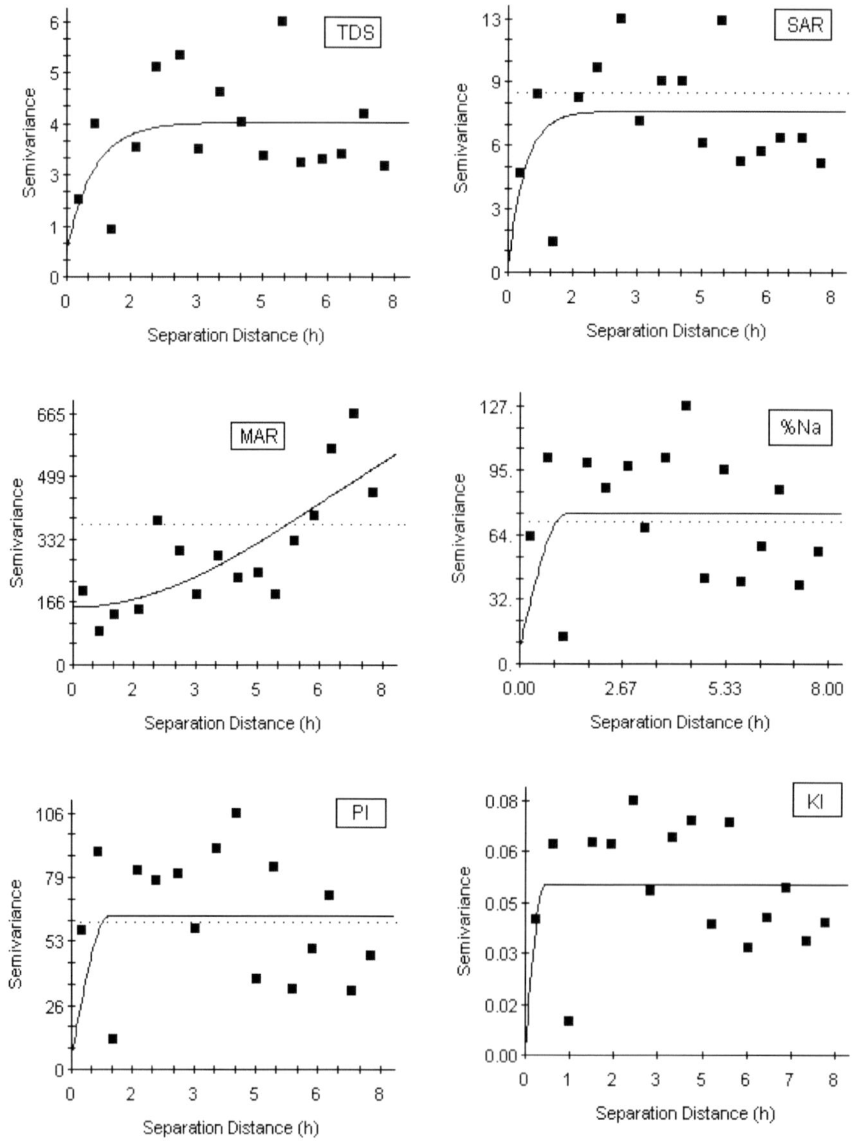

Figure 7. Experimental semi-variograms and fitted models of TDS, SAR, MAR, %Na, PI and KI.

4.4.2. Groundwater Assessment for Irrigation Use

A summary statistics of irrigations parameters values in study area are shown in Table 3. SAR values range from 2.26 to 16.44. MAR showing a high values varied from 21.09 to 94.59. Permeability index values range from 24.73 to 57.63 exceeding in some wells 50%. Kelly's index showing values from 0.26 to 1.29 with a low mean 0.67.

Total Dissolved Salts

TDS is an important criterion of groundwater assessment in agriculture use because increase in dissolved salts in irrigation water affects soil efficiency, growth, and yield of plants.

For long-term irrigation, the TDS should not exceed 2 g/l. High increase in water salinity increases salts amount in soil and leads to salinization problem (Khodapanah et al., 2009).

A kriged map of TDS spatial development in studied area (Figure 8) indicate that only a central part showing a moderate salinity (3 – 4 g/L). However, the coastal area of Jorf peninsula shows the highest values of TDS (> 7 g/L) which indicates that base on TDS categorization; the shallow groundwater from Jorf peninsula is permissible to poor water quality.

Table 3. Basic statistics of parameters of irrigation

	SAR	MAR	Na%	PI	KI
Min	2.26	21.09	21.29	24.73	0.26
Max	16.44	94.59	56.77	57.63	1.29
Mean	6.79	44.41	39.28	41.19	0.67
S.D.	2.98	19.26	8.36	7.81	0.24
Variance	8.89	371.11	69.93	60.93	0.06
kurtosis	1.70	0.77	-0.42	-0.40	0.03

SAR: Sodium Adsorption Ratio
%Na: Sodium Percent
MAR: Magnesium Adsorption Ratio
PI: Permeability Index
KI: Kelly's index
SD: Standard deviation

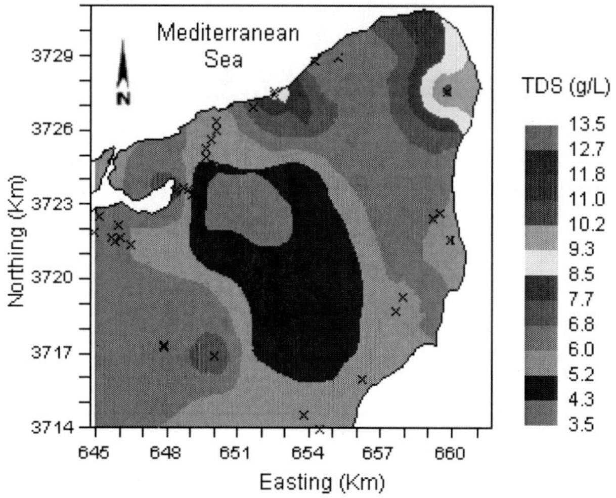

Figure 8. Kriged map of spatial development of TDS (g/L).

Sodium Percentage

The classification of groundwater was grouped based on percent Sodium as excellent (20%), good (20 – 40%), permissible (40 – 60%), doubtful (60 – 80%) and unsuitable (> 80%) (Sadashivaiah et al., 2008).

Out of selected wells based on percent sodium, 16 wells are permissible, 17 are doubtful and 3 are unsuitable. Figure 9 shows that spatial development of sodium percentage was permissible in central region. However, the poor quality is shown in Northeastern part of study area.

Sodium Adsorption Ratio

According to salinity and alkali hazard classification (Table 4) (Richards, 1954), 10 wells are excellent water quality, 14 wells range in good quality and 12 showing a doubtful to unsuitable water quality for irrigation use.

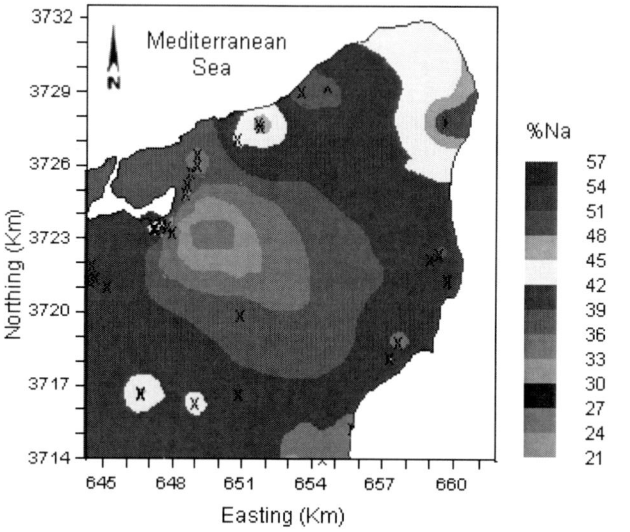

Figure 9. Kriged map of spatial development of Sodium percentage.

Table 4. Salinity and alkali hazard classes (After Richards, 1954)

Quality of water	SAR	Number of wells in study area
Excellent	< 10	10
Good	10 – 18	14
Doubtful	18 – 26	6
Unsuitable	> 26	6

Magnesium Adsorption Ratio (MAR) and Permeability Index (PI)

In the analyzed waters from study area, 14 water samples have MAR values exceeding 50%. Figure 10 shows that the groundwater from Jorf is not doubtful to unsuitable for irrigation in the south and North-Estearn part of peninsula.

Permeability index is an important factor which influences quality of irrigation water, in relation to soil for development in agriculture. Based on permeability index, Doneen (1964) classified the groundwater as class I, class II and class III to find out suitability of groundwater for irrigation purpose. The analysis as per Doneen's classification, has shown that most wells in he study area lie under permeability class I or class II (Figure 11) and in general water may be considered, according to Doneen's classification, permissible for irrigation.

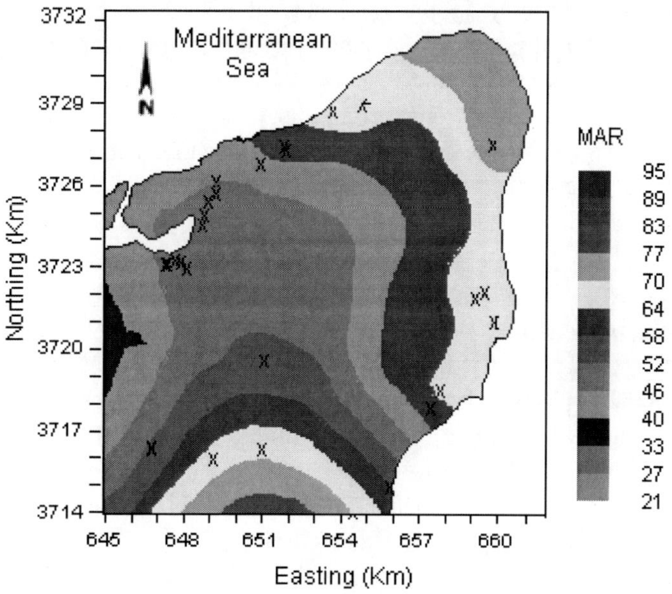

Figure 10. Kriged map of spatial development of MAR.

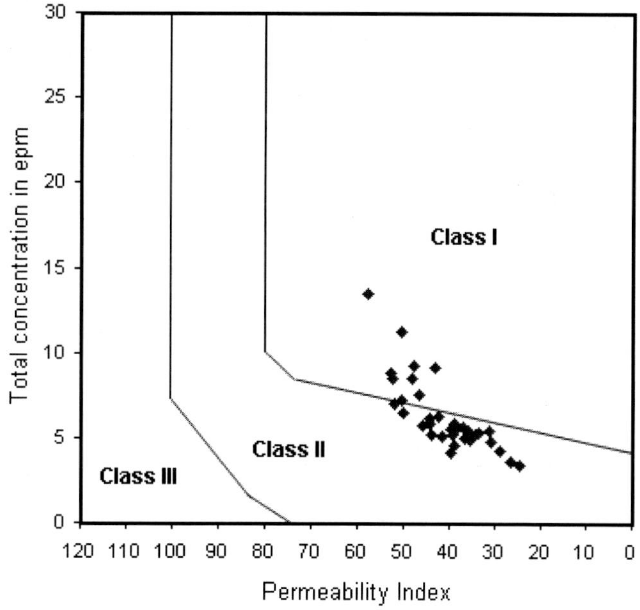

Figure 11. Classification of irrigation water for soils of Medium Permeability (After Doneen, 1964).

Kelly's Index

Based on Kelly's index (KI) groundwater from coastal Jorf aquifer are classified for irrigation. Sodium measured against calcium and magnesium was considered by Kelly (1940) to calculate this parameter.

A Kelly's index of more than one indicates an excess level of sodium in waters. Therefore, waters with a Kelly's index less than one are suitable for irrigation, while those with a ratio more than one are unsuitable.

KI in the sampled groundwater from Jorf aquifer varied between 0.26 and 1.29. Out of 36 water samples 32 shown a KI values <1 and only 4 wells show a values of KI >1. Therefore, according to Kelly's index, the water samples are suitable for irrigation except the four sampled wells in which KI values are more than one.

4.5. Richards Classification

As per the Richards (1954) classification based on SAR values (Figure12), out of 36 water samples 13 samples fall in the low and medium alkali hazard (categories S1 and S2). Fourteen are high alkali hazard (S3) and 9 are considered out of range indicating very high alkali hazard. According to Electrical conductivity in Richards diagram, all water samples fall in C4 and C5 categories suggesting à high water salinity.

According to the Richards (1954) Riverside diagram (Table 5), in which the EC is taken as salinity hazard and SAR as alkali hazard. It shown in table 5 that water samples fall in the category C4S1, C4S2, C5S2, and C5S3. However 25% fall outside the diagram, indicating high alkali hazard and high saline water.

In general, groundwater from coastal Jorf aquifer is a medium to high saline water, suitable for very high salt tolerant crops, organic soils, and coarse textured soils with chemical amendments. More than 60% of water samples are unsuitable for irrigation water.

4.6. Salinity Hazard of Groundwater Resource Degradation

Mapping salinity hazard of study area is essential to indicate hazard level of each area in coastal Jorf aquifer. In this objective, assessment water parameters (SAR, MAR, %Na, PI and KI) were used as indicators of salinity hazard classes.

Table 5. Classification of the groundwater samples (based on Richards Diagram)

Class of water	Number of wells	Number wells (%)	Salinity hazard
C4S1	02	5.6%	Suitable for very high salt tolerant crops
C4S2	02	5.6%	Suitable for organic soils or coarse textured soils
C5S2	09	25%	Groundwater can be used with chemical amendments in open soils and with excessive leaching.
C5S3	14	38.8%	Unsuitable for irrigation water
Out of range	09	25%	Poor water quality

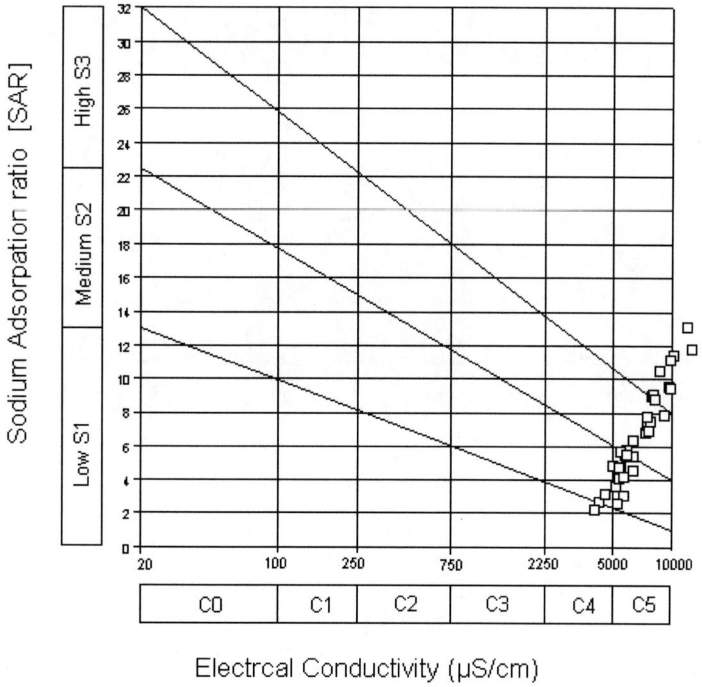

Figure 12. Classification of groundwater from coastal Jorf for irrigation use, based on Richards Method.

Table 6. Class limits and their scores

Indicators	Hazard class limits			
	Low	Slight	Moderate	Severe
SAR	<10	<18	<26	>26
MAR	<30%	<40%	<50%	>50%
%Na	<20%	<40%	<60%	>60%
PI	>75%	>50%	>25%	<25%
KI	<0.6	<0.8	<1	>1
Number of wells	6	11	14	5

Four classes were obtained (Table 6). For each indicator a score is attributed from 1 to 4.

According to hazard classes, groundwater from study area was grouped into four groups: low hazard (17%), slight hazard (31%), moderate hazard (38%) and severe hazard (14%).

Groundwater salinity hazard for irrigation use was mapped based on salinity hazard classes. Ordinary kriging was then used to map the spatial development of groundwater salinity hazard for irrigation use (Figure 13). The map obtained clearly shows that only the central part of Jorf peninsula shows a low salinity hazard. Therefore, a coastal area in which a highest salinity was recorded, due to saltwater intrusion, shows a severe salinity hazard and poor water quality, which is unsuitable for irrigation.

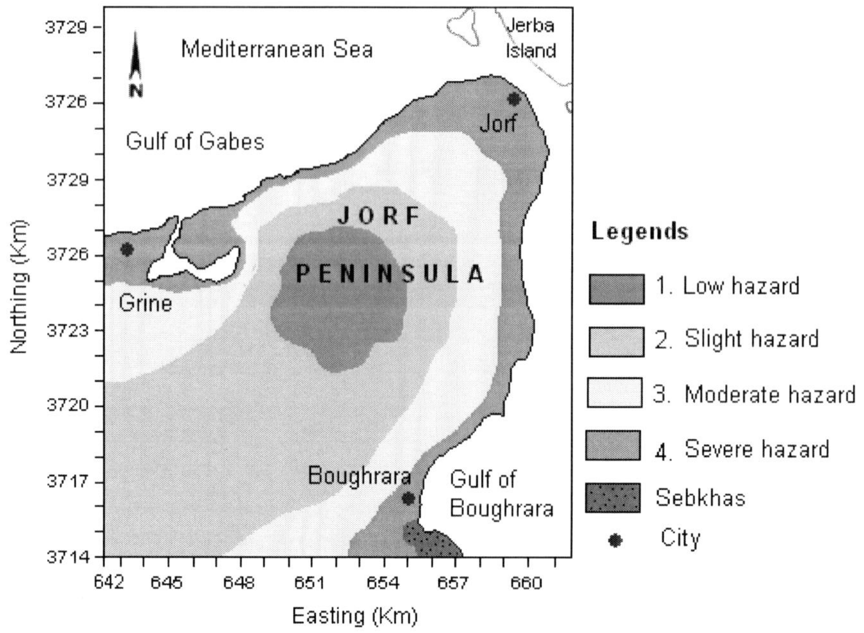

Figure 13. Salinity hazard map of groundwater resource in the coastal Jorf.

CONCLUSION

In the costal Jorf aquifer, groundwater was assessed for irrigation purpose. Geochemical modelling shows that groundwater was classified as highly saline water type rich in Na, Cl, Ca and SO_4 ions. Two processes controlled groundwater mineralization: saltwater intrusion and rock-water interaction.

Several assessment parameters were used to evaluate groundwater quality for irrigation use such us sodium adsorption ratio, magnesium adsorption ration, percentage sodium, permeability index and Kelly's index. Results indicate a permissible to unsuitable groundwater quality. Based on the salinity hazard classes, groundwater samples fall in fours classes salinity hazard: low (17%), slight (31%), moderate (38%) and severe (14%).

According to Richards's classification, only 11% of water samples were considered suitable for irrigation use for high salt tolerant crops. However, 25% can be used with chemical amendments in open soils and with excessive leaching and about 65% of water samples are uunsuitable for irrigation use due to poor water quality.

Geostatistical tools were used to analyse the spatial autocorrelation of groundwater assessment indicators. Results showed that geostatistical techniques can reveal the stochastic structure of groundwater chemical composition and salinity hazard parameters. A spatial delineation map of salinity hazard was obtained by kriging techniques. It clearly shows that coastal zone of study area showing a severe salinity hazard and the central part of peninsula is the lowest hazard.

REFERENCES

Agoubi B., Kharroubi A. and Abida H. (2011). Hydrochemistry of groundwater and its assessment for irrigation purpose in coastal Jeffara Aquifer, southeastern Tunisia. *Arab Journal Geosci.* DOI 10.1007/s12517-011-0409-1

Agoubi, B; Kharroubi, A., Bouri S., & Abida, H. (2010) Contribution of geostatistical modelling to mapping groundwater level and aquifer geometry, case study of Sfax's deep aquifer, Tunisia. *Middle East Journal of Scientific Research,* 6(3), 305-316.

Assaf, H., and M. Saadeh. (2009). Geostatistical Assessment of Groundwater Nitrate Contamination with Reflection on DRASTIC Vulnerability Assessment: The Case of the Upper Litani Basin, Lebanon. *Water Ressources Management* 23, 775–796.

Bouaziz, S., Y. Jedoui, É. Barrier, and J. Angelier. (2003). Néotectonique affectant les dépôts marins tyrrhéniens du littoral sud-est tunisien: implications pour les variations du niveau marin. *Comptes Rendus Geoscience* 335, 247–254

Ceron J. C., Jimenez-Espinosa R. et Pulido-Bosch, A. (2000). Numerical analysis of hydrochemical data: a case study (Alto Guadalentín, southeast Spain). *Appl Geochem* **15,** 1053–1067.

Doneen L.D., (1964). Notes on water quality in agriculture, published as water sciences and engineering paper 4001, Dept. of water, science and engineering, University of California, Davis.

G.D.S (2007) Gamma Design Software, Geostatistics for Environmental Sciences, ver. 7.0 Plainwell, Michigan USA: MI.

Gibbs (1970), mechanisms controlling worlds water chemistry science, 170, 1088-1090.

Gundogdu K. S and I. Guney, (2007). Spatial analyses of groundwater levels using universal kriging, *J. Earth Syst. Sci.* 116, No. 1, pp. 49–55

Hani, A. and M.T. Karimineja. (2010). Toxic metal distribution in Soils of Kaveh Industrial city, Iran. *World Applied Sciences Journal* 8 (11), 1333-1342.

Hem D. (1991), study and interpretation of the chemical characteristics of natural waters, 3 rd edition *US Geol. Survey, water supply paper* 2254.

Issaks E and R. Srivastava, (1989). Applied Geostatistics. Oxford University Press, New York.

Jedoui, Y. (2000). Sédimentologie et géochronologie des dépôts littoraux quaternaires: reconstitution des variations des paléoclimats et du niveau marin dans le sud est tunisien. Thèse de doctorat es-sciences géologiques, université de Tunis II.

Johnson K, V.er Hoef J. M, Krivoruchko K. and Lucas N., (2001), Using ArcGIS Geostatistical Analyst, GIS by ESRI, Redlands, USA.

Journel A., and C.H. Huijbregts. (1978). Mining Geostatistics. Academic Press Inc., New York, p600.

Kelly W. P. (1940). Permissible composition and concentration of irrigated waters. *Proceedings of the ASCF66* (p. 607).

Khodapanah L., W.N.A. Sulaiman, N. Khodapanah (2009). Groundwater Quality Assessment for Different Purposes in Eshtehard District, Tehran, Iran. European *Journal of Scientific Research,* Vol.36 No.4,543-553

Kitanidis P.K, (1997). Introduction to geostatistics, Application in Hydrogeology, *Cambridge Univ. Press,* pp 86-95.

Kirda C. (1997). Options Méditerranéennes, Sér. A /n031, Séminaires Méditerranéens.

Kooveei H.H., J. Ghayomian, and A.S.M. Gieske (2005). Geostatistical assessment of groundwater quality in Sarchahan, (available on line 02/05/2011 in gisdevelopment.net http://www.gisdevelopment.net/ application/nrm/water/ground/ma05168pf.htm

Matheron G., (1962). Traité de géostatistique appliquée, tome I. Mémoire du bureau de recherche géologique et mines, n° 14. Edition technip, Paris.

Matheron G., (1969). Le krigeage universel. Vol. 1. Cahiers du Centre de Morphologie Mathematique, Ecole des Mines de Paris, Fontainebleau.

Matheron G., (1970). Note de géostatistique N° 107, Krigeage universel pour une dérive aléatoire, Cahiers du Centre de Morphologie Mathematique, Ecole des Mines de Paris, *Fontainebleau.* 33p.

Monego M., P. Ramos, and M.V. Neves. (2010). Geostatistical Methods for Polluted Sites Characterization. geoENV VII – Geostatistics for Environmental Applications, Quantitative Geology and Geostatistics 16, 187-198.

O.S.S. (2005). Observatoire du Sahara et du Sahel: Etude hydrogéologique du système aquifère de la Djeffara tuniso-libyenne. *Rapport de synthèse,* p 209. (in Frensh)

Obiefuna G.I., and A. Sheriff (2011). Assessment of Shallow Ground Water Quality of Pindiga Gombe Area,Yola Area, NE, Nigeria for Irrigation and Domestic Purposes. *Research Journal of Environmental and Earth Sciences* 3(2), 132-142.

Ouessar M., M. Sghaier, N. Mahdhi, F. Abdelli, J. de Graaff, and H. Chaieb. (2004). An integrated approach for impact assessment of water harvesting techniques in dry areas: The case of Oued Oum Zessar watershed (Tunisia). *Environmental Monitoring and Assessment* 99, 127–140.

Phocaides A. (2007). Handbook on Pressurized irrigation techniques. Second Edition, Food and Agriculture Organization of the United Nations, Rome, Italy. Available on line, on http://www.fao.org/docrep/010/a1336e/ a1336e00.htm.

Raghunath H.M. (1987). Groundrwater. 2nd Edn, Wiley Eastern Limited. New Delhi, India, p 344-369.

Rakotomalala R., (2008). Tests de normalité: Techniques empiriques et tests statistiques, *Version 2.0.* Université Lumière Lyon 2, 59p

Rentier C., (2002). Methods stochastic delineation of protection zones around water abstraction. *PhD Thesis,* Université de Liège, 230p.

Richards L.A. (1954). Diagnosis and improvement of saline and alkali soils. Agric. *Handbook 60, USDA*, Washington D.C, p 160.

Sadashivaiah C., Ramakrishnaiah C R, Ranganna C., (2008). Hydrochemeical analysis and evaluation of groundwater in Tumkur Taluk, Karnataka state, India. *Intl. J. Envion. Research and Public health,* 5(3): 158-164.

Sanchez-Martos, F., Jiménez-Espinosa, R., Pulido-Bosch, A. (2001) Mapping groundwater quality variables using PCA and geostatistics: a case study of Bajo Andarax, southeastern Spain, *Hydrological Sciences-Journal-des Sciences Hydrologiques,* 46(2), 543-555.

Sarkar A. A., and A. A. Hassan, (2006). Water quality assessment of a groundwater basin in Bangladesh for irrigation. *Pakistan Journal of biological Sciences* 9(9), 1677-1648.

Soares A (2010) Geostatistical methods for polluted sites characteriation. Geostatistics for environment application. geoEenv VII-geostatistics for environmental applications. *Quant Geol Geostat* 16:199–209.

Shi, Z., Y. Li, R. C. Wang, and F. Makeschine, (2005). Assessment of temporal and spatial variability of soil salinity in a coastal saline field. *Environ Geo,* 48, 171-178.

Singh A.K., G.C. Mondal, B. K. Tewary, and A. Sinha, (2008). Major ion chemistry, solute acquisition process and quality assessment of mine water in damodar valley coalfields, India. *Abstracts of the International Mine Water Conference, 19th – 23rd October 2009,* Pretoria, South Africa.

Sundaray, S. K., Nayak, B. B., & Bhatta, D. (2009). Environmental studies on river water quality with reference to suitability for agricultural purposes:Mahanadi river estuarine system, India: a case study. *Environmental Monitoring and Assessment,* 155, 227–243.

Umapathy S., (2011). A study on ground water quality of Neyveli area, Cuddalore district, tamilnadu. Inter. Nation. *Journal of Géomatics and geosciences,*V(2), 49-56.

Wackernagel, H. (2003). Multivariate geostatistics. An introduction with applications, *3e Ed. Springer-Verlag,* Berlin.

Wilcox LV (1955) Classification and use of irrigation waters. *US Department of Agriculture.* Cir. 969, Washington.

Yeh M.S., Lin Y.P, Chang L.-C., (2006). Designing an optimal multivariate geostatistical groundwater quality monitoring network using factorial kriging and genetic algorithms. *Environ Geol* (2006) 50: 101–121. DOI 10.1007/s00254-006-0190-8.

In: Irrigation Management, Technologies ...
Editor: M. Hossain Ali
ISBN: 978-1-62417-862-7

Chapter 7

ASSESSING THE FACTORS INFLUENCING GROUNDWATER IRRIGATION TECHNOLOGY ADOPTION IN GHANA

Victor Owusu[1,]* ***Alberta Vincentia Asante***[1]
and Paul Pavelic[2]
[1]Department of Agricultural Economics, Agribusiness and Extension
Kwame Nkrumah University of Science and Technology, Kumasi, Ghana
[2]International Water Management Institute, Hyderabad, India

ABSTRACT

Groundwater irrigation is believed to have the highest level of water control necessary for smallholder irrigation farmers to overcome household-level poverty and food insecurity. This study analyses the effects of the determinants of groundwater irrigation technology adoption with a data from 240 smallholder irrigation farmers in Ghana. The empirical results show that household size and membership of farmer-based organizations tend to have positive significant impacts on the probability of adopting groundwater manual irrigation whilst extension contacts and farm size have negative impacts. In addition the age, gender and education of the farmer, household size, access to credit, farm size and owner-cultivation tend to have significant positive effects on the probability of adopting groundwater motor pump irrigation technology whilst experience in irrigation farming and extension contacts have negative effects. The probability to adopt surface water motor pump irrigation technology is positively influenced by extension contacts and farm size whilst owner-cultivation has a negative effect. Gross margin analyses indicate higher gross margins for groundwater motor pump irrigators. To promote the groundwater irrigation technology adoption, the update of skills and knowledge of the farmers on the existing irrigation technologies must be facilitated through frequent participation in workshops and training programs on irrigation. Smallholder irrigation farmers must be supported with accessible micro-credit facilities for acquisition of low-cost irrigation pumps.

---

* Email Address: vowusu.agric@knust.edu.gh.

1. INTRODUCTION

Despite the fact that agriculture plays a prominent role in Ghana's socioeconomic development, agricultural production is mainly rainfed and plagued with drought-related risks and unforeseeable weather patterns. Under these conditions, smallholder irrigation agriculture has been advocated as a way out of the persistent household-level food insecurity and poverty trap experienced by most rural households in the country. Irrigation tends to increase the productivity of smallholder farmers (Bhaduri et al., 2009) and stimulates accelerated economic growth (Namara et al., 2010).

As in most parts of sub-Saharan Africa, smallholder irrigation agriculture in Ghana has not been optimal in terms of acreages under irrigation although the potential to expand irrigable area through the adoption of small-scale irrigation technologies exists. Evidence shows that less than two percent of the total cultivatable area in Ghana is irrigated (Gumma et al., 2011). In response to the potential benefits and challenges of irrigation technology adoption, policy makers and other stakeholders are making frantic efforts to promote small-scale irrigation technology adoption among smallholder farmers in Ghana (Namara et al, 2011b). In spite of all these efforts, Ghana's agriculture is still characterized by low rates of irrigation technology adoption (Namara et al., 2010).

Ghana's irrigation water sources are mainly groundwater and surface water. Farmers normally use ropes and buckets and sometimes motorized pumps to lift and distribute the surface or groundwater to irrigate their crops. In some cases, watering cans have been used to lift water from rivers and streams (Adeoti et al., 2007). The motorized pumps which are relatively efficient are more capital intensive. The adoption of these capital intensive micro-irrigation schemes by smallholder farmers who usually operate small farms with relatively small capital has therefore been difficult (Adeoti, 2009). As argued by Brabben and Kay (2000), the lack of simple, affordable and well adapted technologies for irrigation agriculture is a serious handicap to the efforts of achieving food security in sub-Saharan Africa.

A number of potential barriers to the adoption of micro-irrigation technologies have been identified in the literature (He et al., 2007; Namara et al., 2010; Namara et al., 2011a). These include inadequate information, education, and training on the irrigation technology (He et al., 2007). In addition, the information about the technology to the farmers involves uncertainty and risk, which discourage adoption (Foltz, 2003). Lack of access to credit, especially when a significant expenditure is required to purchase the irrigation equipments could also prevent adoption (He et al., 2007; Adeoti, 2009). Absent or unreliable supply of equipments and insufficient transportation or infrastructure could also be major barriers to the adoption of the new irrigation technology.

Previous studies that have investigated the adoption of irrigation technologies in Ghana (Adeoti et al., 2007; Namara et al., 2010; Yidana, 2010; Namara et al., 2011b) have not paid specific attention to the effects of the factors which tend to influence the adoption of groundwater irrigation technology by smallholder farmers. Despite the fact that groundwater sources are believed to have the highest level of water control for irrigation (Sha, 2006; Sha, 2009; Pavelic et al., 2012), rigorous studies on the adoption of these technologies in Ghana are relatively scanty (Adeoti, 2009; Namara et al., 2011a).

Although donors and policymakers express interest in supporting smallholder irrigation technology adoption in Ghana, there are limited information on the opportunities and constraints of the existing irrigation technologies (Namara et al., 2011a). The present study expands the existing literature on irrigation technology adoption by rigorously analyzing the effects of the factors which tend to influence the adoption of groundwater irrigation technologies in Ghana. Moreover, gross margin estimates on the financial performance of the smallholder farmers under the different irrigation technologies are provided.

2. Groundwater Irrigation Potential in Ghana

Although groundwater irrigation in sub-Saharan Africa in general is low (Sha, 2006; Sha, 2009), the utilization of groundwater as the most reliable water resource for irrigation is relatively gaining some prominence amongst smallholder farmers in Ghana (Nicola et al, 2005; Namara et al., 2010; Namara et al., 2011a). This relative rise in agricultural use of groundwater, once mainly for domestic use, is due to the increasing availability of small pumping technologies (Sha, 2006; Namara et al., 2010). In Ghana, several areas abound in groundwater. The most predominant areas are the Upper Volta Basin which encompasses the Upper East Region, Upper West Region and the Northern Region. Groundwater is also available in the areas around the Lower Volta including the Volta Region and the coastal areas of Ghana.

According to Odame-Ababio (2002), the groundwater resources in Ghana are mainly found in the sedimentary rock formation made up of the Voltarian origin (which occupies about 43 percent of the country's total area) and the non-sedimentary formation made up of the crystalline basement complex of pre-Cambrian origin (which occupies 57 percent of the total area of the country with yields of 1.5 m^3/ha to 32.0m^3/ha and 20m to 100m deep). The quality of groundwater resources in Ghana is generally good except for the few pollutions of iron and fluoride, as well as mineralization of dissolved solids usually found in some coastal aquifers (WRC, 2000). In Northern Ghana, aquifers have been located at between 10m and 60m deep with an average of 27m. In the Southern Ghana, boreholes are deeper ranging between 25m and 90m deep, with an average of 42m due to thicker soil cover. Nicola et al. (2005) have reported that in the Volta Basin alone, the annual groundwater production through boreholes, hand dug wells, and piped systems has increased substantially over the past decade to an estimated 88 million m^3/year.

Despite the potentials of groundwater irrigation in Ghana, numerous challenges exist. These include the absence of explicit policy support, lack of access to affordable drilling technology, and cost of energy for abstracting water (Namara et al., 2010; Ghana Irrigation Policy, 2011). Moreover, groundwater irrigation in Ghana has little policy or strategic support. For example, groundwater irrigation is not an important strategy for policy makers in the formulation of agriculture development policies in the country (MoWRWHG, 2007; Ghana Irrigation Policy, 2011; Namara et al., 2011a; MoFA, 2011).

3. Conceptual Framework

The smallholder farmer is assumed to be cultivating crops using the existing irrigation technologies. These include groundwater manual, groundwater motor pump and surface water motor pump irrigation technologies. The farmer decides on the irrigation technology to adopt taken into consideration the perceived net benefits from the technology, as well as the field's characteristics. Given these assumptions and denoting the net benefits from adoption of the irrigation technology by A^*, the farmer adopts the technology if the net benefits from adoption exceeds that of non-adoption (i.e. $A^* > 0$). Let A_{ij}^* represents the underlying latent variable for adoption of the irrigation technology *j* by farmer *i*. Although A_{ij}^* is unobservable; it can be expressed as a function of observable elements in a latent variable model. Denoting adoption of groundwater manual, groundwater motor pump and surface water motor pump irrigation technologies by G_{mn}, G_{mp} and S_{mp} respectively, the latent variable can be related to a set of household-level characteristics and farm-level as in (1):

$$A_{ij}^* = \lambda X_{ij} + \varepsilon_{ij}, \qquad A_{ij} = 1\left[A_{ij}^* > 0\right] \; (j = G_{mn}, G_{mp}, S_{mp}) \tag{1}$$

where A_{ij} is a binary variable indicating 1 if household *i* adopts and 0 otherwise, λ is a vector of parameters to be estimated, X_{ij} is a vector of household and plot-level characteristics and ε_{ij} is an error term assumed to be normally distributed. The probability of adoption of the existing irrigation technology can be represented as:

$$\Pr(A_{ij} = 1) = \Pr(A_{ij}^* > 0) = \Pr(\varepsilon_{ij} > -\lambda X_{ij}) = 1 - H(-\lambda X_{ij}), \tag{2}$$

where $H(\cdot)$ is the cumulative distribution function for ε_{ij}. This study specifically employs the probit model to examine the probability of adoption of an existing irrigation technology by the smallholder farmers (Greene, 2008). The personal and household characteristics of the farmer investigated in the probit model are the age (X_1), gender (X_2), education (X_3), experience (X_4), household size (X_5), extension contact (X_6), membership of farmer's association (X_7) and credit access for irrigation farming (X_8). The plot-level characteristics investigated are the farm size (X_9) and land tenure (owner-cultivation, X_{10}). Given these characteristics of the farmer, and from equation (1), it is hypothesized that $\partial A/\partial X_1 > 0$, $\partial A/\partial X_2 > 0$, $\partial A/\partial X_3 > 0$, $\partial A/\partial X_3 > 0$, $\partial A/\partial X_4 > 0$, $\partial A/\partial X_5 > 0$, $\partial A/\partial X_6 > 0$, $\partial A/\partial X_7 > 0$, $\partial A/\partial X_8 > 0$, $\partial A/\partial X_9 > 0$ and $\partial A/\partial X_{10} > 0$.

4. DATA DESCRIPTION

The survey for the data collection was carried out between January and March in 2011 amongst 240 smallholder farmers in 3 administrative districts of Ghana. These districts are the Akuapem South Municipality in the Eastern Region, the Keta Municipality in the Volta

Region and the Kasena Nankana District in the Upper East Region of Ghana. A multi-stage sampling procedure was employed. The 3 districts were purposely selected because of the prevailing irrigation farming in the districts. About 5 communities were randomly selected from each of the 3 districts. The communities randomly sampled from the Akuapem South Municipality were Pakro, Nyamebekyere, Krokese, Akraman and Afumkrom. The randomly selected communities from the Keta Municipality were Anloga, Whuti, Woe, Srogbe and Tegbi. In the Kasena Nankana district, Nania, Kazugu, Doba, Naghalkinia and Pungu were randomly sampled for the study. In the last stage of the sampling, about 16 farm households were randomly sampled from the 15 sampled communities making a total sample size of 240. The sample specifically contains 60 rainfed farmers who have not adopted any irrigation technology, 60 groundwater manual irrigators, 46 groundwater motor pump irrigators and 74 surface water motor pump irrigators.

The sources of groundwater for irrigation farming were shallow wells, deep wells and boreholes. The surface water sources were rivers and streams. The lists of farmers under the existing irrigation technologies were obtained from the various farmer-based organizations in the sampled communities. Carefully designed questionnaires were used to solicit relevant primary data on the personal and household-level characteristics as well as plot-level characteristics of the smallholder farmers. Information on production, sales and irrigation technologies were also collected.

5. Discussion of Results

The descriptive statistics of the variables used in the probit models reveal some differences in the adopters and non-adopters of the existing irrigation technologies (Table 1). The maximum likelihood estimates provided in Table 2 are from the three probit models explaining respectively, the probability to adopt groundwater manual irrigation technology, groundwater motor pump irrigation technology and surface water motor pump irrigation technology. The models predicted correctly 23.6%, 28.7% and 72.3% respectively for the sample, the probability to adopt groundwater manual irrigation, groundwater motor pump irrigation and surface water motor pump irrigation. The value of the likelihood ratio test statistic in each specification is highly significant indicating a rejection of the null hypothesis that the parameters are jointly equal to zero.

Generally, the coefficients of age, gender (male) and education of the farmer, as well as household size, farm size, being an owner-cultivator and a member of farmer-based association and having access to credit all have the expected signs in the probit model explaining the probability to adopt groundwater motor pump irrigation technology (Table 2). The coefficients for household size, experience in irrigation farming, being a member of farmer-based association and having access to credit exhibited the hypothesized positive signs in the model for the adoption of groundwater manual irrigation technology.

Table 1. Descriptive statistics of variables used in the probit models

Variable	Variable definition	Groundwater manual (G_{mn})		Groundwater motor pump (G_{mp})		Surface water motor pump (S_{mp})	
		Adopters (N=60)	Non-adopters (N=180)	Adopters (N=46)	Non-adopters (N=194)	Adopters (N=74)	Non-adopters (N=166)
AGE	Age of the farmer (years)	42.30 (12.08)	40.23 (12.61)	43.65 (10.68)	40.06 (12.81)	35.49 (11.78)	43.10 (12.11)
GEND	1 if farmer is a male, 0 otherwise	0.77 (0.43)	0.91 (0.29)	0.96 (0.21)	0.85 (0.36)	0.93 (0.25)	0.84 (0.36)
EDUC	Number of years of schooling	6.97 (5.06)	6.83 (5.11)	9.37 (5.38)	6.27 (4.84)	5.91 (4.78)	7.30 (5.17)
HHSIZE	Household size	6.42 (2.35)	5.44 (2.30)	6.70 (2.36)	5.45 (2.29)	5.03 (2.42)	5.98 (2.26)
EXPER	Number of years in irrigation farming	16.25 (9.19)	15.52 (10.02)	15.41 (9.57)	15.77 (9.89)	13.14 (10.31)	16.84 (9.38)
FASSO	1 if member of farmer-based association, 0 otherwise	0.62 (0.49)	0.42 (0.49)	0.61 (0.49)	0.43 (0.50)	0.36 (0.48)	0.51 (0.50)
CREDIT	1 if farmer received credit for farming in 2010, 0 otherwise	0.35 (0.48)	0.26 (0.44)	0.41 (0.50)	0.25 (0.43)	0.22 (0.41)	0.31 (0.46)
EXTEN	1 if farmer extension contact, 0 otherwise	0.47 (0.50)	0.67 (0.47)	0.41 (0.50)	0.67 (0.47)	0.78 (0.41)	0.55 (0.50)
FSIZE	Farm size (Ha)	2.56 (1.90)	4.57 (3.38)	4.44 (2.92)	3.98 (3.26)	5.38 (4.13)	3.48 (2.48)
LTEN	1 if farmer is owner-cultivator, 0 otherwise	0.77 (0.43)	0.61 (0.49)	0.89 (0.31)	0.59 (0.49)	0.11 (0.31)	0.89 (0.32)

Note: Figures in parenthesis are the standard deviations.

Also the coefficients for gender, household size, experience in irrigation farming, being a member of farmer-based association farm size, having access to credit, extension contact and farm size all exhibited the correct a priori signs in the model explaining the probability to adopt surface water motor pump irrigation technology by the smallholder farmers. Also shown in Table 3 are the marginal effects of the explanatory variables for which adoption of the irrigation technologies may have policy impacts. The positive marginal effect of age for groundwater motor pump irrigation and the negative marginal effect for surface motor pump irrigation suggest that the probability to adopt groundwater motor pump irrigation is higher for older farmers but lower for young adopters of surface water motor pump irrigation. The marginal effect of 0.1239 for the gender variable indicates that being a male increases ones probability to adopt groundwater motor irrigation technology by 12.4%. The positive significant marginal effect of education suggests a 1.0% higher probability to adopt groundwater motor pump irrigation technology as the farmer's number of years of schooling increases. Education is believed to permit a more critical evaluation of the productive characteristics and costs of adoption innovations.

Table 2. Maximum likelihood estimates on the adoption of irrigation technologies

Variable	Groundwater manual (G_{mn})	Groundwater motor pump (G_{mp})	Surface water motor pump (S_{mp})
CONSTANT	0.0025 (0.00)	-3.4930*** (-5.12)	0.6900 (0.87)
AGE	-0.0067 (-0.58)	0.0306** (2.45)	-0.0397** (-2.40)
GEND	-0.3171 (-1.05)	0.7721** (2.05)	0.1766 (0.33)
EDUC	-0.0270 (-0.77)	0.0544** (1.97)	-0.0285 (-0.80)
HHSIZE	0.1591*** (3.22)	0.0927** (1.92)	0.0982 (1.51)
EXPER	0.0090 (0.59)	-0.0462*** (-2.90)	0.0016 (0.06)
FASSO	0.6560*** (2.78)	0.1930 (0.77)	0.1712 (0.55)
CREDIT	0.2775 (1.18)	0.7282*** (2.73)	0.1141 (0.26)
EXTEN	-0.7574*** (-3.45)	-0.9569*** (-4.06)	2.1686*** (4.37)
FSIZE	-0.2790*** (-4.73)	0.0633* (1.74)	0.1154** (2.18)
LTEN	-0.0787 (-0.34)	0.8832*** (3.05)	-3.3410*** (-7.35)
Number. of observations	240	240	240
Log- likelihood	-103.1616	-82.5495	-41.3128
Wald $\chi^2(10)$	63.56***	60.66***	200.29***
Pseudo- R^2	0.2356	0.2873	0.7228

***, **, * are 1%, 5% and 10% significant levels respectively.
Z-values are in parentheses.
Source: Authors' computation based on field survey data.

The marginal effects of 0.0398 and 0.0173 for household size indicate that as the household size of the farmer increases, there are corresponding increases of 3.98% and 1.73% respectively in the probability to adopt groundwater manual irrigation and groundwater motor pump irrigation. Household size was used as a proxy for the availability of family labor at the household-level for irrigation farming, since some irrigation technologies are labor intensive. Using Ghanaian data, Adeoti (2009) found a positive influence of availability of labor on the probability to adopt micro-irrigation technologies. The negative significant marginal effect of the experience variable suggests a 0.9% decrease in the probability to adopt groundwater motor pump irrigation technology as the number of years in irrigation farming increases. Compared to the "new" irrigation technologies, the smallholder farmers may have been become used to the "traditional" technologies of irrigation farming (Namara et al, 2007). The results also concur with the hypothesis that the longer a farmer gets used to an irrigation technology the less probable would he be willing to adopt a new technology.

Table 3. Marginal effects of the probit estimates

Variable	Groundwater manual (G_{mn})	Groundwater motor pump (G_{mp})	Surface water motor pump (S_{mp})
AGE	-0.0017 (-0.58)	0.0057** (2.40)	-0.0086** (-2.02)
GEND	-0.0881 (-0.95)	0.1239*** (3.38)	0.0347 (0.36)
EDUC	-0.0070 (-1.25)	0.0101** (2.24)	-0.0060 (-0.80)
HHSIZE	0.0398*** (3.33)	0.0173* (1.91)	0.0208 (1.51)
EXPER	0.0023 (0.59)	-0.0086*** (-2.87)	0.0003 (0.06)
FASSO	0.1671*** (2.74)	0.0391 (0.76)	0.0461 (0.54)
CREDIT	0.0734 (1.14)	0.1639** (2.38)	0.0249 (0.25)
EXTEN	-0.2048*** (-3.33)	-0.2060*** (-3.65)	0.3751*** (4.88)
FSIZE	-0.0697*** (-5.34)	0.0118* (1.72)	0.0245** (2.14)
LTEN	-0.0197 (-0.33)	0.1472*** (3.31)	-0.7085*** (-5.00)

***, **, * are 1%, 5% and 10% significant levels respectively.
Z-values are in parentheses.
Source: Authors' computation based on field survey data.

The empirical results also indicate a 16.7% higher probability to adopt groundwater manual irrigation as more irrigation farmers join farmer-based organizations. The probability to adopt groundwater motor pump irrigation technology increases by 16.4% as the farmers' access to credit increases. Access to capital in the form of either accumulated savings or capital markets is necessary in financing the adoption of new agricultural technologies (Adeoti, 2009). Farmers' contacts with Agricultural Extension Agents (AEAs) serve as a source of information on the relevance and appropriate use of new agricultural technologies. The empirical results show that the probability to adopt surface water motor pump irrigation technology increases by 37.5% but the adoption of groundwater manual irrigation and groundwater motor pump irrigation decreases by margins of 20.5% and 20.6% respectively when farmers contact with extension agents increases. The result on the surface water motor pump irrigation is consistent with studies on agricultural technology adoption (Adeoti, 2009).

Table 4. Gross margin estimates

	Groundwater manual (G_{mn})	Groundwater motor pump (G_{mp})	Surface water motor pump (S_{mp})	Rainfed
Variable costs (GH¢/Ha)				
Land preparation		100.45	57.5	65.5
Sowing of seeds		79	36.82	63.86
Weed control		85.91	67.73	68.25
Harvesting		59	31	
Seeds	24.58	22.44	16.22	11.55
Insecticide	17.35	15.42	15.55	7.5
Herbicide	16.14	15.89	15.25	6
Fungicide	29	17.17	21.5	6
Neem	30	42.57	48.75	0
NPK	29.97	32.93	38.97	27.23
Sulphate of Ammonia	23.55	24.75	40.73	19.18
Urea	29.73	30.9	50.45	2.88
Animal manure	41	13.43	45	30
Green manure	37.5	47.5	45	0
Compost	50	110	20	0
Water	40	9.17	120	0
Fuel	7	7	5.8	0
Total variable cost (GH¢/Ha)	375.82	713.53	676.27	307.95
Revenue (GH¢/Ha)				
Crop 1	701.4	876.8	765.81	618.94
Crop 2	617.87	976.8	1071.46	655.74
Crop 3	916.67	1056.25	668.56	607.3
Crop 4	232.21	625	649.5	219.7
Crop 5	989.98	784.31	696.25	302.5
Total Revenue(GH¢/Ha)	3458.13	4319.16	3851.58	2404.18
Gross margin (GH¢/Ha)	3082.31	3605.63	3175.31	2096.23

[1] US Dollar (US$) = 1.5053 Ghana Cedi (GH¢) in 2011.

The negative impacts of the extension variable on the adoptions of groundwater manual irrigation and groundwater motor pump irrigation suggest that although extension contact may be high amongst farmers, the failure of AEAs to focus on issues that could help improve the adoption of irrigation may discourage farmers from adopting the technology. The farm size variable exhibited positive marginal effects for the probability to adopt groundwater motor pump and surface water motor pump irrigation technologies but a negative effect for surface water motor pump irrigation. The traditional manual irrigation technology where ropes and buckets are used to transport irrigation water from dug-out wells is relatively more tedious, *ceteris paribus* discouraging expansion of acreages under cultivation. On the other hand, the use of motor pumps with flexible rubber pipes and tubes could easily convey both surface and groundwater to the farm for irrigation, *ceteris paribus*, resulting in expansion in area under cultivation. The variable representing owner-cultivation has a positive marginal effect of 0.1472 for groundwater motor pump irrigation and a negative marginal effect of -0.7085 for surface water motor pump. These suggest a 14.72% higher probability to adopt groundwater motor pump irrigation for farmers with owned plots and 70.85% lower probability to adopt surface water motor pump irrigation for farmers with owned plots. Other things being equal, we expect owner-cultivators to invest more in permanent irrigation structures such as dug-out wells and motor pumps compared to farmers with temporary rented plots. The empirical results concur with a study by Namara et al. (2007) with Indian data.

The gross margin on the relative financial performance of each irrigation technology was computed as the total revenue less the total variable costs. The revenues represent returns from combination of crops cultivated under each technology. The gross margin assumes away the fixed capital costs associated with each irrigation technology. As Table 4 shows, the groundwater motor pump irrigators tend to have higher gross margins despite their relatively large variable costs (or fixed costs incurred in the investment of groundwater motor pumps). The high revenue from groundwater motor pump irrigation may come from the level of water control for irrigation almost twice or thrice a year depending on the resources and labor availability of the farmer. The policy implications of these findings are that smallholder irrigation farmers would be better-off if they were supported with subsidies and easily accessible micro-credits to invest in motor pumps, rather than resorting to the traditional rainfed farming which hinges on the vagaries of the weather.

Conclusion and Policy Recommendations

This chapter has investigated the adoption of groundwater irrigation technologies in Ghana with a cross-sectional data from 240 smallholder farmers collected in 2011. The field survey was carried out in the Akuapem South Municipality in the Eastern Region, the Keta Municipality in the Volta Region and the Kasena Nankana District in the Upper East Region of Ghana.

The factors which tend to influence the probability of smallholder farmers to adopt groundwater manual irrigation, groundwater motor pump irrigation and surface water motor pump irrigation technologies were analyzed with the probit models. The empirical results show that household size and membership of farmer-based organization tend to have positive significant impacts on the probability to adopt groundwater manual irrigation whilst extension contacts and farm size have negative impacts. Also the age, gender and education of the farmer, household size, access to credit, farm size and owner-cultivation tend to have significant positive impacts on the adoption of groundwater irrigation technologies whilst experience in irrigation farming and extension contacts have negative effects. Also extension contacts and farm size have significant positive effects on the probability to adopt surface water motor pump irrigation whilst owner-cultivation has a negative effect. Gross margin analyses indicate higher gross margins for groundwater motor pump irrigators.

To promote the groundwater technology adoption, update of skills and knowledge of the farmers on the existing irrigation technologies must be facilitated through frequent participation in workshops and training programmes on irrigation. In addition, smallholder irrigation farmers must be supported with accessible micro-credit facilities for acquisition of low-cost irrigation pumps. The Government of Ghana (GoG), NGOs and other stakeholders should step up the efforts in providing financial assistance to smallholder irrigation farmers. Also efficient water management and good agronomic practices (GAP) that assist farmers to produce crops at the minimum cost should be promoted by Agricultural extension Agents (AEAs) and other stakeholders to make the adoption of small-scale irrigation technologies more economically and financially viable to farmers.

REFERENCES

Adeoti, A., Barry, B., Namara, R., Kamara, A. and A. Titiati (2007). *Treadle pump irrigation and poverty in Ghana.* IWMI Research Report 117. Colombo, Sri Lanka: International Water Management Institute. pp. 30.

Adeoti, A. I. (2009). Factors influencing irrigation technology adoption and its impact on household poverty in Ghana. *Journal of Agriculture and Rural Development in the Tropics and Subtropics* 109(1):51–63.

Bhaduri, A., Amarasinghe, U.A. and T. Shah (2009). Groundwater expansion in Indian agriculture: past trends and future opportunities. In: Amarasinghe, U.A., Shah, T., Malik, R.P.S. (Eds.), India's Water Future: Scenarios and Issues. International Water Management Institute, Colombo, Sri Lanka, pp. 181–196.

Brabben, T. and M. Kay (2000). Treadle pumps – A viable option for Africa? GRID, Issue 16. IPTRID Network Magazine. International Program for Technology and Research in Irrigation and Drainage/Food and Agriculture Organization of the United Nations.

Foltz, J. D. (2003). The economics of water-conserving technology adoption in Tunisia: An empirical estimation of farmer technology choice. *Economic Development and Cultural Change* 51 (2): 359-73.

Ghana Irrigation Development Policy (2011). National Irrigation Policy, Strategies and Regulatory Measures. Ministry of Food and Agriculture (MoFA) and Ghana Irrigation Development Authority (GIDA).

Greene, W. H. (2008). Econometric analysis (6th ed.). Upper Saddle River, NJ: Prentice Hall.

Gumma, M. K., Thenkabail, P. S., Hideto, F., Nelson, A., Dheeravath, V., Busia, D. and A. Rala (2011). Mapping Irrigated Areas of Ghana Using Fusion of 30 m and 250 m Resolution Remote-Sensing Data. *Remote Sensing* 3: 816-835.

He, X.-F., Cao, H., and F.-M. Li (2007). Econometric analysis of the determinants of adoption of rainwater harvesting and supplementary irrigation technology (RHSIT) in the semiarid Loess Plateau of China. *Agricultural Water Management* 89(3): 243-250.

MoWRWHG (2007). Ministry of Water Resources, Works and Housing, Ghana (MoWRWH).

MoFA (2011). Ministry of Food and Agriculture, Ghana (MoFA).

Namara, R. E., Nagar R. K. and B. Upadhyay (2007). Economics, adoption determinants, and impacts of micro-irrigation technologies: Empirical results from India. *Irrigation science* 25:283–297.

Namara, R. E., Horowitz, L., Kolavalli, S., Kranjac-Berisavljevic, G., Dawuni, B. N., Barry, B.

M. Giordano (2010). *Typology of irrigation systems in Ghana.* Colombo, Sri Lanka: International Water Management Institute. 35p. (IWMI Working Paper 142).

Namara, R. E., Awuni, J. A., Barry, B., Giordano, M., Hope, L., Owusu, E. S. and G. Forkuor (2011a). *Smallholder shallow groundwater irrigation development in the upper east region of Ghana*. Colombo, Sri Lanka: International Water Management Institute. 35p. (IWMI Research Report 143). www.iwmi.org/Publications/IWMI_Research_Reports/index.aspx

Namara, R.E., Horowitz, L., Nyamadi, B. and B. Barry (2011b). Irrigation Development in Ghana: Past experiences, emerging opportunities, and future directions. Ghana Strategy Support Program (GSSP) Working Paper No. 0026. Accra, Ghana: International Food Policy Research Institute (IFPRI).

Nicola, M. and N. Van De Giesen (2005). Spatial distribution of groundwater production and development potential in the Volta River basin of Ghana and Burkina Faso. *Water International* 30 (2): 239–249.

Odame-Ababio, K. (2002). Integrated water resources management, environment and equity in national water laws: Ghana's perspective. A paper presented at the Workshop on Water Governance: Legal and Institutional Aspects of Water Resources Management in West Africa. Organized in Ouagadougou 25–27 September 2002 by IUCN-West Africa Office and IUCN Center for Environmental Law.

Pavelic, P., Patankar, U., Acharya, S., Jella, K. and M.K. Gumma (2012). Role of groundwater in buffering irrigation production against climate variability at the basin scale in South-West India. *Agricultural Water Management* 103:78– 87.

Shah, T., Singh, O. P. and A. Mukherji (2006). Some aspects of South Asia's groundwater irrigation economy: analyses from a survey in India, Pakistan, Nepal Terai and Bangladesh. *Hydrogeology Journal* 14: 286–309.

Shah, T. (2009). *Taming the anarchy: Groundwater governance in South Asia*. Washington, DC, USA: Resources for the Future (RFF); Colombo, Sri Lanka: International Water Management Institute (IWMI).

Yidana, S.M. (2010). Groundwater classification using multivariate statistical methods: Birimian Basin, Ghana. *Journal of Environmental Engineering* 136(12):1379-1388.

WRC (2000). *Water Resources Management Problems Identification, Analysis and Prioritization*. Final Report. Water Resources Commission (WRC), Accra, Ghana.

In: Irrigation Management, Technologies …
Editor: M. Hossain Ali
ISBN: 978-1-62417-862-7

Chapter 8

IRRIGATION MANAGEMENT: EFFECTS OF SOYBEAN DISEASES ON SEED COMPOSITION IN GENOTYPES DIFFERING IN DISEASE RESISTANCE UNDER IRRIGATED AND NONIRRIGATED CONDITIONS

Nacer Bellaloui[1,*], Alemu Mengistu[2], Daniel K. Fisher[3], L. H. S. Zobiole[4] and Hamed K. Abbas[5]

[1]Crop Genetics Research Unit, USDA-ARS, Stoneville, MS, US
[2]Crop Genetics Research Unit, USDA-ARS, Jackson, TN, US
[3]Crop Production Systems Research Unit, USDA-ARS, Stoneville, MS, US
[4]Crop Protection, R&D, Dow AgroSciences – Brazil, Cascavel, Paraná, Brazil
[5]Biological Control of Pests Research Unit, USDA-ARS, Stoneville, MS, US

ABSTRACT

Soybean seed is a major source of protein and oil in the world. Nutritional qualities of soybean seed are determined by the quantity and quality of seed composition components (protein, oil, fatty acids, isoflavones, and minerals). Charcoal rot is a disease caused by the fungus *Macrophomina phaseolina* (Tassi) Goid and results in yield loss and poor seed quality, especially under drought conditions. Phomopsis, caused by *Phomopsis longicolla* Hobbs, is also one of the major soybean seed decay pathogens, resulting in poor seed quality, especially under humid, warm conditions. Disease infection is influenced by irrigation management. Since there are no commercial soybean cultivars resistant to either charcoal rot or *Phomopsis*, irrigation management to minimize charcoal rot and phomopsis infection to maintain high seed quality is crucial. Therefore, the objective of this research was to evaluate the effect of charcoal rot and *Phomopsis* on seed protein, oil, fatty acids, sugars, minerals, and isoflavones under irrigated (IR) and nonirrigated (NIR) conditions. Field experiments were conducted using soybean

* Email: Nacer.Bellaloui@ars.usda.gov.

genotypes susceptible (S) and moderately resistant (MR) to charcoal rot and *Phomopsis*. This research evaluates the effect of charcoal rot on seed composition by infesting the soil with charcoal rot (infested soil conditions, INF) or control (noninfested soil conditions, NINF). The effects of *Phomopsis* on seed composition were achieved by infecting soybean plants and assessing seed composition components at harvest maturity and 15 days after harvest maturity. Our research demonstrated that seed composition components were significantly altered by irrigation treatment (IR or NIR) and disease (charcoal rot or *Phomopsis*), and the degree of alteration of these components was dependent on the level of resistance of the genotype to the disease. The research highlights the significance of irrigation management and soybean varietal selection in maintaining high seed nutritional qualities. A special section is devoted to soybean irrigation and current irrigation techniques.

1. INTRODUCTION

Soybean is a major crop in the world, and the soybean seed is an essential source of protein and oil (Belewu & Belewu, 2007), minerals and sugars (Hou et al., 2009; Zobiole et al., 2010; Bellaloui et al., 2010 b; Bellaloui et al., 2011), and phenolics (Sakthivelu et al., 2008; Bellaloui et al., 2012a, b). Soybean seeds contain about 42.1% protein, 19.5 % oil (Wilson, 2004), 33% carbohydrates (Hymowitz & Collins, 1974), and 5% ash (Wikipedia, 2012). Soybean seeds also contain unsaturated fatty acids (oleic acid, 24%; linoleic acid, 54%; linolenic acid, 8%) (Schnebly & Fehr, 1993) and saturated fatty acids (palmitic acid, 10-12%; stearic acid, 2.2-7.2%) (Cherry et al., 1985). Seed carbohydrates include sucrose (4-5%), raffinose (2%), and stachyose (3.5-4.5%) (Wilson et al., 1995). Soybean seeds contain secondary metabolite phenolic compounds such as phenol, lignin, and isoflavones (Sakthivelu et al., 2008; Bellaloui et al., 2012a, b). The beneficial effects of soybean seed nutrients on human health have been previously reported (Sakthivelu et al., 2008; Messina et al., 1994; Potter et al., 1998; Wikipedia, 2011). From the soybean industry perspectives, high oleic and low linolenic acids are desirable because they contribute to the stability of oil. High sucrose and low raffinose and stachyose are also desirable traits because sucrose is digestible and contributes to taste, but raffinose and stachyose are indigestible, causing flatulence or diarrhea in human and monogastric animals such as swine and chicken (Liu, 1997; Obendorf et al., 1998). Biotic stress factors such as disease, genotype, maturity group and abiotic stress factors such as temperature, irrigation/drought, and soil type have been reported to affect soybean seed protein, oil, fatty acids, sugars, minerals, and phenolics (Maestri et al., 1998; Piper & Boote 1999; Tsukamoto et al., 1995; Lee et al., 2003a, b; Dardanelli et al., 2006; Bellaloui et al., 2009, 2010, 2011, 2012). These variations in seed constituents were explained as an interaction between gene (s) and environment (Bellaloui et al., 2009, 2010, 2011). It was reported that the contribution of the interaction between the genetic background and environment can reach up to 39% for protein and 48% for oil (Bellaloui et al., 2009). Bellaloui et al. demonstrated, using near-isogenic lines for maturity genes, that maturity genes, temperature, and photoperiod significantly contributed to the variability of seed composition constituents (Bellaloui et al., 2009). In addition, the effect of drought, temperature, cultivars/genotypes, maturity group, diseases, and agricultural practices (such as irrigation, crop rotation, fertilizers, soil mineral status) on seed composition and mineral nutrition have been studied (Dardanelli et al., 2006; Maestri at al., 1998; Piper & Boote 1999;

Lee et al., 2003a, 2003b, Tsukamoto et al., 1995; Bellaloui et al., 2009, 2010,2011, 2012a, b), but conclusive results have been not yet achieved. Charcoal rot and *Phomopsis* are among the major diseases in the Early Soybean Production System (ESPS) in the midsouth USA. Charcoal rot is widespread in the midsouth in Mississippi, Arkansa, Missouri, and in the central part of the Midwest, especially in Kansas, Nebraska, and parts of Missouri. Charcoal rot is a root disease caused by the soil-borne fungus *Macrophomina phaseolina* (Tassi) Goid. The disease can cause more than 70% yield loss, poor seed quality, and under severe infection of high heat and severe drought, can kill the crop (Figure 1a-c). Charcoal rot becomes severe under hot, dry weather conditions, or when the crop grows under environmental stress conditions. Charcoal rot infects the root and the lower part of the stem, leading to early maturation, chlorosis, and incomplete pod and seed fill (Figure 1a-c and Figure 2a-c). *Phomopsis* is another major disease in ESPS, but it is more virulent under high temperature and humidity conditions. Although ESPS has shown yield benefits in Arkansas (Taylor, 1999), Mississippi (Heatherly, 1999), Texas (Savoy et al., 1992), and Missouri (Wrather et al., 1996), seed quality in the ESPS is a major concern due to phomopsis seed decay.

a. Irrigation effect: Infested field with charcoal rot under nonirrigated conditions.

b. Irrigation effect: Irrigation removes drought stress, reducing charcoal rot severity.

Figure 1. (Continued)

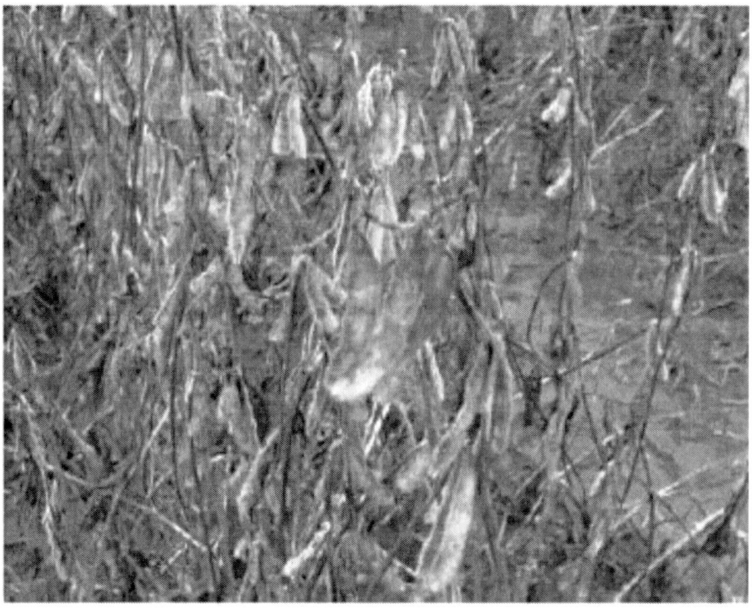

c. Seed-fill is severly affected by charcoal rot under nonirrigated conditions.

d. Seed-fill is less affected by charcoal rot under irrigated conditions.

Figure 1. Charcoal rot infestation in the field and how it is affected by irrigation.

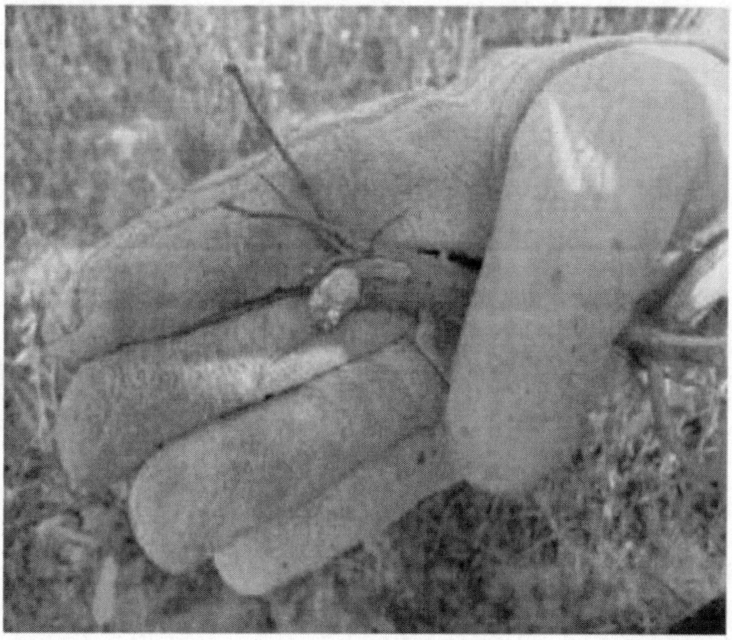

a. Charcoal rot infection to the lower stem (crown) of soybean.

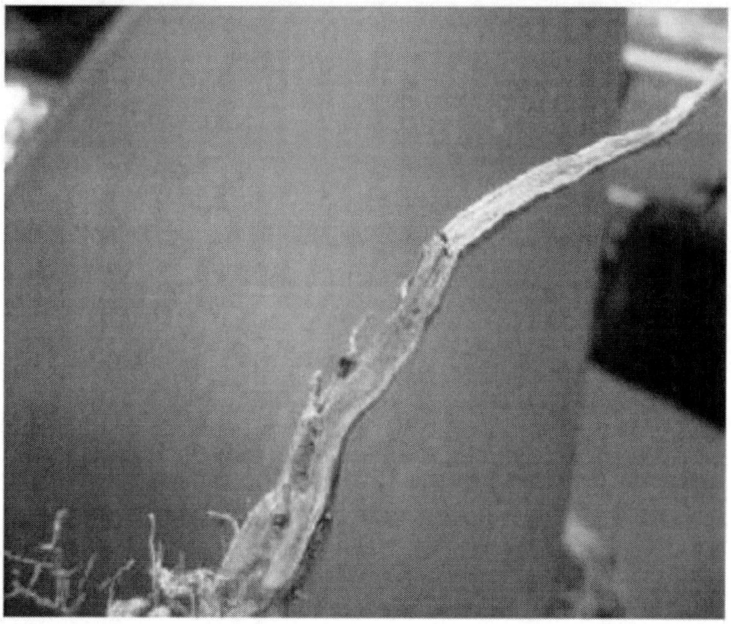

b. Microsclerotia, gray speckles or dots within roots.

Figure 2. (Continued)

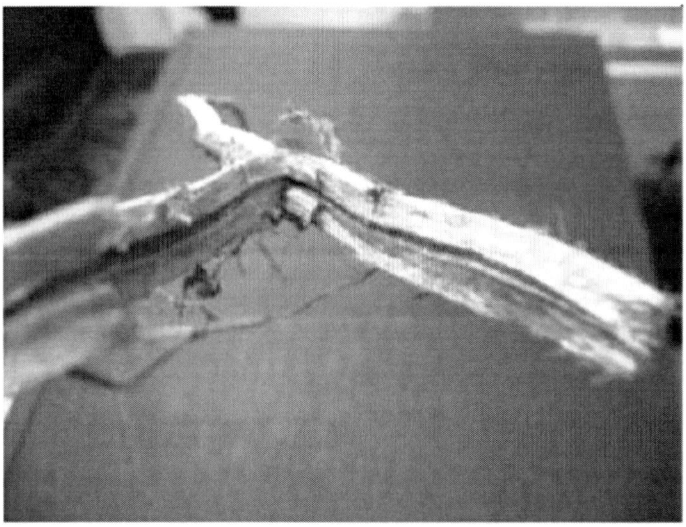

c. Charcoal rot infection to the conductive tissues and red/brown lesions.

d. Spotty infection of charcoal rot to the field. Some areas infected and others not.

Figure 2. Charcoal rot infection and symptoms on stem and root.

The poor seed quality was due to high temperature and humidity between August and September when the soybean matures (TeKrony et al., 1996; Mengistu & Heatherly 2006; Smith et al., 2008). Currently, there are no phomopsis seed decay resistant soybean cultivars available on the market (Mengistu et al., 2010), although resistant germplasm genotypes such as MO/PSD- 0259 (Minor et al., 1993), and SS 93-6012 and SS 93-6181 (Pathan et al., 2009) have been developed. When wet weather delays normal harvesting, a reduction of 10 % in seed weight and 50 % or more in seed germination are found in susceptible cultivars. Soybean seeds having more than 2% damage are reduced in grade, thus lowering the price per bushel and profit to the grower. Seeds infected with the fungi of this complex produce low-quality oil and flour and other value-added products, and are often unsuitable for vegetable or other food and feed uses. Liver damage in chickens, and possibly in other animals, can be caused by toxic compounds produced by these fungi. The fungus survives the winter as dormant mycelia within infected seed and plant materials (Kulik & Sinclair, 1999). Mycelia within pycnidia and perithecia produce asexual or sexual spores during the growing season, and the microscopic spores germinate in 4 to 18 hours when moisture becomes available and penetrate the tissue. The fungus penetrates immature, senescent, or wounded tissues directly. Pods can be infected any time during their development and lead to seed decay. Seeds usually become invaded by the fungus during or after infected pods turn yellow. Seed infection starts on lower nodes, and then spreads throughout the plant, especially in delayed harvesting as is the case in ESPS where early rain usually occurs, delaying harvesting. Phomopsis seed decay affects seed integrity, resulting in lowering nutritional qualities and poor germination.

Mechanisms of the effect of *Phomopsis* on seed quality and composition are still not understood. However, physiological explanations, due to fungal damage, were reported to be related to: a) the disruption of cellular membranes (principally lecithins) (List et al., 1992); b) inhibition of uptake and assimilation of Ca, K, B, Cu, Fe, Co, Zn, and Mn, that are involved in seed coat resistance against diseases; c) disturbances in phenol metabolism and lignin biosynthesis and their effects on membrane permeability and integrity (Marschner, 1995); and d) alteration in cell walls and cell membranes integrity, affecting fungal penetration (Marschner, 1995).

Since there are no commercial soybean cultivars resistant to either charcoal rot or *Phomopsis*, irrigation management to minimize charcoal rot and *Phomopsis* effects to maintain high seed quality is crucial. Therefore, the objective of the current research was to investigate the impact of the relationships between irrigation management and charcoal rot and *Phomopsis* on soybean seed protein, oil, fatty acids, sugars, isoflavones, and minerals in the Early Soybean Production System in the midsouth USA. Susceptible (S) and moderately resistant (MR) soybean genotypes to charcoal rot and *Phomopsis* were used to understand the defense mechanism and response dynamics of these genotypes to these diseases under irrigated and nonirrigated conditions. Since this research is dealing with the effect of diseases on soybean seed composition under irrigated and nonirrigated conditions, soybean irrigation and irrigation techniques will be addressed as well.

2. Soybean Irrigation and Irrigation Techniques

2.1. Water Use

Water use of soybean varies throughout the growing season, and can vary considerably from one day to the next. Environmental factors, including temperature, solar radiation, and wind determine the evaporative demand of the atmosphere, while the growth stage of the crop affects the plants' water requirements. Seasonal water requirements for irrigated soybeans vary considerably, depending on factors such as variety and location, and can range from 450 to 700 mm per year (FAO, 2010). Daily water use can exceed 8 mm (about 0.31 inches) per day during peak developmental stages (R1-R7). Water requirements vary with growth stage, and optimal yields are obtained by irrigating regularly throughout the season. There are two critical stages, however, where insufficient water can lead to significant yield loss (Doss et al., 1974; Eck et al., 1987; Foroud et al., 1993; Kranz et al., 2005). During early reproductive stages (R3-R4), pod development occurs, and adequate water availability is needed to establish maximum pod set. Sufficient water during later reproductive stages (R5-R6) is needed to ensure optimal seed-fill and maximum seed weight.

Therefore, if water resources are limited, soybean should be adequately irrigated during these critical stages. In extreme water-limited conditions, a single irrigation was shown to slightly improve yield and quality, but not enough to justify the irrigation practice (Sweeney et al., 2003). While an adequate water supply is needed for optimal growth, an excess of water can be detrimental to plant health and yield. Flooding from excess rainfall or irrigation can affect the soil microclimate, negatively impacting growing conditions and yield (Sullivan et al., 2001). Therefore, in order to avoid extended periods of waterlogged conditions, maintaining adequate surface drainage is critical, especially in high-rainfall regions (Linkemer et al., 1998).

2.2. Scheduling

Proper timing of irrigations is critical in maintaining good crop health to achieve high seed quality and maximum yield. To minimize drought stress, it is necessary to determine when an irrigation is needed, and to ensure that the entire field receives water in a timely manner. Irrigation decisions have often been made based on visual observation, such as the appearance of the crop or the apparent dryness of the soil surface. These observations can be misleading because they can indicate that stress conditions are already occurring, and that irrigation should have been initiated earlier to avoid yield loss.

2.3. Scientific Methods of Scheduling Irrigations

A number of scientific methods of scheduling irrigations are available which can aid the producer in making better-informed decisions. Estimating the daily crop-water use or maintaining a soil-water balance are two critical components used to monitor soil moisture levels. Knowing these two components will allow the producer to better manage the timely application of water and forecast irrigation needs. The following are few techniques used for soil moisture monitoring.

2.3.1. Tensiometers

Tensiometers have been used extensively and provided accurate measurements of soil-water potential status. Tensiometers are simple to install and use, and provide a measure of the soil-water potential or the amount of energy plant roots must expend to extract water from the soil (Alam & Rogers, 1997).

However, tensiometers operate only over a limited range of moisture conditions, require periodic maintenance, and are usually read manually, requiring a fair amount of time and labor.

Guidelines for using tensiometers in scheduling irrigations of soybean have been developed by various researchers (e.g. Heatherly, 1999; Shock et al., 2005). It has been recommend that tensiometers be placed at a depth of 12 inch (304.8 mm) in a clay soil and 6 inch in silt loam and sandy soils, and that irrigation should commence when tensiometer readings reach -40 to - 60 kPa.

2.3.2. Granular Matrix Sensors

Granular matrix sensors offer measurements similar to those of tensiometers and have several desirable features. The sensors are solid-state electrical devices, require no maintenance, and operate over a wider range of soil-water potentials than tensiometers do (Figure 3a-b).

While sensor measurements are often made manually by visiting each sensor location and attaching a hand-held electronic meter, measurements can also be automated using dataloggers, allowing frequent measurements to be collected with little labor requirement (Irmak et al., 2006). Automated measurements allow a more complete picture of soil-water status and crop-water use, giving the producer more information to help in making irrigation decisions (Fisher, 2007).

2.3.3. The Waterbalance Method

Another common approach to scheduling irrigations is to maintain a daily accounting of soil-water reserves, and to initiate an irrigation when a preset deficit level is reached. A daily water balance is calculated using measurements or estimates of the amounts of water deposited and withdrawn from the field.

Water is deposited in the form of precipitation or irrigation, and withdrawn as the crop uses water through evapotranspiration. Reference evapotranspiration, which gives a measure of the evaporative demand of the atmosphere, is estimated from weather data. This is then adjusted by a crop coefficient function, which takes the particular crop's physiological characteristics into account to determine the crop evapotranspiration.

A standardized methodology for estimating reference and crop evapotranspiration was detailed by Allen et al. (1998), and was tested by Ortega-Farias et al. (2004) on a soybean crop. The water-balance method has been incorporated into a number of computer-based models for scheduling irrigations.

Many scheduling programs have been developed in different regions of the US to simplify the scheduling process by automatically retrieving weather data from nearby weather stations, including crop coefficient functions for a variety of crops, and offering guidance on entering additional information and interpreting model output.

The models are available as stand-alone programs, such as the Arkansas Irrigation Scheduler (http://www.aragriculture.org/computer_programs/ irrigation_ scheduling/), spreadsheet models, such as Kansas State University's KanSched (http://www.ksre. ksu.edu/mil/) and Michigan State University's Irrigation Cookbook (http://www.agweather. geo.msu.edu/mawn/irrigation/), or Internet web-based applications, such as the North Dakota Agricultural Weather Network (NDAWN) (http://ndawn.ndsu. nodak.edu/) and the California Irrigation Management Information System (CIMIS) (http://www cimis.water.ca.gov).

a. Soil water potential and temperature sensors.

b. Placement of the sensors in the field for data collection during the growing season.

Figure 3. Irrigation monitoring using sensors.

2.4. Improving Efficiency

Shortages of water and an increase in concern over the amounts of water used for agriculture make it necessary to improve irrigation efficiency. Improvements can be realized through increases in management effort and through application of scientific modeling and advances in technology (Greenwood et al., 2010). Surface irrigation systems are among the least efficient due to non-uniformities in water application and distribution across the field. The use of flexible plastic tubing (polypipe) is increasing due to its low cost, ease of installation, and increased control over water distribution (Enciso and Peries, 2005). Adjustable outlets enable flow rates to be controlled in individual furrows, allowing more uniform flow of water across the field (Figure 4). As water nears the end of the field, flow rates can be cut back to reduce runoff while maintaining wetted furrows until the desired amount of water is applied. The PHAUCET program, developed by the Natural Resources Conservation Service, is a computer program that can evaluate existing surface irrigation systems and recommend improvements to increase efficiency (NRCS, 2001). The program takes field dimensions and irrigation system characteristics into account to define pipe and outlet-size combinations to improve distribution uniformity, provide more consistent infiltration times, and decrease required pumping times. Accurate water measurements are necessary to ensure effective water management and proper irrigations. Flow-meters are used to measure both the total amount of water applied during an irrigation and the rate at which the water is applied. Measuring the amount of water applied, and knowing the dimensions of the field, the depth of water can be determined to ensure that it is appropriate (Boman & Shukla, 2009). The rate of application is important in order to avoid excessive soil erosion, set system run times, and adjust field and system operating practices to maximize efficiencies (Rogers, 1995).

a. Irrigation control by gate opening.

b. Soybean in irrigated and nonirrigated plots.

Figure 4. (Continued)

c. Irrigating cracked clay soil is a challenge.

d. Inefficiency in irrigating cracked clay soils.

Figure 4. Example of type of soils, in this case, clay soil, and how irrigation management is crucial to increase irrigation water use efficiency.

Sprinkler irrigation systems offer advantages over surface irrigation systems, especially automated, center pivot devices. Labor and maintenance requirements are reduced, the equipment is simple to operate, application uniformity is higher, and runoff is generally easier to avoid. Proper nozzle sizing and operation are critical, and periodic evaluation is needed to ensure proper functioning of the system for optimal uniformity (Harrison & Perry, 2010). Variable-rate technology has the potential to further increase uniformity and reduce water use by varying the application rate at any point in the field. Water can be applied precisely to account for crop and soil variability, as well as withheld if the center pivot passes over non-cropped areas (Harrison & Perry, 2010).

3. Materials and Methods

3.1. Charcoal Rot Experiment

3.1.1. Growth Conditions

A field experiment was conducted in 2004 and 2005 at the Delta Research and Extension Center, Stoneville, MS (33°26'N). Susceptible (S) DK 3964 and moderately resistant (MR) AG 3905 maturity group (MG) III soybean genotypes were selected based on their susceptibility to charcoal rot disease as described by Mengistu et al. (2011). Before planting, the seeds were treated with mefenoxam (R)-2-{2, 6-(dimethylphenyl)-methoxyacetylamino}-propionic acid to avoid a stand loss due to *Pythium* spp. Plots were fumigated using a mix of methyl bromide in combination with chloropicrin in the proportions 0.895 and 0.105, respectively, at a depth of 254 mm at a broadcast rate of 504 kg/ha. Row spacing was 0.5 m and seeding rate was 16–20 seeds/m of row. Plots were 23 m long. Plots were either infested with charcoal rot inoculum (INF) or noninfested (NINF), and either irrigated (IR) or nonirrigated (NIR). Planting dates were by end of April in 2004 and 2005, and harvesting dates were in mid-September. Irrigation of the plots were applied whenever water potential reached -50 KPa at 300 mm depth, and monitored using tensiometers according to Heatherly et al. (2001) and Bellaloui & Mengistu (2008). Irrigation was applied beginning from emergence (VC) to the end of seed fill (R6) growth stage (Fehr and Caviness 1977). Inoculation and disease assessments were detailed in Mengistu et al. (2007). The scale for root and stem severity was assessed as detailed in Paris et al. (2006) and Mengistu et al. (2007).

3.1.2. Experimental Design and Statistical Analysis

The design of the experiment was a Randomized Complete Block (RCB) with split-plot arrangement of the treatments, with irrigation as the main plot, infestation as subplot, and cultivar as sub-subplot. Three replicates were used. Differences in means were considered significant at $P \leq 0.05$. Results were analyzed by SAS, using Proc Mixed procedure (SAS, 2001). The detailed design of this experiment was previously reported (Bellaloui & Mengistu, 2008; Mengistu et al., 2011).

3.2. Phomopsis Experiment

3.2.1. Growth Conditions

Field experiments were conducted in 2003 and 2005 at the Delta Research and Extension Center, Stoneville, MS (33° 26′ N, 90° 91′W). One experiment was overhead irrigated (IR) and the other was nonirrigated (NIR). Details for field irrigation, design, and field conditions were previously detailed by Mengistu et al. (2010) and Bellaloui et al. (2012a, b). Seed were harvested at two times (maturity harvest at R8 and 15 days after maturity harvest). Harvesting times were considered treatments and nested within genotypes and maturity group (MG). Genotypes were planted in a randomized complete block design with three replications. Irrigation and MG were not replicated. In each year, the same genotypes and same experimental location were used. Seeds were planted 12 May 2003 and 10 May 2005, and planting details were detailed by Mengistu et al. (2012 a, b) and Bellaloui et al. (2012a, b). Susceptible (S) and moderately resistant (MR) genotypes were used and classified as indicated by Mengistu et al. (2010). Soybean genotypes used were MG IV susceptible PI 360835 and MG IV susceptible PI 93055, and MG IV moderately resistant PI 417479 and MG IV moderately resistant PI 82264. Seeds were harvested at harvest maturity (R8) (Fehr and Caviness 1977) and 15 days after harvest maturity (delayed harvesting). Harvesting dates for MG IV were 1 September in 2003 and 4 September in 2005 under the non-irrigated conditions. Harvesting dates under the irrigated conditions were 6 September in 2003 and 10 September in 2005. Irrigation was applied based on using tensiometer and moisture sensor readings. The soil water potential under irrigated conditions was kept between 0 and −17 kPa until harvest. For non-irrigated treatment, soil water potentials ranged from −76 to −184 kPa.

3.2.2. Inoculation and Disease Rating

Inoculation and disease ratings were performed according to Mengistu et al. (2010). A conidial suspension (106 conidia mL^{-1}) was sprayed onto the plants, and the plants were sprayed at dusk. Plastic shields were placed between plots of each MG category to prevent spore drift. Percent seed infection index (PSII) was used for each genotype for measuring resistance. The PSII was calculated as reported by Mengistu et al. (2010); Bellaloui et al. (2012a, b). The PSII was proven to be more consistent than percent infection seed (Mengistu et al. 2007; Mengistu et al., 2010). Genotypes were classified as resistant = 1 to 10%, moderately resistant = 11 to 20%, moderately susceptible=21 to 30%, and susceptible >30% (Mengistu et al. 2010). The full spectrum of MG genotypes for phomopsis evaluation and data analysis was presented in Mengistu et al. (2010). Here, we selected a subset of MG IV genotypes from those trials and analyzed for seed protein, oil, fatty acids, sugars, and minerals.

3.2.3. Data Analysis

Analysis of variance was performed as previously described by Mengistu et al. (2010) using the Proc Mixed Model procedure in SAS (SAS Institute, 2001) and means were separated using Fisher's Least Significant Difference test at $P \leq 0.05$. Since irrigation was not replicated (Mengistu et al., 2010) and since year and genotype interacted for some seed composition components, the results we presented for each irrigation treatment.

3.3. General Materials and Methods for Charcoal Rot and Phomopsis Experiments

For the charcoal rot experiment, seeds were collected for analysis at harvest maturity (R8). For the Phomopsis experiment, seed were collected for analysis at harvest maturity and 15 days after harvest maturity. Seeds were analyzed for protein, oil, fatty acids, sugars, isoflavones, and minerals as described below.

3.3.1. Seed Analysis for Protein, Oil, and Fatty Acids

About 25 g of seed was ground using a Laboratory Mill 3600 (Perten, Springfield, IL). Seed analyses were conducted by near infrared reflectance (Wilcox & Shibles, 2001;

Bellaloui et al., 2009) using a diode array feed analyzer, model AD 7200 (Perten, Springfield, IL). Calibrations were developed by the University of Minnesota, using Perten's Thermo Galactic Grams PLS IQ software. The calibration curve was established according to AOAC methods (AOAC, 1990a, b). Analyses of protein and oil were performed based on a seed dry matter basis (Wilcox & Shibles, 2001; Boydak et al., 2002), while fatty acids were analyzed on an oil basis.

3.3.2. Seed Analysis for Sucrose, Raffinose, and Stachyose

About 25 g of seed from each plot were ground using a Laboratory Mill 3600 (Perten, Springfield, IL). Analyses were conducted by near infrared reflectance (NIR) (Wilcox & Shibles 2001; Bellaloui et al., 2010b) using the AD 7200 array feed analyzer (Perten, Springfield, IL). Analyses of sugars were performed based on a seed dry matter basis (Wilcox & Shibles 2001; Boydak et al., 2002; Bellaloui et al., 2010b).

3.3.3. Analysis of Isoflavones

Seeds were collected from each plot and ground to pass through a 1-mm sieve using a Laboratory Mill 3600 (Perten, Springfield, IL). Seed concentrations of the isoflavones daidzein, genistein, and glycitein were measured similar to those of Morrison et al. (2008) and Sato et al. (2008) using a Perten near-infrared (NIR) reflectance diode array feed analyzer (Perten, Spring Field, IL, U.S.A.). Initial calibration equations were developed by the Perten company and University of Minnesota using Thermo Galactic Grams PLS IQ. The analysis was conducted on a dry matter basis. We refer to the total/combined isoflavones as the sum of the major individual isoflavones daidzein, genistein, and glycitein.

3.3.4. Seed Mineral Composition

Seeds at harvest maturity and 15 days after harvest maturity were collected from each plot and ground to pass through a 1-mm sieve using a Perten Laboratory Mill 3600 (Perten, Springfield, IL). Seed mineral concentrations were analyzed at the University of Georgia's Soil, Plant, and Water Laboratory, Athens, GA. Seed K, Ca, Zn, Mn, and Cu concentrations were analyzed by digesting 0.5 g of dried ground seed in HNO_3 in a microwave digestion system, and values were then determined using inductively coupled plasma spectrometry. Seed P and B concentrations were determined as indicated in the following sections.

3.3.5. Boron Measurement

The concentration of boron in seeds at harvest maturity and 15 days after harvest maturity was measured with the Azomethine-H method (Lohse, 1982). Detailed description of the method was previously reported elsewhere (Dordas, 2006; Dordas et al., 2007; Bellaloui et al., 2010a). Briefly, one gram of ground seed was and ashed at 500 ºC, extracted with 20 ml of 2 M HCl at 90 ºC for 10 min, and then filtered. The filtered mixture was transferred to plastic vials, a 2 ml sample of the solution was then added to 4 ml of buffer solution (containing 25% ammonium acetate, 1.5% EDTA, and 12.5% acetic acid), and to 4 ml of freshly prepared azomethine-H solution (0.45% azomethine-H and 1% of ascorbic acid) (John et al., 1975). After 45 minutes of color development, seed boron concentration was measured using a Beckman Coulter DU 800 spectrophotometer (Fullerton, California, U.S.A.) at 420 nm.

3.3.6. Phosphorus Measurement

The concentration of phosphorus in seed at harvest maturity and after 15 days after harvest maturity was measured spectrophotometrically as the yellow phospho-vanado-molybdate complex (Analytical Methods Committee, 1959). Two grams of dry seed were ground, ashed, 10 ml of 6 M HCl was added, and then the samples were placed in a water bath at 70°C to evaporate and dry. After drying, the samples were kept under heat, 2 ml of 36% m/m HCl was added, and the mixture was gently boiled. Then, 10 ml of water was added and the solution was carefully boiled for about 1 min. The samples were diluted to 50 ml in a volumetric flask, and after the first 2 ml were discarded, the sample solution was then filtered and kept for P analysis.

A sample of 5 ml was taken, and 5 ml of 5 M HCl and 5 ml of ammonium molybdate-ammonium metavanadate (a solution of ammonium molybdate, $(NH_4)_2$ MoO_4 (25 g/500 ml water), and ammonium metavanadate, NH_4VO_3) (1.25 g/500 ml water) reagent were added, and the mixture was diluted to 50 ml. A standard curve was created by preparing a range of phosphorus concentrations (from 0-50 µg/ml) using dihydrogen orthophosphate dissolved in both water and 36% m/m HCl.

After 30 minutes for color development at ambient temperature, samples were read at 400 nm using a Beckman Coulter DU 800 spectrophotometer at 400 nm (Fullerton, California, U.S.A.).

4. RESULTS

4.1. Charcoal Rot

4.1.1. Effects on Seed Protein, Oil, Fatty Acids, and Sugars under Irrigated Conditions

Seed protein and oleic acid concentrations were higher and oil was lower under infested (INF) in S and MR genotypes than under noninfested (NINF) (Table 1). There were no changes in linoleic acid. Linolenic acid was lower under INF in S genotypes, but no consistent pattern of linolenic acid was observed in MR genotypes. Sucrose concentration was lower under INF in S genotypes, but no consistent pattern of sucrose was observed in MR genotypes. Stachyose concentration was higher under INF in S and MR genotypes. No consistency in raffinose concentration pattern was observed in S or MR genotypes (Table 1).

4.1.2. Effects on Seed Minerals and Isoflavones under Irrigated Conditions

Seed Ca and K percentages were lower in S genotypes under INF than under NINF. The opposite trend was shown in MR genotypes where K and Ca percentage were higher under INF than under NINF in both years. Potassium and Ca percentages were higher in MR genotypes than in S genotypes in both years. Boron concentration was lower in S genotypes than in NINF under INF than under NINF in both years. No differences in B concentration in MR genotypes were observed between INF and NINF. Copper concentration did not show any consistency in S or MR genotypes. Isoflavone daidzein concentration was higher under INF than under NINF in S genotypes. Isoflavone genistein concentration was higher under INF than in NINF in both S and MR genotypes. There was no consistency in isoflavone glycitein concentration in S or MR genotypes under INF or NINF (Table 2).

Table 1. Effect of charcoal rot (*Macrophomina phaseolina*) infestation on seed protein, oil, fatty acids, and sugar concentrations of irrigated soybean genotypes differing in resistance to charcoal rot disease in 2004 and 2005. Genotypes were sensitive (S) or moderately resistant (MR). The experiment was conducted at the Delta Research and Extension Center, Stoneville, MS[1]

2004 Sugars									
Genotype (sensitivity to charcoal rot)	Infestation condition	Protein (g/kg)	Oil (g/kg)	Oleic (g/kg)	Linoleic (g/kg)	Linolenic (g/kg)	Sucrose (mg/g)	Raffinose (mg/g)	Stachyose (mg/g)
DK 3964 (S)	Infested	418 a	193 b	245 a	534 a	74 b	13.2 b	3.2 a	63.3 a
	Non-infested	383 b	206 a	212 b	556 a	89 a	16.2 a	3.4 a	55.1 b
AG 3905 (MR)	Infested	406 a	205 b	278 a	543 a	45 b	22.0 a	3.5 a	87.4 a
	Non-infested	382 b	218 a	205 b	521 a	75 a	21.7 a	3.8 a	64.2 b
2005									
DK 3964 (S)	Infested	436 a	181 b	283 a	553 a	64 b	10.4 b	4.2 a	75.0 a
	Non-infested	402 b	209 a	212 b	575 a	76 a	17.3 a	3.3 b	64.4 b
AG 3905 (MR)	Infested	431 a	204 b	293 a	536 a	87 a	21.1 a	4.8 a	89.3 a
	Non-infested	420 b	212 a	245 b	568 a	73 b	20.9 a	4.2 a	71.1 b

[1] Means within a column for each infestation condition and in each year separately followed by the same letter are not significantly different at the 5% level as determined by Fisher's LSD test. Values are means of three replicates.

Table 2. Effect of charcoal rot (*Macrophomina phaseolina)* infestation on seed minerals and isoflavone concentrations of irrigated soybean genotypes differing in resistance to charcoal rot disease in 2004 and 2005. Genotypes were sensitive (S) or moderately resistant (MR). The experiment was conducted at the Delta Research and Extension Center, Stoneville, MS[1]

2004 Isoflavones									
Genotype (sensitivity to charcoal rot)	Infestation Condition	P (%)	K (%)	Ca (mg/kg)	B (mg/kg)	Cu (mg/kg)	Daidzein (µg/g)	Genistein (µg/g)	Glycitein (µg/g)
DK 3964 (S)	Infested	0.31 a	1.2 b	0.32 b	27 b	3.2 a	953 a	865 a	756 a
	Non-infested	0.35 a	1.6 a	0.46 a	37 a	3.1 a	730 b	732 b	739 a
AG 3905 (MR)	Infested	0.51 a	1.9 a	0.65 a	49 a	4.6 a	876 a	747 a	687 b
	Non-infested	0.47 a	1.3 b	0.62 a	45 a	4.1 b	789 b	653 b	867 a
2005									
DK 3964 (S)	Infested	0.27 a	1.0 b	0.31b	34 b	3.0 a	806 a	787 a	671 b
	Non-infested	0.30 a	1.7 a	0.51 a	42 a	2.3 b	715 b	533 b	723 a
AG 3905 (MR)	Infested	0.37 a	2.3 a	0.57 a	45 a	4.1 a	976 a	577 a	604 b
	Non-infested	0.41 a	1.7 b	0.61 a	42 a	4.4 a	943 a	470 b	785 a

[1] Means within a column for each infestation condition and in each year separately followed by the same letter are not significantly different at the 5% level as determined by Fisher's LSD test. Values are means of three replicates.

Table 3. Effect of charcoal rot (*Macrophomina phaseolina*) infestation on seed protein, oil, fatty acids, and sugar concentrations of nonirrigated soybean genotypes differing in resistance to charcoal rot disease in 2004 and 2005. Genotypes were sensitive (S) or moderately resistant (MR). The experiment was conducted at the Delta Research and Extension Center, Stoneville, MS[1]

2004 Sugars									
Genotype (sensitivity to charcoal rot)	Infestation condition	Protein (g/kg)	Oil (g/kg)	Oleic (g/kg)	Linoleic (g/kg)	Linolenic (g/kg)	Sucrose (mg/g)	Raffinose (mg/g)	Stachyose (mg/g)
DK 3964 (S)	Infested	432 a	183 b	283 a	523 a	65 b	12.0 b	3.0 b	63.3 a
	Non-infested	394 b	201 a	241 b	536 a	82 a	18.5 a	3.6 a	56.0 b
AG 3905 (MR)	Infested	427 a	193 b	294 a	565 a	45 a	16.6 b	3.2 a	73 .1a
	Non-infested	430 a	209 a	253 b	542 a	59 a	20.1 a	2.1 b	50.4 b
2005									
DK 3964 (S)	Infested	436 a	176 b	315 a	526 a	76 a	11.8 b	3.7 a	77.2 a
	Non-infested	402 b	215 a	267 b	561 a	64 a	19.2 a	3.8 a	51.0 b
AG 3905 (MR)	Infested	429 a	185 b	326 a	568 a	47 a	12.9 b	3.7 a	72.4 a
	Non-infested	435 a	194 a	277 b	571 a	50 a	17.5 a	3.9 a	51.2 b

[1] Means within a column for each infestation condition and in each year separately followed by the same letter are not significantly different at the 5% level as determined by Fisher's LSD test. Values are means of three replicates.

Table 4. Effect of charcoal rot (*Macrophomina phaseolina*) infestation on seed minerals and isoflavone concentrations of nonirrigated soybean genotypes differing in their resistance to charcoal rot disease in 2004 and 2005. Genotypes were sensitive (S) or moderately resistant (MR). The experiment was conducted at the Delta Research and Extension Center, Stoneville, MS[1]

2004 Isoflavones									
Genotype (sensitivity to charcoal rot)	Infestation conditions	P (%)	K (%)	Ca (mg/kg)	B (mg/kg)	Cu (mg/kg)	Daidzein (µg/g)	Genistein (µg/g)	Glycitein (µg/g)
DK 3964 (S)	Infested	0.42 a	1.0 b	0.21 b	22 b	3.1 a	652 a	530 a	652 a
	Non-infested	0.45 a	1.5 a	0.35 a	28 a	2.9 a	654 a	473 b	578 b
AG 3905 (MR)	Infested	0.33 a	1.7 a	0.42 a	41 a	3.3 a	534 a	655 a	561 b
	Non-infested	0.36 a	1.1 b	0.43 a	39 a	3.1 a	523 a	553 b	743 a
2005									
DK 3964 (S)	Infested	0.40 a	0.07 b	0.21 b	20 b	2.5 b	512 a	686 a	670 a
	Non-infested	0.29 b	1.5 a	0.34 a	27 a	3.1 a	476 b	430 b	422 b
AG 3905 (MR)	Infested	0.46 a	2.1 a	0.51 a	38 a	3.5 b	531 a	613 a	572 b
	Non-infested	0.30 b	1.3 b	0.49 a	41 a	4.0 a	467 b	575 b	662 a

[1] Means within a column for each infestation condition and in each year separately followed by the same letter are not significantly different at the 5% level as determined by Fisher's LSD test. Values are means of three replicates.

4.1.3. Effects on Seed Protein, Oil, Fatty Acids, and Sugars under Nonirrigated Conditions

Protein concentration was higher in S genotypes under INF than under NINF. There was no change in protein concentration in MR genotypes under INF or NINF. Oil concentration was lower under INF than under NINF in S and MR genotypes. Oleic acid concentration was higher under INF than under NINF in S and MR. Linolenic acid concentration did not show a consistent pattern in both S and MR genotypes. Sucrose concentration was lower in S and MR under INF than under NINF, but stachyose concentration was higher in S and MR genotypes under INF than under NINF. There was no consistent pattern in raffinose concentration in both S or MR genotypes under INF or NINF (Table 3).

4.1.4. Effects on Seed Minerals and Isoflavones under Nonirrigated Conditions

Potassium concentration was lower in S genotypes under INF than under NINF, but higher under INF than NINF in MR genotypes. Calcium concentration was lower in S genotypes under INF than under NINF. No differences in Ca concentration in MR genotypes under either infestation treatment were observed.

Concentration of B was lower in S genotypes under INF than under NINF, but no differences were observed in MR genotypes. Potassium, Ca, and B levels in MR genotypes were higher than S genotypes. Genistein isoflavone concentration was higher in S and MR genotypes under INF than under NINF. Glycitein concentration was higher in S under INF than under NINF, but glycitein was higher under NINF than under INF in MR genotypes (Table 4).

4.2. Phomopsis Effects

4.2.1. Effects on Seed Protein, Oil, Fatty Acids, and Sugars under Irrigated Conditions

At harvest maturity, seed protein and oil concentrations were higher in MR than in S genotypes, but oleic acid was higher in S genotypes than in MR genotypes. Linoleic acid concentration was lower in S than in MR genotypes, but linolenic acid was higher in S than in MR genotypes (Table 5).

Sucrose concentration was higher in MR genotypes than in S genotypes. Raffinose concentration was higher in MR genotypes than in S genotypes, but stachyose concentration was higher in S genotypes than in MR genotypes (Table 5).

The general pattern was repeated at 15 days after harvest maturity, except that protein and sucrose concentrations decreased further compared with those at harvest maturity in both years.

4.2.2. Effects on Seed Minerals and Isoflavones under Irrigated Conditions

At harvest maturity, P concentration showed no consistent pattern in either S or MR genotypes (**Table 6**). The concentrations of K, Ca, B, and Cu were higher in MR than in S genotypes (**Table 6**), but differed among S genotypes and MR genotypes.

A similar pattern was shown at 15 days after harvest maturity in both years, except that K concentration was further decreased even in PI 82264 (MR), and Ca concentration was further decreased in both susceptible PI 360835 and PI 93055 genotypes.

Table 5. Effect of phomopsis (*Phomopsis longicolla*) at harvest maturity (HM) and 15 days after harvest maturity (AHM) on seed protein, oil, fatty acids, and sugar concentrations in maturity group IV soybean genotypes differing in susceptibility to phomopsis resistance under irrigated environments in 2003 and 2005. Genotypes were sensitive (S) or moderately resistant (MR). The experiment was conducted at the Delta Research and Extension Center, Stoneville, MS[1]

HM 2003 Sugars								
Genotype (sensitivity to Phomopsis)	Protein (g/kg)	Oil (g/kg)	Oleic (g/kg)	Linoleic (g/kg)	Linolenic (g/kg)	Sucrose (mg/g)	Raffinose (mg/g)	Stachyose (mg/g)
PI 360835 (S)	444 c	191 c	303 a	443 b	74 a	35 b	5.4 c	40 b
PI 93055 (S)	431 d	183 d	285 a	463 b	82 a	38 b	4.6 d	58 a
PI 417479 (MR)	466 a	205 b	253 b	512 a	63 b	64 a	7.5 b	32 c
PI 82264 (MR)	458 b	213 a	247 b	536 a	58 b	64 a	9.6 a	31 c
HM 2005								
PI 360835 (S)	435 d	183 b	294 a	432 b	74 a	31 d	5.3 c	52 a
PI 93055 (S)	454 c	194 c	312 a	427 b	82 a	42 c	4.5 d	41 b
PI 417479 (MR)	463 b	214 a	254 b	563 a	56 b	76 a	8.5 a	33 c
PI 82264 (MR)	474 a	201 b	242 b	571 a	53 b	67 b	7.5 b	30 c
AHM 2003	Protein (g/kg)	Oil (g/kg)	Oleic (g/kg)	Linoleic (g/kg)	Linolenic (g/kg)	Sucrose (mg/g)	Raffinose (mg/g)	Stachyose (mg/g)
PI 360835 (S)	421 c	183 c	243 b	472 b	73 a	23 b	5.4 c	52 a
PI 93055 (S)	403 d	172 d	232 b	486 b	81 a	22 b	5.0 c	51 a
PI 417479 (MR)	454 a	212 a	276 a	532 a	54 b	54 a	6.5 b	30 b
PI 82264 (MR)	443 b	202 b	254 a	566 a	60 b	57 a	7.6 a	31 b
AHM 2005								
PI 360835 (S)	402 d	181 d	254 b	478 b	85 a	22 b	4.6 d	51 a
PI 93055 (S)	414 c	192 c	243 b	482 b	89 a	21 b	5.3 c	49 b
PI 417479 (MR)	445 b	208 b	283 a	535 a	47 b	52 a	8.1 b	34 c
PI 82264 (MR)	450 a	223 a	292 a	562 a	53 b	53 a	9.3 a	33 c

[1] Means within a column in each year separately followed by the same letter are not significantly different at the 5% level as determined by Fisher's LSD test. Values are means of three replicates.

Table 6. Effect of phomopsis (*Phomopsis longicolla*) at harvest maturity (HM) and 15 days after harvest maturity (AHM) on seed minerals and isoflavone concentrations in maturity group IV soybean genotypes differing in susceptibility to phomopsis resistance under irrigated environments in 2003 and 2005. Genotypes were sensitive (S) or moderately resistant (MR). The experiment was conducted at the Delta Research and Extension Center, Stoneville, MS[1]

HM 2003 Isoflavones								
Genotype (sensitivity to Phomopsis)	P (%)	K (%)	Ca (%)	B (mg/kg)	Cu (mg/kg)	Daidzein (µg/g)	Genistein (µg/g)	Glycitein (µg/g)
PI 360835 (S)	0.31 c	1.1 c	0.21 c	24 b	3.1 d	453 c	445 d	743 b
PI 93055 (S)	0.42 a	1.0 c	0.20 c	27 b	4.2 c	334 d	578 c	645 c
PI 417479 (MR)	0.43 a	1.6 b	0.45 b	42 a	6.5 a	463 b	694 b	834 a
PI 82264 (MR)	0.37 b	2.1 a	0.53 a	39 a	5.3 b	642 a	783 a	892 a
HM 2005								
PI 360835 (S)	0.25 c	1.0 c	0.21 d	21 c	2.1 d	542 c	554 d	721 d
PI 93055 (S)	0.32 b	1.2 c	0.32 c	22 c	3.6 c	436 d	631 c	830 c
PI 417479 (MR)	0.52 a	1.7 b	0.53 b	43 a	5.1 a	571 b	736 b	972 b
PI 82264 (MR)	0.30 b	2.5 a	0.60 a	34 b	4.7 b	831 a	861 a	1002 a
AHM 2003	P (%)	K (%)	Ca (%)	B (mg/kg)	Cu (mg/kg)	Daidzein (µg/g)	Genistein (µg/g)	Glycitein (µg/g)
PI 360835 (S)	0.24 b	0.9 c	0.29 b	21 d	4.3 c	354 b	363 d	431 c
PI 93055 (S)	0.32 a	1.3 b	0.26 b	25 c	3.5 d	342 b	426 c	503 b
PI 417479 (MR)	0.37 a	1.8 a	0.53 a	47 a	5.2 b	532 a	564 b	871 a
PI 82264 (MR)	0.27 b	1.7 a	0.56 a	37 b	5.8 a	573 a	634 a	862 a
AHM 2005								
PI 360835 (S)	0.21 c	1.0 b	0.21 d	19 c	2.1 c	426 b	332 d	639 b
PI 93055 (S)	0.32 b	1.0 b	0.27 c	20 c	3.6 b	532 a	477 c	579 c
PI 417479 (MR)	0.53 a	1.5 a	0.53 b	41 a	5.0 a	562 a	631 b	793 a
PI 82264 (MR)	0.37 b	1.4 a	0.67 a	35 b	5.1 a	551 a	801 a	830 a

[1] Means within a column in each year separately followed by the same letter are not significantly different at the 5% level as determined by Fisher's LSD test. Values are means of three replicates.

Table 7. Effect of phomopsis (*Phomopsis longicolla*) at harvest maturity (HM) and 15 days after harvest maturity AHM on seed protein, oil, fatty acids, and sugar concentrations in maturity group IV soybean genotypes differ in susceptibility to phomopsis resistance under nonirrigated environment in 2003 and 2005. Genotypes were sensitive (S) or moderately resistant (MR). The experiment was conducted at the Delta Research and Extension Center, Stoneville, MS[1]

HM 2003 Sugars								
Genotype (sensitivity to Phomopsis)	Protein (g/kg)	Oil (g/kg)	Oleic (g/kg)	Linoleic (g/kg)	Linolenic (g/kg)	Sucrose (mg/g)	Raffinose (mg/g)	Stachyose (mg/g)
PI 360835 (S)	421 c	184 b	263 a	502 b	71 a	21 b	4.5 d	56 a
PI 93055 (S)	413 d	174 c	274 a	543 a	76 a	23 b	5.7 c	53 a
PI 417479 (MR)	437 b	201 a	286 a	556 a	58 b	57 a	6.5 b	42 b
PI 82264 (MR)	446 a	207 a	273 a	565 a	61 b	50 a	7.1 a	44 b
HM 2005								
PI 360835 (S)	416 c	191 c	254 b	562 a	85 a	38 c	5.8 c	51 a
PI 93055 (S)	402 d	186 d	262 b	503 b	70 b	33 c	4.3 d	50 a
PI 417479 (MR)	434 b	203 a	297 a	496 c	60 c	56 b	6.7 b	40 c
PI 82264 (MR)	454 a	214 b	313 a	529 b	53 d	61 a	7.1 a	46 b
AHM 2003	Protein (g/kg)	Oil (g/kg)	Oleic (g/kg)	Linoleic (g/kg)	Linolenic (g/kg)	Sucrose (mg/g)	Raffinose (mg/g)	Stachyose (mg/g)
PI 360835 (S)	393 c	197 c	226 b	523 a	87 a	19 c	4.6 c	53 b
PI 93055 (S)	382 d	184 d	237 b	545 a	83 a	17 c	4.3 c	61 a
PI 417479 (MR)	416 a	206 b	264 a	555 a	51 c	46 a	5.8 a	47 c
PI 82264 (MR)	407 b	215 a	277 a	531 a	66 b	34 b	5.2 b	41 d
AHM 2005								
PI 360835 (S)	386 c	191 c	237 b	571 a	83 a	25 c	4.8 a	55 b
PI 93055 (S)	406 b	180 d	213 c	524 b	75 b	16 d	3.3 c	67 a
PI 417479 (MR)	425 a	203 b	269 a	557 a	57 c	46 a	4.7 a	42 c
PI 82264 (MR)	421 a	217 a	273 a	578 a	43 d	32 b	3.9 b	40 c

[1] Means within a column in each year separately followed by the same letter are not significantly different at the 5% level as determined by Fisher's LSD test. Values are means of three replicates.

Table 8. Effect of phomopsis (*Phomopsis longicolla*) at harvest maturity and 15 days after harvest maturity on seed minerals and isoflavone concentprations in maturity group IV soybean genotypes differ in susceptibility to phomopsis resistance under nonirrigated environment in 2003 and 2005. Genotypes were sensitive (S) or moderately resistant (MR). The experiment was conducted at the Delta Research and Extension Center, Stoneville, MS[1]

HM 2003 Isoflavones								
Genotype (sensitivity to Phomopsis)	P (%)	K (%)	Ca (mg/kg)	B (mg/kg)	Cu (mg/kg)	Daidzein (µg/g)	Genistein (µg/g)	Glycitein (µg/g)
PI 360835 (S)	0.23 c	1.0 b	0.23 b	22 b	3.1 b	453 c	435 b	873 b
PI 93055 (S)	0.36 b	1.2 b	0.21 b	20 b	4.2 a	354 d	481 b	794 c
PI 417479 (MR)	0.37 b	2.1 a	0.43 a	35 a	3.5 b	531 b	601 a	932 a
PI 82264 (MR)	0.42 a	2.0 a	0.47 a	31 a	4.5 a	786 a	643 a	993 a
HM 2005								
PI 360835 (S)	0.31 b	1.2 b	0.31 b	19 c	3.3 b	436 c	577 c	768 d
PI 93055 (S)	0.30 b	1.1 b	0.25 b	21 c	3.0 b	388 d	465 d	864 c
PI 417479 (MR)	0.46 a	2.9 a	0.49 a	32 b	4.3 a	530 b	689 b	1133 a
PI 82264 (MR)	0.33 b	1.8 a	0.51 a	41 a	3.2 b	798 a	733 a	1002 b
AHM 2003	P (%)	K (%)	Ca (%)	B (mg/kg)	Cu (mg/kg)	Daidzein (µg/g)	Genistein (µg/g)	Glycitein (µg/g)
PI 360835 (S)	0.23 c	1.1 b	0.20 c	19 c	2.4 d	354 c	487 c	805 c
PI 93055 (S)	0.20 c	1.0 b	0.20 c	22 c	4.0 b	255 d	325 d	863 c
PI 417479 (MR)	0.33 b	2.3 a	0.35 a	33 b	4.7 a	453 b	508 b	927 b
PI 82264 (MR)	0.47 a	2.5 a	0.39 b	40 a	3.1 c	687 a	626 a	1011 a
2005								
PI 360835 (S)	0.26 b	1.0 b	0.26 c	20 b	3.0 b	357 c	433 c	802 b
PI 93055 (S)	0.21 b	0.9 b	0.20 c	26 b	3.8 a	356 c	345 d	872 b
PI 417479 (MR)	0.49 a	2.1 a	0.31 b	37 a	3.2 b	450 b	651 b	1034 a
PI 82264 (MR)	0.47 a	1.5 a	0.45 a	39 a	4.2 a	672 a	754 a	1017 a

[1] Means within a column in each year separately followed by the same letter are not significantly different at the 5% level as determined by Fisher's LSD test. Values are means of three replicates.

Individual isoflavones concentration responded differently, depending on the genotype and its level of resistance to *Phomopsis*. For example, genistein and glycitein concentrations were lower in S genotypes than in MR genotypes, but there was no consistency in daidzein concentration in S or MR genotypes. Although individual isoflavone concentrations significantly differed in each genotype, concentration were higher in MR genotypes than in S genotypes.

4.2.3. Effects on Seed Protein, Oil, Fatty Acids, and Sugars under Nonirrigated Conditions

At harvest maturity, protein and oil concentrations were lower in S genotypes than in MR genotypes. Linolenic acid concentration was higher in S genotypes than in MR genotypes (Table 7), but there were no consistency in oleic and linoleic acids. Sucrose concentration was lower, but stachyose concentration was higher in S genotypes than in MR genotypes. No clear pattern was observed for raffinose concentration.

Results were similar at 15 days after harvest maturity, with further protein and oleic acid decreases in both S and MR genotypes (Table 7).

4.2.4. Effects on Seed Minerals and Isoflavones under Nonirrigated Conditions

At harvest maturity, K, B, and Ca concentrations were lower in S genotypes than in MR genotypes, while P and Cu had no consistent pattern. The levels of Ca, K, and B were higher in MR than in S genotypes, but genotypes accumulated significantly differed levels of K, Ca, B, and Cu. There was a clear pattern that isoflavone concentrations were lower in S genotypes compared to MR genotypes, and the concentrations of isoflavones between S genotypes and MR genotypes were different (Table 8).

5. DISCUSSION

5.1. Charcoal Rot Effects on Seed Composition

The higher concentration of seed protein and oleic acid under INF in S and MR genotypes (Table 1 and Table 3) indicated that all genotypes responded similarly to the INF conditions regardless of the level of resistance, suggesting that increase in protein and oleic acid could be used as a tolerance mechanism to charcoal rot infestation, especially when charcoal rot infestation level is moderate (Bellaloui et al., 2008). The increase of protein and oleic acid as a response to biotic and abiotic stress has been previously reported. For example, it was shown that increasing infection with *S. sclerotiorum* resulted in higher seed protein in one cultivar, but not in other soybean cultivars (Hoffman et al., 1998). Other researchers found that soybean infected with bean pod mottle virus resulted in higher seed protein and decreased oil (Ziems et al., 2007). The decrease of seed oil concentration was due to the inverse relationship between protein and oil, as indicated by others (Brim & Burton, 1979; Burton, 1985; Wilson, 2004). The mechanism controlling this inverse relationship is still not known, however, this inverse relationship was thought to be genetically controlled, related to gene× environment interactions, and depended on energy cost partitioning between protein and oil syntheses (Egli & Bruening, 2007). It was also reported that the response of soybean genotypes to charcoal rot infestation and their effects on seed composition cannot be generalized, and a combination of charcoal rot infection, genotype, and environmental factors could all be involved.

The decrease of seed sucrose concentrations under INF conditions may indicate the negative impact of charcoal rot on seed sucrose. The higher stachyose under INF conditions support the suggestion that the plant may tolerate biotic (pathogens) and abiotic (drought) stresses by increasing stachyose concentrations to prevent the pathogen from penetrating the seed. The lower sucrose and higher stachyose concentrations were previously reported with other pathogens such as phomopsis (Bellaloui et al., 2012a, b). The increase of K, Ca, and B concentrations in MR than in S genotypes supports previous observations that these nutrients may be involved in cell walls and membrane integrity and disease resistance (Marschner, 1995). The increase of genistein concentrations under INF conditions in both S and MR genotypes may indicate that plants may increase this fraction of isoflavones as a defense mechanism. It was thought that phenolics, including isoflavones may have a role in plant disease resistance and defense mechanisms (Paxton, 1980; Graham, 1983; Graham & Webb, 1991; Dixon, 1990). The higher sucrose concentrations in MR genotypes than S genotypes regardless the charcoal rot infestation and drought conditions indicated that MR genotypes were able to maintain sucrose level, and this may be due to their resistance to charcoal rot. The further decrease in sucrose and genistein under drought conditions, especially in S genotypes indicated the negative combined effects of charcoal rot infestation and drought.

5.2. Phomopsis Effects on Seed Composition

The higher protein and oil concentrations in MR genotypes may be due to resistance of these genotypes to *Phomopsis*. It was previously reported that the lower linolenic acid in S may be due to the inverse relationship between oleic and linolenic acids, but this relationship was not observed between oleic and linolenic in S genotypes, but was observed in MR genotypes. This pattern was also observed at 15 days after harvesting maturity. It was reported that a significant positive correlation between seed infected with *Phomopsis* spp. and the concentration of seed protein or oil was found, but the existence of this correlation depended on geographic location and severity of the disease infection (Bradley et al., 2002). It was also found that increased infested seed resulted in higher concentration of linolenic acid (Bradley et al., 2002). Higher linolenic acid in seed is undesirable because its degradation leads to poor flavor and undesirable odors in soybean oil (Beare-Rogers, 1995). The poor quality oil resulting from phomopsis infection could be due to the damaging effects of *Phomopsis* on seed coat tissue and the exposure of the underlying tissue to air and consequent oxidation of lipids (Hepperly & Sinclair, 1978). Other researchers also reported that oil, meal, and flour derived from infected seeds had lower quality than those from non-infected seeds (Roy, 1976), and this was possibly due to the degradation of seed coat proteins by fungal released enzymes (Ryley, 2004).

Our results support previous findings that higher linolenic acid concentrations were observed in S genotypes than in MR genotypes, suggesting that the source of poor quality oil in S genotypes could be due to high linolenic acid oxidation, especially at 15 days after maturity harvest. High protein and oil concentrations in MR genotypes could be due to higher resistance in MR genotypes than in S genotypes. The higher level of oleic acid concentrations in both S and MR genotypes, especially under NIR could be due to the possible role of oleic acid as stress compound indicator under *Phomopsis* and drought stress (Bellaloui et al., 2012a, b). The further decrease of protein and oil in some genotypes 15 days after harvest maturity under NIR conditions could be due to the severe negative effects of both *Phomopsis* and drought stress on protein and oil. It was reported that the severity of *Phomopsis spp.* infection on seed composition is dependent on heat, drought, genotypes, and level of phomopsis resistance (Bellaloui et al., 2012a, b).

The lower sucrose concentrations in S genotypes than in MR genotypes, especially 15 days after harvest maturity under NIR conditions, could be related to the effect of phomopsis resistance level on sucrose mobility from leaves to seed or due to sucrose synthesis inhibition during seed-fill stage. The higher stachyose concentrations in S genotypes than in MR genotypes, especially 15 days after harvest maturity under NIR conditions, may indicate that stachyose may be involved in tolerance mechanisms to *Phomopsis* and drought. Previous research reported that sucrose decreased and stachyose increased under *Phomopsis* and drought conditions (Bellaloui et al., 2012a, b). The decrease of sucrose by phomopsis infection is supported by others who found that fungus infected soybean seed resulted in lower carbohydrates (Pathan et al., 1989). In addition, our results showed that there was a redistribution of carbohydrate profile in that sucrose concentration decreased, but stachyose concentration increased, especially in S genotypes under *Phomopsis* and drought conditions, supporting previous research (Bellaloui et al., 2012). Previous research showed that other sugar fractions such as raffinose and galactinol may have a role in protecting the plants under abiotic stress such as drought (Taji et al., 2002) or diseases (Bellaloui et al., 2012). The lower sucrose in S genotypes may indicate that the activity of sucrose synthase decreased under drought conditions (Streeter, 2003), suggesting that sucrose is more sensitive to biotic and abiotic stress than raffinose and stachyose (Bellaloui et al., 2012a, b).

The higher concentrations of K, Ca, and B in MR than in S genotypes under IR and NIR conditions may indicate that these minerals are related to resistance to diseases, but it remains unclear whether this involvement is direct or indirect. Although previous reports indicated involvement of Cu in disease resistance (Marschner, 1995; Bellaloui et al., 2012), the inconsistency of P and Cu concentration pattern indicated further research is needed before conclusive remarks are made. Previous research indicated that *Phomopsis* may inhibit the uptake and assimilation of nutrients involved in seed coat resistance (Bellaloui et al., 2012a, b) or be involved in membrane permeability and integrity (Ca and B), or involved in metabolism and lignin biosynthesis (B and Cu) (Marschner, 1995). These nutrients could be important to cell walls and cell membrane integrity to prevent disease from penetration (Marschner, 1995).

The higher concentrations of isoflavones genistein and glycitein in MR genotypes than in S genotypes under IR and NIR also indicated the negative impact of diseases on isoflavones. The further decrease in genistien under NIR conditions indicated that genistein may be more sensitive to diseases and drought than daidzein and glycitein.

The current research demonstrated that both charcoal rot and *Phomopsis* affected seed composition, especially protein, oil, oleic and linolenic acids, sucrose, stachyose, genistein isoflavone, K, Ca, and B. For charcoal rot, S genotypes tend to increase protein and oleic acid concentrations as a response mechanism to the charcoal rot infestation. The increase in protein concentration resulted in a decrease in oil concentration. Under NINF, MR genotypes had higher protein concentrations than S genotypes. Sucrose, K, Ca, and B were lower in S genotypes compared with MR genotypes, indicating a negative impact of charcoal rot on these nutrients, especially under drought conditions. Genistein and stachyose concentrations may be involved in disease resistance mechanisms as these compounds increased in both S and MR genotypes under charcoal rot infestation, especially under drought conditions. For *Phomopsis,* MR genotypes showed superiority in protein and oil concentrations than S genotypes. However, S genotypes had higher oleic acid concentration than MR genotypes, could be a result of stress response to phomopsis infection. Lower concentrations of seed sucrose, K, Ca, and B in S genotypes indicated the negative impact of phomopsis infection on these nutrients as in case of charcoal rot. The further decrease in these nutrients 15 days after harvest maturity indicated that delaying harvest leads to further deterioration of seed quality, partially due to the involvement of these nutrients in cell walls and membrane, and seed coat integrity. Phomopsis infection negatively impacted all isoflavones, especially genistein 15 days after harvest maturity, indicating delaying harvesting also affects isoflavones, desirable seed quality components. Therefore, availability of charcoal rot and phomopsis resistant soybean cultivars is urgently needed.

ACKNOWLEDGMENTS

Authors would like to thank Dr. Vijay Nandula and Dr. Manju Pande for reviewing the manuscript. Thanks to Dr. Anne Gillen for helping taking pictures. We also thank Debbie Boykin for statistical support and analysis, and Sandra Mosley for seed composition analysis and references list check.

REFERENCES

Alam, M. & Rogers, D. H. (1997). Tensiometer use in scheduling irrigation. Publication L-796, Kansas State University Agricultural Experiment Station and Cooperative Extension Service, Manhattan, Kansas, USA.

Allen, R. G., Pereira, L. S., Raes, D. & Smith, M. (1998). Crop evapotranspiration: guidelines for computing crop water requirements. *FAO Irrigation and Drainage Paper #56, Food and Agriculture Organization of the United Nations,* Rome.

Analytical Methods Committee. (1959). *Analyst,* London, 84, 214.

AOAC. (1990a). Method 988.05. In: Official methods of analysis, 15th edn., edited by Helrich, K. The Association of Official Analytical Chemists, Inc., Arlington

AOAC. (1990b). Method 920.39. In: Official methods of analysis, 15th edn., edited by Helrich, K. The Association of Official Analytical Chemists, Inc., Arlington.

Beare-Rogers, J. (1995). Food fats and fatty acids in human nutrition. In: Development and processing vegetable oils for human nutrition, edited by Przybylski, R. & McDonald, B. E., AOCS, 1-7, Champaign, IL.

Belewu, M. A. & Belewu, K. Y. (2007). Comparative physico-chemical evaluation of tiger-nut, soybean and coconut milk sources. *Int. J. Agri. Biol.,* 9, 785–787.

Bellaloui, N. & Mengistu, A. (2008). Seed composition is influenced by irrigation regimes and cultivar differences in soybean. *Irrig. Sci.,* 26, 261–268.

Bellaloui, N., Mengistu, A. & Paris, R. L. (2008). Soybean seed composition in cultivars differing in resistance to charcoal rot (*Macrophomina phaseolina*). *J. Agric. Sci.* 146, 667-675.

Bellaloui, N., Mengistu, A. & Paris, R. L. (2008). Soybean seed composition in cultivars differing in resistance to charcoal rot (*Macrophomina phaseolina*). *J. Agric. Sci.,* 146, 667-675.

Bellaloui, N., Mengistu, A. & Zobiole, L. H. S. (2012b). Phomopsis seed infection effects on soybean seed phenol, lignin, and isoflavones in maturity group V genotypes differing in phomopsis resistance. *J. Crop Imp.* In press.

Bellaloui, N., Mengistu, A. Fisher, D. K. & Abel, C. A. (2012a). Soybean seed composition constituents as affected by drought and *Phomopsis* in phomopsis susceptible and resistant genotypes. *J. Crop Imp.* In press.

Bellaloui, N., Reddy, K. N., Gillen, A. M. & Abel, C. A. (2010b). Nitrogen metabolism and seed composition as influenced by foliar boron application in soybean. *Plant Soil,* 336, 143-155.

Bellaloui, N., Smith, J. R., Gillen, A. M. & Ray, J. D. (2010a). Effect of maturity on seed sugars as measured on near-isogenic soybean (*Glycine max*) lines. *Crop Sci.,* 50, 1978-1987.

Bellaloui, N., Smith, J. R., Gillen, A. M. & Ray, J. D. (2011). Effects of maturity, genotypic background, and temperature on seed mineral composition in near-isogenic soybean lines in the Early Soybean Production System. *Crop Sci.,* 51, 1161-1171.

Bellaloui, N., Smith, J. R., Ray, J. D. & Gillen, A. M. (2009). Effect of maturity on seed composition in the Early Soybean Production System as measured on near-isogenic soybean lines. *Crop Sci.,* 49, 608-620.

Boman, B. & Shukla, S. (2009). Water measurement for agricultural irrigation and drainage systems. Circular 1495, Florida Cooperative Extension Service, Institute of Food and Agricultural Sciences, University of Florida.

Boydak, E., Alpaslan, M.,Hayta, M., Gercek, S. & Simsek, M. (2002). Seed composition of soybeans grown in the Harran region of Turkey as affected by row spacing and irrigation. *J. Agric. Food Chem.,* 50, 4718-4720.

Bradley, C. A., Hartman, G. L., Wax, L. M. & Pedersen, W. L. (2002). Quality of harvested seed associated with soybean cultivars and herbicides under weed-free conditions. *Plant Dis.,* 86, 1036-1042.

Brim, C.A. & Burton, J.W. (1979). Recurrent selection in soybeans. II. Selection for increased percent protein in seeds. *Crop Sci.,* 19, 494–498.

Burton, J.W. (1985). Breeding soybean for improved protein quantity and quality. In: World Soybean Research Conference III: Proceedings, edited by Shibles, R., Boulder, CO: Westview Press, pp. 361–367.

Cherry, J. H., Bishop, L., Hasegawa, P. M. & Leffler, H. R. (1985). Differences in the fatty acid composition of soybean seed produced in northern and southern areas of the U.S.A. *Phytochemistry,* 24, 237–241.

Dardanelli, J. L., Balzarini, M., Martinez, M. J., Cuniberti, M., Resnik, S., Ramunda, S. F., Herrero, R. & Baigorri, H. (2006). Soybean maturity groups, environments, and their interaction define mega-environments for seed composition in Argentina. *Crop Sci.,* 46, 1939–1947.

Dixon, R.A. (1990). Isoflavonoids: biochemistry, molecular biology and biological functions. In: Comprehensive natural products chemistry, edited by Sankawa, U, Amsterdam: *Elsevier,* 773-823.

Dordas, C. (2006). Foliar boron application improves seed set, seed yield, and seed quality of alfalfa. *Agron. J.,* 98, 907–913.

Dordas, C., Apostolides, G. & Goundra, O. (2007). Boron application affects seed yield and seed quality of sugar beets. *J. Agric. Sci.,* 145, 377-384.

Doss, B. D., Pearson, R. W. & Rogers, Howard T. (1974). Effect of soil water stress at various growth stages on soybean yield. *Agron J.,* 66, 297-299.

Eck, H. V., Mathers, A. C. & Musick, J. T. (1987). Plant water stress at various growth stages and growth and yield of soybeans. *Field Crop Res.,* 17, 1-16.

Egli, D.B. & Bruening, W. P. (2007). Accumulation of nitrogen and dry matter by soybean seeds with genetic differences in protein concentration. *Crop Sci.* 47, 359–366.

Enciso, J. & Peries, X. (2005). Using flexible pipe (polypipe) with surface irrigation. Publication L-5469, Texas Cooperative Extension Service, The Texas A&M University System.

FAO. (2010). Crop Water Information: Soybean. United Nations Food and Agriculture Organization, Rome. http://www.fao.org/nr/water/cropinfo_ soybean.html

Fehr, W. R. & Caviness, C. E. (1977). Stages of soybean development. Spec. Rep.80, *Iowa Agric. Home Econ. Exp. Stn.,* Iowa State Univ., Ames.

Fisher, D. K. (2007). Automated collection of soil-moisture data with a low-cost microcontroller circuit. *Appl. Eng. Agric.,* 23, 493-500.

Foroud, N., Miindel, H. H., Saindon, G. & Entz, T. (1993). Effect of level and timing of moisture stress on soybean plant development and yield components. *Irrig. Sci.,* 13, 149-155.

Graham, N. & Graham, M.Y. (1991). Gyceollinelicitors induce major but distinctly different shifts in isoflavonoid metabolism in proximal and distal soybean cell populations. *Mol. Plant Microbe In.act,* 4, 60-68.

Graham, N., Kim, I. E. & Graham, M. Y. (1990). Role of constitutive isoflavone conjugates in the accumulation of glyceollin in soybean infected with *Phytaphthora megasperma. Mol. Plant Microbe* In., 3, 157-166.

Greenwood, D. J., Zhang, K., Hilton, H. W. & Thompson, A. J. (2010). Opportunities for improving irrigation efficiency with quantitative models, soil water sensors and wireless technology. *J. Agric. Sci.,* 148, 1–16.

Harrison, K. & Perry, C. (2010). Evaluating and Interpreting Application Uniformity of Center Pivot Irrigation Systems. *Circular* 911, The University of Georgia Cooperative Extension.

Heatherley, L. G., Elmore, C. D., Wesley, R.A. & Spurlock, S. R. (2001). Row spacing and weed management systems for non-irrigated early soybean production system plantings in the midsouthern USA. *Crop Sci.,* 41, 784–791.

Heatherly, L. G. (1999). Early soybean production system (ESPS). In: *Soybean Production in the Midsouth,* edited by Heatherly, L. G. & Hodges, H. F., 103-118. CRC Press, Boca Raton, FL.

Hepperly, P. R. & Sinclair, J. B. (1978). Quality losses in Phomopsis-infected soybean seeds. *Phytopathology, 68,* 1684-1687.

Hoffman, D. D., Hartman, G. L., Mueller, D. S., Leitz, R. A., Nickell, C. D. & Pedersen, W. L. (1998). Yield and seed quality of soybean cultivars infected with *Sclerotinia sclerotiorum*. *Plant Dis*., 82, 826-829.

Hou, A., Chen, P., Alloatti, J., Li, D., Mozzoni, L., Zhang, B. & Shi. A. (2009). Genetic variability of seed sugar content in worldwide soybean germplasm collections. *Crop Sci.,* 49, 903–912.

Hymowitz, T. & Collins, F. I. (1974). Variability of sugar content of seed of *Glycine max* (L.) Merr. and *G. soja* Serb. and Zucco. *Agron. J.*, 66, 239–240.

Irmak, S., Payero, J. O., Eisenhauer, D. E., Kranz, W. L., Martin, D. L., Zoubek, G. L., Rees, J. M., VanDeWalle, B., Christiansen, A. P. & Leininger, D. (2006). Watermark granular matrix sensor to measure soil matric potential for irrigation management. Publication EC783, University of Nebraska Extension, Lincoln, NE USA.

John, M. K., Chuah, H. H. & Neufeld, J. H. (1975). Application of improved azomethine-H method to the determination of boron in soils and plants. *Anal. Lett.,* 8, 559-568.

Kranz, W. L., Elmore, R. W. & Specht, J. E. (2005). Irrigating soybean. Publication G1367, University of Nebraska-Lincoln Extension, Lincoln Nebraska USA.

Kulik, M. M. & Sinclair, J. B. (1999). Phomposis seed decay. In: Compendium of soybean diseases, 4th edition., edited by Hartman, G. L., Sinclair, J. B. & Rupe, J. C). APS Press, The American Phytopathological Society, St. Paul MN, USA, pp. 31-32.

Lee, S. J., Ahn, J. K., Kim, S. H., Kim, J. T., Han, S. J. & Jung, M. Y. (2003a). Variation in isoflavone of soybean cultivars with location and storage duration. *J. Agric. Food. Chem.,* 51, 3383-3389.

Lee, S. J., Yan, W. K., Ahn, J. K. & Chung, I. M. (2003b). Effects of year, site, genotype and their interactions on various soybean isoflavones. *Field Crop Res.,* 81, 181–192.

Linkemer, G., Board, J. E. & Musgrave, M. E. (1998). Waterlogging effect on growth and yield components of late-planted soybean. *Crop Sci.,* 38, 1576–1584.

List, G. R., Mounts, T. L. & Lanser, A. C. (1992). Factors promoting the formation of nonhydratable soybean phosphatides. *J. Am. Oil Chem. Soc.,* 69, 443.

Liu, K. (1997). Soybeans chemistry, technology, and utilization. *Chapman & Hall,* New York.

Lohse, G. (1982). Microanalytical azomethine-H method for boron determination in plant tissue. *Comm. Soil Sci. Plant Anal.,* 13, 127–134.

Maestri, D. M., Labuckas, D. O., Meriles, J. M., Lamarques, A. L., Zygadlo, J. A. & Guzman, C. A. (1998). Seed composition of soybean cultivars evaluated in different environmental regions. *J. Sci. Food Agric.,* 77, 494–498.

Marschner, H. (1995). Mineral nutrition of higher plants. Academic Press, San Diego, CA, 379-396.

Mengistu, A. & Heatherly, L. G. (2006). Planting date, irrigation, maturity group, year, and environment effects on *Phomopsis longicolla*, seed germination, and seed health rating of soybean in the early soybean production system of the midsouthern USA. *Crop Prot.,* 25, 310-317.

Mengistu, A., Ray, J. D., Smith, J. R. & Paris, R. L. (2007). Charcoal rot disease assessment of soybean genotypes using a colony-forming unit index. *Crop Sci.,* 47, 2453-2461.

Mengistu, A., Smith, J. R., Bellaloui, N., Paris, R. L. & Wrather, J. A. (2010). Irrigation and time of harvest effects on evaluation of selected soybean accessions against *Phomopsis longicolla*, *Crop Sci.,* 50, 2055-2064.

Mengistu, A., Smith, J. R., Ray, J. D. & Bellaloui, N. (2011). Seasonal progress of charcoal rot and its impact on soybean productivity. *Plant Dis.,* 59, 1159-1166.

Messina, M. J., Persky, V., Setchell, K. D. R. & Barnes, S. (1994). Soy intake and cancer risks: a review of the in vitro and in vivo data. Nutr. *Cancer,* 21, 113–131.

Minor, H. C., Brown, E. A., Doupnik Jr., B., Elmore, R. W. & Zimmerman, M. S. (1993). Registration of Phomopsis seed decay resistant soybean germplasm MO/PSD-0259. *Crop Sci.,* 33, 1105.

Morrison, M. J., Cober, E. R., Saleem, M. F., McLaughlin, N. B., Fregeau-Reid, J. Ma, B. L., Yan, W. & Woodrow, L. (2008). Changes in isoflavone concentration with 58 years of genetic improvement of short-season soybean cultivars in Canada. *Crop Sci.,* 48, 2201–2208.

NRCS. (2001). PHAUCET. http://www.wsi.nrcs.usda.gov/products/W2Q/ water_mgt/ Irrigation/irrig-mgt-models.html

Obendorf, R. L., Horbowicz, M,, Dickerman, A. M., Brenac, P., & Smith, M. E. (1998). Soluble oligosaccharides and galactosyl cyclitols in maturing soybean seeds in planta and in vitro. *Crop Sci.*, 38, 78–84.

Ortega-Farias, S., Olioso, A., Antonioletti, R. & Brisson, N. (2004). Evaluation of the Penman-Monteith model for estimating soybean evapotranspiration. *Irrig. Sci.,* 23, 1–9.

Paris, R. L., Mengistu, A., Tyler, J. M. & Smith, J. R. (2006). Registration of soybean germplasm line DT97-4290 with moderate resistance to charcoal rot. *Crop Sci.,* 46, 2324-2325.

Pathan, M. S., Clark, K. M., Wrather, J. A., Sciumbato, G. L., Shannon, J. G., Nguyen, H. T. & Sleper, D. A. (2009). Registration of soybean germplasm SS93-6012 and SS93-6181 resistant to phomopsis seed decay. *J. Plant Reg.,* 3, 91- 93.

Paxton, J. D. (1980). A new working definition of the term “phytoalexin”. *Plant Dis,* 64, 734.

Piper, E. L., & Boote, K.J. (1999). Temperature and cultivar effects on soybean seed oil and protein concentrations. *JAOCS.,* 76, 1233–1242.

Potter, S. M., Baum, J. A., Teng, H. Y., Stillman, R. J., Shay, N. F. & Erdman, J. W. (1998). Soy protein and isoflavones: their effects on blood lipids and bone density in postmenopausal women. *Am. J. Clin. Nutr.,* 68, 1375s–1379s.

Rogers, D. H. (1995). Managing furrow irrigation systems. Publication L-913. Cooperative Extension Service, Kansas State University, Manhattan, KS USA.

Roy, K. W. (1976). The mycoflora of soybean reproductive structures. Ph.D. thesis, Purdue University, West Lafayette, Indiana. 255 [cited by Abney, T. S. & Ploper, L. D. (1988). Seed diseases. In: Soybean diseases of the north central region (T. D. Wylie, & D. H. Scott, (Eds.), 3-6. APS Press, St. Paul MN, USA.

Ryley, M. (2004). Effects of some diseases on the quality of culinary soybean seed. http://www.docstoc.com/docs/32751896/Influence-of-diseases-on-the-quality-of-soybean-seed Verified (4/11/2010).

Sakthivelu, G., Devi, M. K. A., Giridhar, P., Rajasekaran, T., Ravishankar, G. A., Nikolova, M. T., Angelov, G. B., Todorova, R. M. & Kosturkova, G. P. (2008). Isoflavone composition, phenol content, and antioxidant activity of soybean seeds from India and Bulgaria. *J. Agric. Food Chem.,* 56, 2090–2095.

SAS. (2001). SAS 9.1 TS LeVel 1M3, Windows Version 5.1.2600, SAS Institute, Cary.

Sato, T., Eguchi, K., Hatano, T. & Nishiba, Y. (2008). Use of near-infra red reflectance spectroscopy for the estimation of isoflavone contents of soybean seeds. *Plant Prod. Sci.*, 11, 481-486.

Savoy, B. R., Cothran, J. T. & Shumway, C. R. (1992). Early-season production systems utilizing indeterminate soybean. *Agron. J.*, 84, 394-398.

Schnebly, S. R., & Fehr, W. R. (1993). Effect of years and planting dates on fatty acid composition of soybean genotypes. *Crop Sci.*, 33, 716–719.

Shock, C. C., Flock, R. J., Feibert, E. B. G., Shock, C. A., Pereira, A. B. & Jensen, L. B. (2005). Irrigation monitoring using soil water tension. EM 8900. Corvallis, OR: Oregon State University Extension Service. http://extension.oregonstate.edu/catalog/ pdf/em/em8900.pdf.

Smith, J. R., Mengistu, A., Nelson, R. L. & Paris, R. L. (2008). Identification of soybean accessions with high germinability in high-temperature environments. *Crop Sci.*, 48, 2279-2288.

Streeter, J. G. (2003). Effects of drought on nitrogen fixation in soybean root nodules. *Plant Cell Environ.*, 26, 1199–1204.

Sullivan, M., VanToai, T., Fausey, N., Beuerlein, J., Parkinson, R. & Soboyejo, A. (2001). Evaluating on-farm flooding impacts on soybean. *Crop Sci.*, 41, 93–100.

Sweeney, D. W., Long, J. H. & Kirkham, M. B. (2003). A single irrigation to improve early maturing soybean yield and quality. *Soil Science Society of America Journal*, 67, 235–240.

Taji, T., Ohsumi, C., Iuchi, S., Seki, M., Kasuga, M., Kobayashi, M., Yamaguchi-Shinozaki, K. & Shinozaki, K. (2002). Important roles of drought- and cold-inducible genes for galactinol synthase in stress tolerance in Arabidopsis thaliana. *Plant J.* 29: 417-426.

Taylor, O. (1999). ESPS method moves north. *Soybean Dig.*, 59, 26-28.

TeKrony, D. M., Grabau, L. J., DeLacy, M. & Kane, M. (1996). Early planting of early-maturing soybean: Effects on seed germination and Phomopsis infection. *Agron. J.*, 88, 428-433.

Tsukamoto, C., Shimada, S., Igita, K., Kudou, S., Kokubun, M. & Okubo, K. (1995). Factors affecting isoflavone content in soybean seeds: changes in isoflavones, saponins, and composition of fatty-acids at different temperatures during seed development. *J. Agric. Food Chem.*, 43, 1184–1192.

Wikipedia, natural Phenol. http://en.wikipedia.org/wiki/Natural_phenol.

Wikipedia, soybean. http://en.wikipedia.org/wiki/Soybean#Chemical_ composition_of_ the_seed verified 4-18-2012.

Wilcox, J. R., & Shibles, R. M. (2001). Interrelationships among seed quality attributes in soybean. *Crop Sci.*, 41, 11–14.

Wilson, R. F. (2004). Seed composition. In: Soybeans: Improvement, production, and uses, edited by Boerma H., J.E. Specht, 3rd ed. *ASA, CSSA, and SSSA, Madison, WI*, pp. 621–668.

Wilson, R. F., Novitzky, W. P. & Fenner, G. P. (1995). Effect of fungal damage on seed composition and quality of soybeans. *JOACS*, 72, 1425-1429.

Wrather, J. A., Kendig, S. R., Wiebold, W. J. & Riggs, R. D. (1996). Cultivar and planting date effects on soybean stand, yield, and Phomopsis sp. seed infection. *Plant Dis.*, 80, 622-624.

Ziems, A. D., Giesler, L. J., Graff, G. L., Redinnaugh, M. G., Vacha, J. L., Berry, S. A., Madden, L.V. & Dorrance, A. E. (2007). Response of soybean cultivars to bean pod mottle virus infection. *Plant Dis.* 91, 719–726.

Zobiole, L. H. S., Oliveira Jr., R. S., Visentainer, J. V., Kremer, R. J., Bellaloui, N. & Yamada, T. (2010). Glyphosate affects seed composition in glyphosate-resistant soybean. *J. Agric. Food Chem.,* 58, 4517–4522.

In: Irrigation Management, Technologies …
Editor: M. Hossain Ali

ISBN: 978-1-62417-862-7

Chapter 9

Micro-Irrigation Scheduling Using Clean Energy

Ahmed Al-Busaidi**[1] and Tahei Yamamoto[2]
[1]College of Agricultural & Marine Sciences, Department of Soils, Water and Agricultural Engineering, Sultan Qaboos University, Muscat, Oman
[2]Arid Land Research Center, Tottori University, Hamasaka, Tottori, Japan

Abstract

A micro-irrigation system is classified as a smallscale irrigation system that can be applied for thousands square meters of agricultural fields. With limited water resources for crop production, there is a need to reduce micro-irrigation facilities while maximizing the efficiency of local environmental conditions such as harvested rain water that can be used for field irrigation. Furthermore, sustainable irrigated agriculture using clean energy should be promoted by applying a micro-irrigation system supported with a tank and apron, which resulting in saving water and energy in crop cultivation.

This study aims to develop a series of program packages of simulation models on irrigation scheduling for two cases of small scale irrigation systems, which are supported by a rain harvesting catchment apron, one or two water storage tanks and a groundwater pump operated by wind or solar energy. The first case comprises an apron, a tank and fresh water-well for a non-saline field. The second case is an apron, a tank, a well and field for saline irrigation. The computer program in this study uses the Microsoft 2008 Visual Basic language. The selected fields were distributed in three locations; Mu Us Shamo desert in Inner Mongolia, China (Mu Us), Shavour Soil Institute Station in the Khuzestan Province, Iran (Shavour) and Arid Land Research Center of Tottori University, Japan. By using soil, water, plants and meteorological data measured previously in the field, initial conditions of simulation models were defined as factors of irrigation scheduling, such as effective rainfall (PEF), evapotranspiration (ET), total readily available moisture (TRAM), irrigation water requirement (Iw), irrigation interval (*Ii*), etc. It is expected that simulation results will be used as a guide to design sustainable micro-irrigation projects.

* E-mail: albusaidiahmed@yahoo.com.

1. Micro-Irrigation Scheduling for Harvested Rainwater and Pumped Groundwater Using Clean Energy

1.1. Introduction

Irrigated agriculture has played an important role in food production since ancient civilization for more than four millenniums, and mustbe the main driver to feed the world population in the near future. However, it is also one of the factors that could increase soil degradation. In order to improve the efficiency of the irrigated agriculture under arid and humid climates, sustainable use of rainfall resource is crucial. Micro-irrigation systems consisting of catchment and water storage tanks may be a good choice. A micro-irrigation system is classified as small scale irrigation which can be applied for thousands square meters of agricultural fields (Yamamoto et al., 1993, 2003). With limited water resources for crop production, there is a need to introduce micro-irrigation systemsto get benefits from natural resources such as harvested rain water. Furthermore, sustainable agriculture using clean energy should be promoted by applying a micro-irrigation systems supported by catchment and storage tanks that will help in power and water saving.

This study aims to develop a series of program packages of simulation models on irrigation scheduling for two cases (fresh and saline groundwater) of small scale irrigation systems supported by catchment to harvest rainwater, water storage tanks and groundwater pumps operated by wind or solar energy. The computer programming used in this study is in the Microsoft 2008 Visual Basic language. The selected fields were distributed in three locations: Mu Us Shamo desert in the Inner Mongolia Autonomous region of China (Mu Us), Shavour Soil Institute Station in the Khuzestan Province of Iran (Shavour) and Arid Land Research Center of Tottori University, Japan.

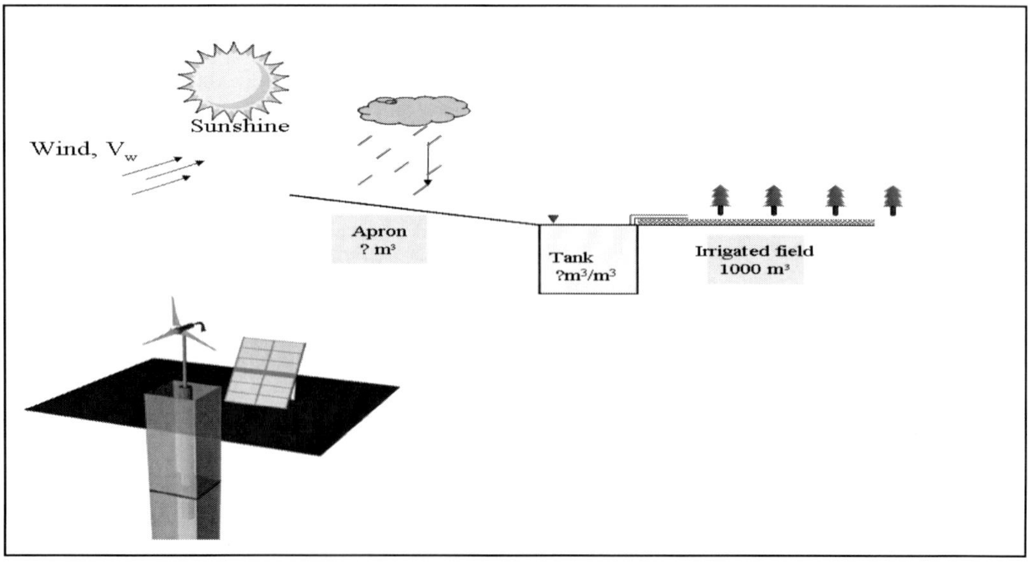

Figure 1. Schematic of small scale irrigation facility operated by clean energy.

Byusing soil, water, plant and meteorological data measured by previous field works, initial conditions of simulation models were defined as factors of irrigation scheduling, such as effective rainfall (PEF), evapotranspiration (ET), total readily available moisture (TRAM), ready available moisture (RE) within the TRAM, supplement irrigation water requirement (Iw) and irrigation interval (*Ii*), etc. It is expected that simulation results will be used as a guide to design sustainable micro-irrigation projects.

1.2. Micro-irrigation Scheduling under Harvested Rainwater

1.2.1. Simulation Model of Small Scale Irrigation Systems

A small scale irrigation system was designed to suit the environments shown in Figure 1. In this system, rain water will be harvested and stored in underground tank. Stored rain water and groundwater will be used as a source of irrigation water. These kinds of irrigation systems using small tanks have played an important role in traditional paddy cultivation in Japanese humid agricultures since ancient times. However, to apply this approach in arid and semiarid areas where rainfall is very scarce and ground water is the only source of irrigation water. Many studies should be conducted and evaluated before formulating any irrigation model.

For this study, simulation of the small scale irrigation system was conducted under the following irrigation parameters:

(1) The system was classified as a micro-irrigation with frequent intervals and for traditional irrigation with intermitted intervals,
(2) The area of irrigated fields was 1,000 m^2,
(3) The catchment was built on an inclined land, and the tank and irrigated fields were located in a flat land. Therefore, rainfall was efficiently harvested and irrigation water was applied to crops uniformly. The following assumptions were made: rainfall harvesting efficiency and water application efficiency were up to 80%, and storage tank capacity of 1,000 m^3 was full.
(4) The pumping systems were constructed near the tank. The ground water was pumped by using a power source from wind or solar during the windy and shiny days,
(5) The suitable size of catchment or tank was defined as the minimum size of the catchment or tank that storage water would not be empty by rain water or/and groundwater during simulation periods, and was determined as follow: If the tank was depleted before the end of the simulation period, either catchment size or tank capacity was increased to ensure larger storage water, andthen the same calculation were repeated with the initial catchment increment or the initial tank increment until the tank didn't depleted throughout simulation periods,
(6) Evaporation from the storage tank was assumed to be zero.

1.3. Theory of Simulation Models

1.3.1. Pumping Water by Solar or Wind Power

The calculation was done using the flow chart of the simulation model shown in Figure 2. Water volume in the storage tank can be pumped using clean energy. Due to sunshine hours, pumping time from the well (SOLRQ (j)) is estimated as follow:

$$SOLRQ\ (j) = Se * Sq * (SOLR\ (i) - 1.0) / Smax \quad (1)$$

SOLRQ (j): Daily water pumped by solar pump on that day (m^3/day).
SOLR (i): Daily sunshine duration onthat day (hrs).
Smax: Maximum sunshine duration to operate solar pump (hrs).
Sq: Daily water volume pumped by electromotive force of 300 watt when SOLR (j) ≥ 1.0 hour.
Sq = 14.4 m^3/day at Smax = 12.0 hrs.
Se: Efficiency of electromotive force, Se = 0.5

Equation (1) can be simplified as follow:

$$SOLRQ(j) = 0.6 * (SOLR\ (j) - 1) \text{ where } 1 < SOLR(j) < 12.0 \quad (2)$$

Due to wind velocity, pumped water from the well (WINQ (j)) is estimated as follow:

$$WINQ\ (j) = Wq * (WIND\ (j) - 2.5) / Wmax \quad (3)$$

WINQ (j): Daily water volume pumped by windmill (m^3/day).
WIND (j): Mean of daily wind velocity (m/sec).
Wq: Daily water volume (m^3/day) pumped by electromotive force of 300 watt when WINT (j) ≥ 2.5 m/s, Wq = 14.4
Wmax: Maximum working hours of windmill per day, Wmax = 10.0 (hrs). For safety reasons, windmill stops when WIND (j) ≥ 12.5 (m/s).

Equation (3) can be simplified as follow:

$$WINQ(j) = 1.44 * (WIND(j) - 2.5) \text{ where } 2.5 < WIND(j) < 12.5 \quad (4)$$

SOLRQ (j) or WINQ (j) will be zero if the storage tank is full.

1.3.2. Water Balance in the Storage Tank

Volume of the stored water (VOL (j)) depends on water balance equation between rainwater or/and pumped water as input and irrigation water as output as follow:

$$VOL(j) = VVOL + APRONQ\ (j) - IwQ\ (j) \quad (5)$$

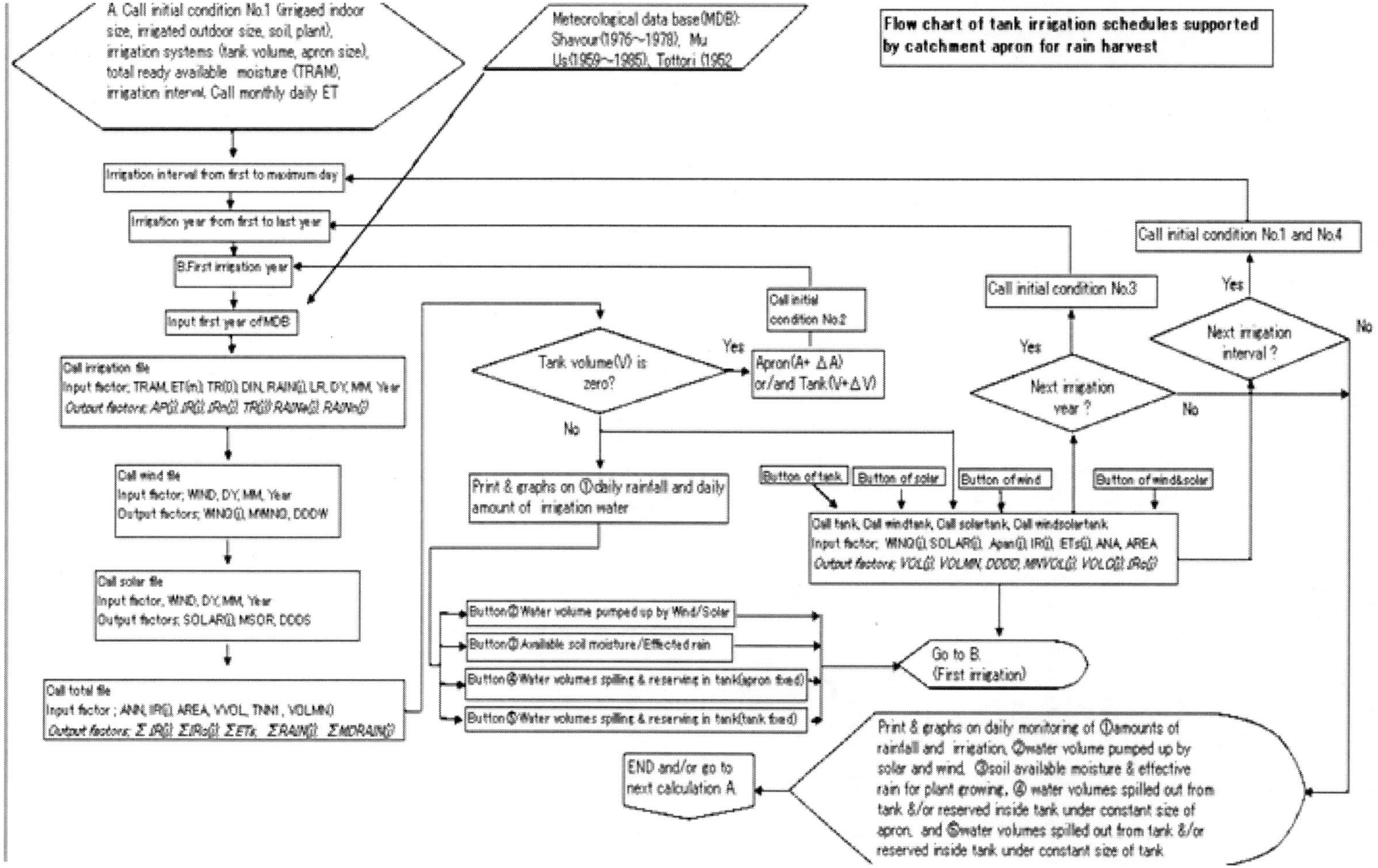

Figure 2. Flow chart of simulation model.

If solar pump is added the VOL(j) will be equal to:

$$VOL(j) = VVOL + SOLAR(j) + APRONQ(j) - IwQ(j) \quad (6)$$

If windmill is used, the VOL(j) will be equal to:

$$VOL(j) = VVOL + WINQ(j) + APRONQ(j) - IwQ(j) \quad (7)$$

$$APRONQ(j) = Rc * Rain(j) / 1000 * ANN \quad (8)$$

$$IwQ(j) = IR(j) * AREA / (1000 * Irrc) \quad (9)$$

WINQ (j): Daily water volume pumped from groundwater by windmill pump.
SOLARQ (j): Daily water volume pumped from groundwater by solar pump.
AREA: Irrigated area (m^2).
ANN: Catchment (apron) size for harvestedrain water (m^2).
VVOL: Water volume stored in (j - 1) thday (m^3).
APRONQ (j): Daily runoff water by volume from apron into tank on jth day (m^3).
IwQ (j): Daily supplemented irrigation water requirement by volume on jth day (m^3).
Rc: Runoff coefficient = 0.8
Irrc: Irrigation application efficiency = 0.8

1.3.3. Accuracy of Simulation Model of One Tank Irrigation System

In order to evaluate the accuracy of simulation modelof one tank irrigation system, the annual balance equation of ready available soil moisture content (RE) and water volume in the tank(VOL) is applied using the input and output factors of the soil root depth and the recharge tank. The water balance equations are:

$$REac(y) = 100 * (1 - (\Delta RE + \sum PEF(j) + \sum Iw(j) - \sum ET(j) / TRAM)) \quad (10)$$

$$VOLac(y) = 100 * (1 - (\Delta VOL + \sum SOLRQ(j) + \sum WINQ(j) + \sum APRONQ(j) - \sum NVOL(j) / TNN) \quad (11)$$

REac (y): Accuracy of ready available soil moisture content yearly consumed and appliedto the plant root depth on yth year (%).
VOLac (y): Accuracy of annual water volume consumed and applied in the tankon yth year (%).
ΔRE: Difference of ready available moisture content reserved in TRAM between first and final day of the irrigation period (mm).
$\sum$PEF(j): Total amount of effective rain during irrigation period per year (mm).
$\sum$ET(j): Total amount of evapotranspiration during irrigation period per year (mm).
TRAM: Total readily available moisture.
ΔVOL: Difference in water volume reserved in the tank between first and final day of irrigation periods (m^3).
$\sum$SOLRQ(j): Total volume of groundwater pumped by solar pump during irrigation period per year (m^3).

∑WINQ(j): Total volume of groundwater pumped by wind pump during irrigation period per year (m^3).
∑APRONQ (j):Total volume of runoff water to the tank from apron during irrigation period per year (m^3).
∑Iw(j): Total volume of irrigation water pumped to the field from tank during irrigation period per year (m^3).
∑NVOL(j): Total volume of spilled water from the tank during irrigation period per year (m^3).
TNN: Volume of water filled the tank (m^3).

1.4. Climatic and Agricultural Characteristics of the Studied Areas

1.4.1. China (Mu Us Shamo Desert)

The Mu Us Shamo desert stretched over four million hectares in the South Ordos Plain of the eastern part of the Chinese desert. Desertification was increasing in a rate of approximately 1,000 km^2 per year since 1950. Adverse weather conditions, soil erosion, soil salinization and over grazing were the main reasons for desertification. Main field works were carried out in 1986-1992 and 1995. Previous mean of meteorological data for 30 years showed: annual mean air temperatures 6.4°C, precipitation 362 mm during June to August, wind velocity 3.5 m/s and evaporation 2,235 mm/yr. The maximum and minimum air temperatures were 34°C in July and -28°C in December. Ground freezing commence in October and last until it thawed in April. Approximately 40% of the land is sand dunes, about 50 % was stabilized sand dunes and about 10% were yellow soils including salt lakes and marshes. Severe wind erosion in the sand dunesand water erosion in the yellow soils were common. The stabilized dunes comprised of windbreak forests of 30-40%, rangelands of 40-45% and 5-10% of arable lands. The arable lands comprised of 1-2 hectares farms under irrigation, and were cultivated for corn, sorghum, nursery trees and vegetables. A very shallow groundwater was available as a source for irrigation with low salt concentration, between 500 to 600 ppm (Yamamoto et al., 1993).

1.4.2. Iran (Esfahan, Fars and KhuzestanProvinces)

Field works were carried out in 1975-1978 and 1995 in Provinces of Esfahan, Fars and Khuzestan of Iran, aiming at managingriver water for sustainable agriculture. Each irrigation project was in a scale of hundreds thousands hectares in alluvial plains. The three Provinces were under arid to semi-arid climates with annual rainfall of 100 to 400 mm during December to March. Farmersusually have a field of 3-5 hectares for wheat, corn, sugar beet, orchard and vegetablecrops. Rice wasthe main crop cultivated in the vicinity of river and swam. Water deficiency conditions were more sever in the Provinces of Esfahan and Fars where groundwater was used as a supplement for irrigation. Whereas, in Khuzestan Province where river water from the Zagros Mountains was used for irrigation. The Khuzestan forms a flat alluvial plain located at the south-west of Iranand has played an important role in petro-chemistry as well as agriculture. The north province was a non-saline light alluvial soil, while, the other showed saline heavy alluvial soil. Main rivers had enough water with good quality for irrigation; however, the downstream of the rivers had saline water which was not suitable for irrigation. Due to drainage canal and tile drain, ECe of soil salinity was reclaimed to be less than 4 dS/m in the main fields by controlling the depth of groundwater below 2-3 m from the soil surface. Salinity of the river water was around 1.35 dS/m and SAR=3, which was evaluated as C3-S1 according to the USDA classification table (USDA, 1954). The crop production in the irrigated field was 2-3 times higher than those of rain-fed fields (Yamamoto et al., 1998).

1.4.3. Japan (Tottori Prefecture)

For Tottori upland fields, metrological data from 1952 to 1982 was measured at Arid Land Research Center, Tottori University. The ET during irrigation seasons was 700-800 mm while the mean rainfall was900-1000 mm. The rainfall decreased to 62-71% in the drought years with more frequency of 1/10 and to 53-54% in that of 1/50. Although Tottori is considered to have relatively high rainfall, whereas drought conditions appeared in the irrigation seasons from May to September. The water quality of irrigation is relatively good and almost suits all irrigation projects. However, salts could come when river water or groundwater is used for irrigation in the vicinity of sea lines, and it increased to high salinity especially under drought seasons. As an advantage of saline irrigation in the Tottori fields, salt leaching might be more useful compared to Shavour or Mu Us, because of its relatively high rainfall characteristics under monsoon climate.

1.5. Model Application: Case Study from Japan

1.5.1. Initial Conditions of Soil, Plant, Climate and Irrigation Water

Initial conditions of soil, plant, climate and irrigation water required for simulation are shown in Figure 3. As for Tottori, simulation period started in year of 1952 to1982. Initial sizes of irrigated field and catchment were 1,000 m^2 and 150 m^2 respectively. At the beginning, water tank was filled to a volume of 1,000 m^3. The irrigation periods were between April and September. Maximum irrigation interval (Max of *Ii*) was 5 days. It was estimated by using TRAM = 32.5 mm and maximum monthly evapotranspiration (Max. of ET) of 6.5 mm with the equation of the Max. *Ii* = TRAM / Max of ET. The input of solar/wind operation index (1/1) means that the system will use both solar and wind pumps. The index=0 means that solar and wind power will not be used. Also, apron/tank changing index (1/0) means that a suitable catchment was estimated with constant capacity for storage tank, which mean that water tank should not be empty during the simulation.

By filling all data and pushing the button of data input, the table of the same data appeared in the textbox as shown in Figure 3. To run the model, select compute, and then new page will open for daily variation in NVOL (j) and VOL (j) during 1952 to 1982 (Figure 4). On the right side of this class form, there are six buttons. Each one is optional and opens different class form, such as the button of Next Interval indicates NVOL (j) and VOL (j) in case of *Ii* was selected between 1 to 5 days. Similarly, each button related to pump operation indicates the source of the power (wind or solar). The buttons of Next Tank or Increment apron indicate the need for rescaling the tank volume or apron size. The suitable size of apron and tank is decided by selecting the right value for the buttons related to interval, tank, apron and pump. The accuracy button indicates REac (y) and VOLac (y) of simulation model.

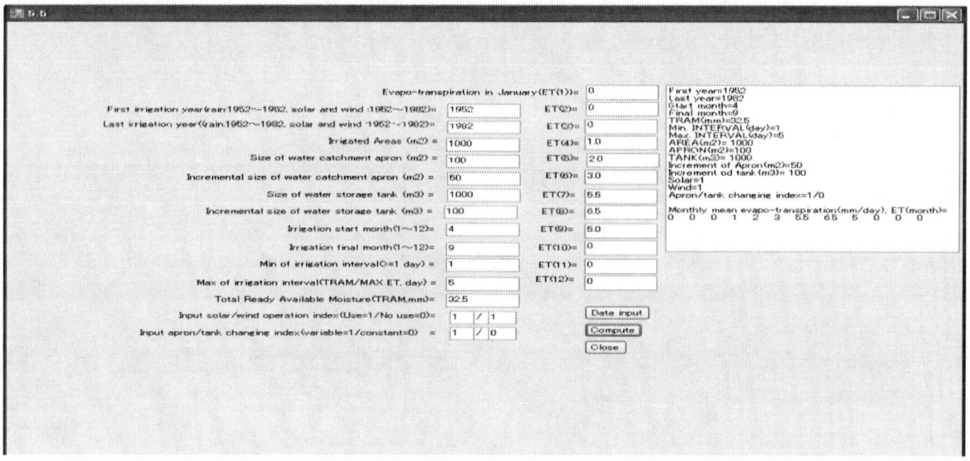

Figure 3. Initial condition of the Tottori fields in Japan.

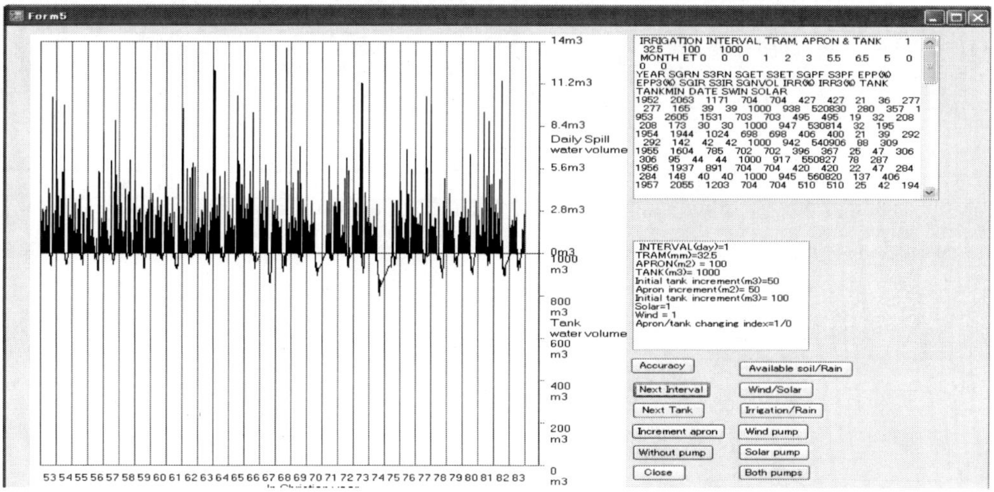

Figure 4. Daily variation of spilled and stored water in 1,000 m^3 tank with *Ii* of 1day in Tottori.

1.5.2. Effective Rainfall and Supplement of Water Requirement

The Ministry of Agriculture in Japan implements the model of equal evapotranspiration (ET), in which ET assumed to be same under both micro-irrigation and traditional irrigation systems such as sprinkler, as shown in Figure 5 (Japan, JIID, 1990). In crops fields, irrigation water is distributed on soil surface with strips of wet zones bounded by strips of dry soil. The ETa under micro-irrigation is defined as ET/p; where p is the wetted fraction in the field and ET is crops evapotranspiration over the entire field under sprinkler irrigation (L). Asa result, the ET in the wet portions of the micro-irrigation field (ETa) is higher than sprinkler irrigation field (ET). If micro-irrigation field contains both dry and wet zones, this means there is no difference between total crops grown under both irrigation methods. TRAM under sprinkler irrigation covers the whole crop area. Whereas, TRAM under micro-irrigation (MTRAM) is confined to uniformly distributed cones surrounding the emitter. In order to relate this to the entire crop area, MTRAM is transformed to the converted micro-irrigation total readily available moisture (CMTRAM) as shown in Figure 6.

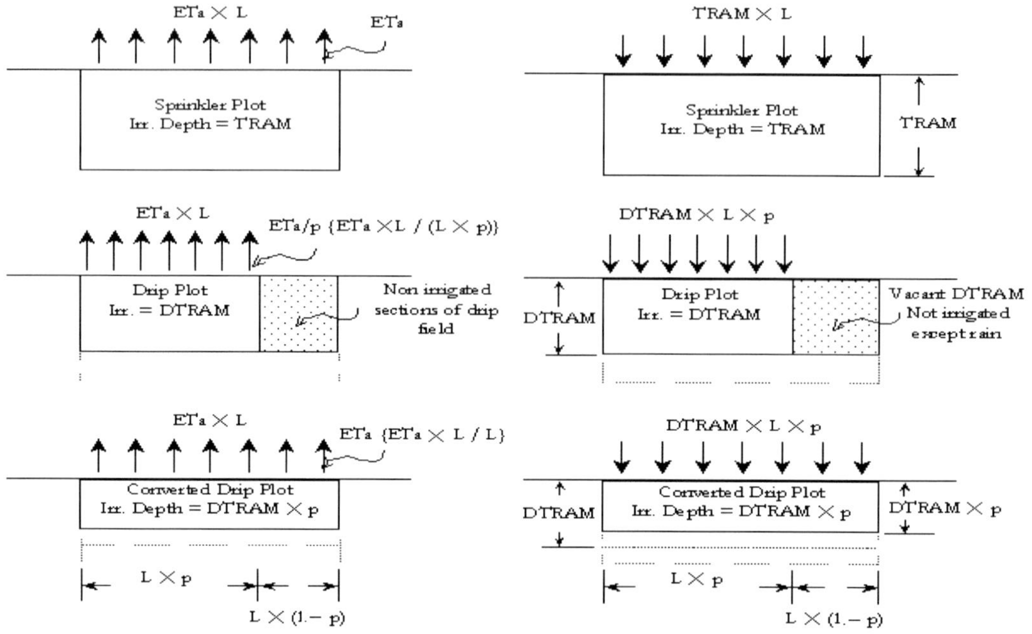

Figure 5. Equal evapotranspiration model for crops under micro and sprinkle rirrigation methods.

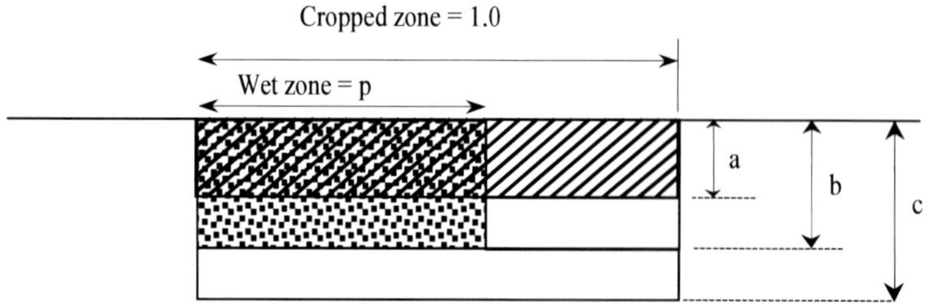

a: Converted Microirrigation Total Ready Available Moisture, CMTRAM (mm)

b: MTRAM (mm)

c: Total Ready Available Moisture in sprinkler irrigation (mm). If the crop coefficient is 0, b = c.

b - a: Apparent void in MTRAM (mm)

p: Fraction of wetted zone due to microirrigation

: Wet zone under microirrigation (no rainfall effect)

Figure 6. Soil moisture fraction model under micro and sprinkler irrigation methods.

The irrigation scheduling factors, as described above, have been obtained by long-term field works carried out in the studied areas; such as monthly mean ET and TRAM were measured using lysimeter with crop cultivated under drip irrigation. From the collected data, TRAM was 32.5 mm in Tottori, 35.0 mm in Mu Us and 150 mm in Shavour. In addition, soil moisture extraction pattern (SMEP) were measured using tensiometers with a layer of 0-40cm in Tottori, 0-100 cm in Mu Us and 0-120 cm in Shavour. In all flow charts of simulation model in Figure 2, the Iw(j) and PEF(j) were estimated from daily calculation of ET (j), TRAM and RE(j) for different irrigation interval (Ii), as shown in Figure 7.

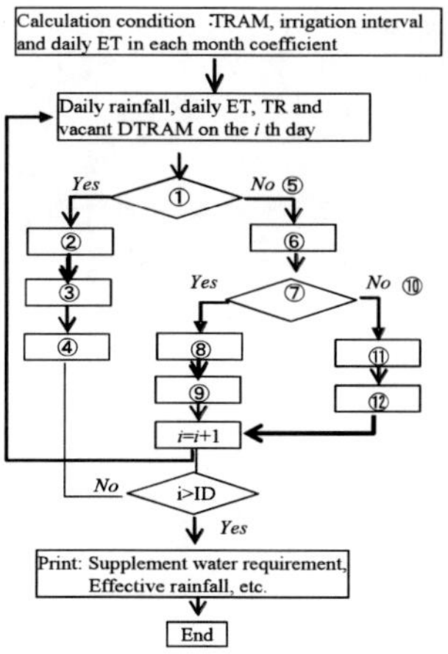

daily rainfall>5mm
R_0 = DTRAM – TR_{i-1}
daily rainfall ≧ R_0 ①
Effective rainfall = R_0 ②
Supplement water requirement=0 ③
TR_i=Tr_{i-1}+ effective rainfall – daily ET ④

0.8 × daily rainfall < R_0 ⑤
Effective rainfall = 0.8 × daily rainfall ⑥
Irrigation interval × daily ET > (TR_i + effective rainfall) ⑦
Supplement water requirement = irrigation interval × daily ET –(TR_i- daily ET) - effective rainfall) ⑧
TR_i= Tr_{i-1} + effective rainfall + supplement water requirement – daily ET ⑨
Irrigation interval × daily ET< (TR_i + effective rainfall) ⑩
Supplement water requirement = 0 ⑪

TR_i=TR_{i-1}+ effective rainfall – daily ET ⑫

Figure 7. Detailed flow chart to estimate water requirement using effective rainfall.

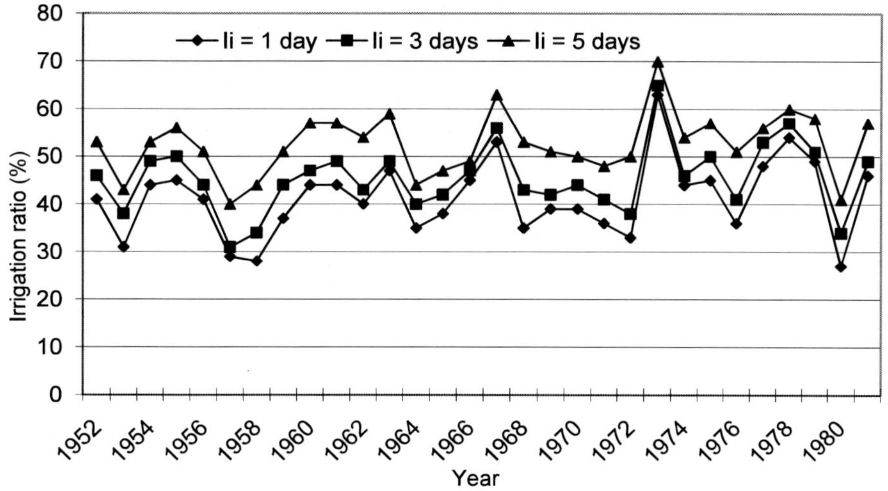

Figure 8. Annual variation of irrigation ratio in Tottori.

Variation in irrigation ratio (ΣIw (j) / ΣET (j)) and rainfall efficiency (ΣPEF (j) / ΣRain (j)) were shown with *Ii* for Tottori in Figures 8 and 9, and for Mu Us in Figures 10 and 11. ∑Iw(j) and ∑PEF(j) were estimated by integrating one year data for Iw (j) and PEF (j). As for the Tottori fields where the ET and rainfall during May to September were 670 and 860 mm, the irrigation ratio and rainfall efficiency were 41.2-52.7% and 38.5-45.3% during 1952-1982, respectively. For the Mu Us fields where ET and rainfall during May to September were 920 and 310 mm, the irrigation ratio and rainfall efficiency were 79.8-82.3%

and 49.1-56.3 % during 1959-1985, respectively. Whereas, for the Shavour fields where ET and rainfall were 2,650 and 260 mm per year, the irrigation ratio and rainfall efficiency were 95 and 66 %, respectively.

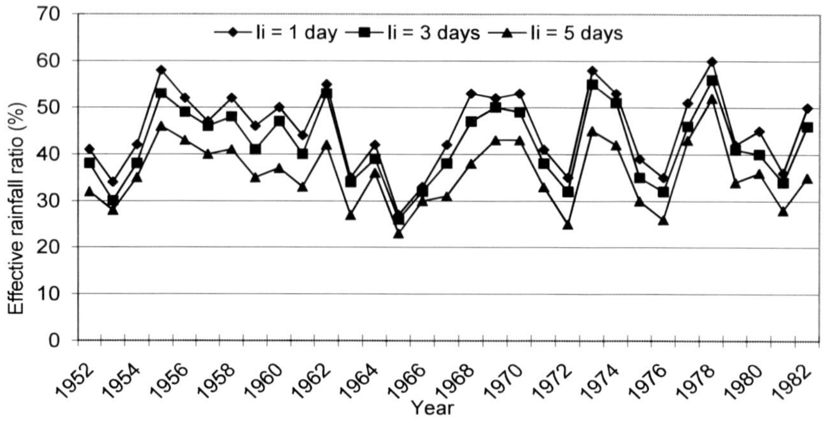

Figure 9. Annual variation of rainfall efficiency in Tottori.

As for Tottori, the *Ii* of 1 day showed 4.3-11.5 % lower irrigation ratio than *Ii* of 3-5 days. Under monsoon climate, a decrease of the *Ii* showed a tendency of high rainfall and low irrigation ratio. This tendency became more apparent when the TRAM value increase. This finding was also concluded by Yamamoto et al. (1999). As for MuUs and Shavour, differencesin the irrigation ratio or rainfall efficiency between *Ii* of daily and maximum values were smaller than Tottori and that happened because of low rainfall in dry areas. In dry zones such as arid or hyper arid lands and because of high evaporation, it is recommended to use wetted fraction model in which ETa under micro-irrigation system is low and Iw is small (Amu-Mensah*et al.*, 2001).

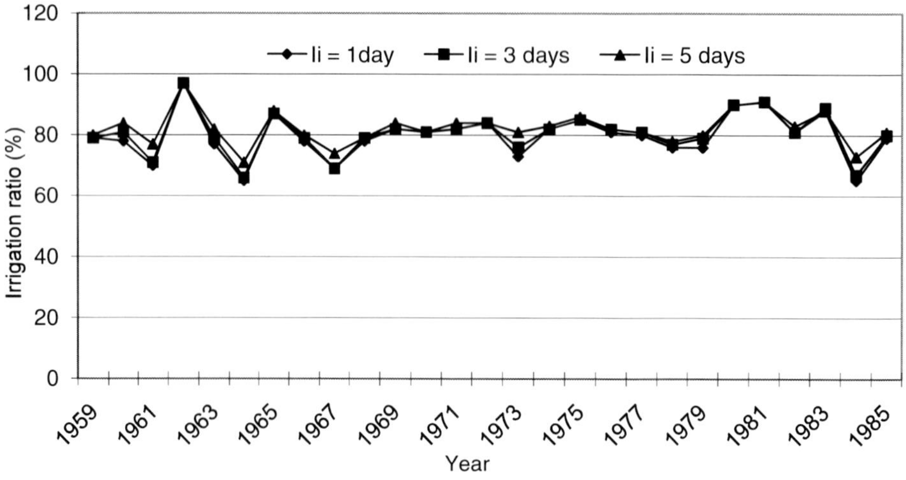

Figure 10. Annual variation of irrigation ratio in Mu Us.

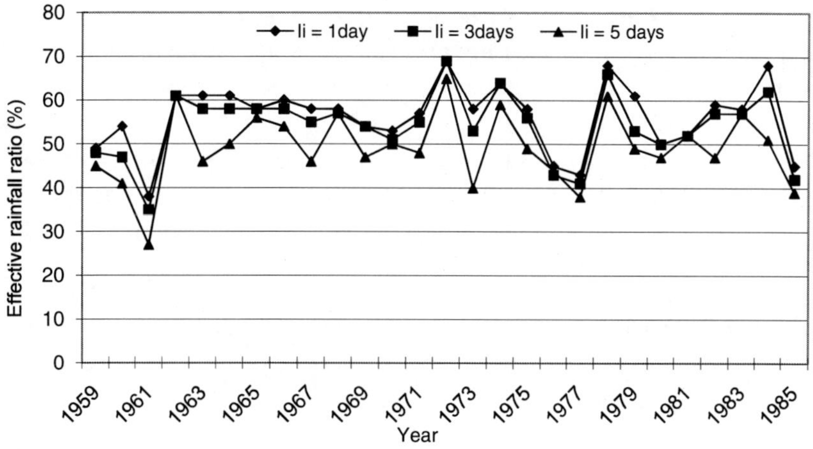

Figure 11. Annual variation of rainfall efficiency in Mu Us.

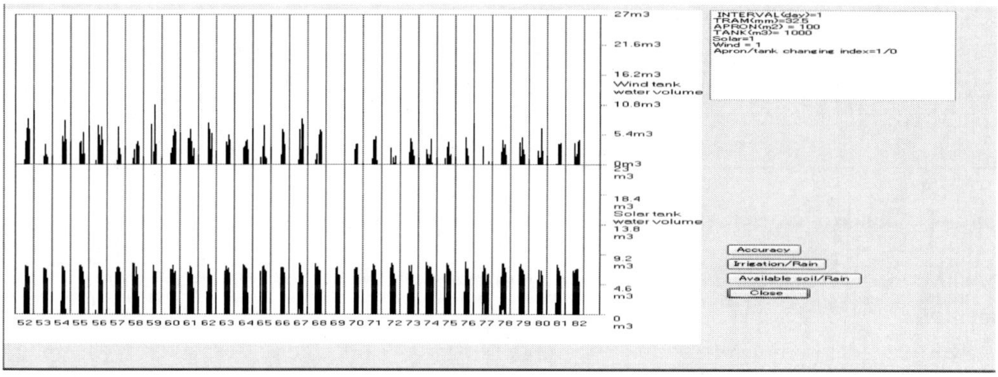

Figure 12. Volume of groundwater pumped by wind and solar energy in Tottori.

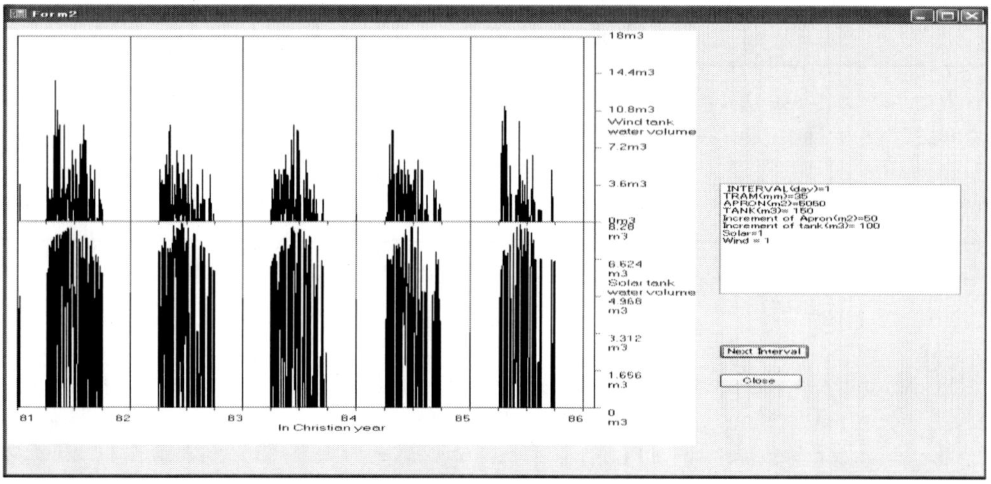

Figure 13. Volume of groundwater pumped by wind and solar energy in Mu Us.

1.5.3. Irrigation Water Pumped by Wind and Solar Energy

Volume of groundwater pumped daily by wind and solar energy is shown in Figures12-14. The volume was selected based on daily sunshine hours or mean wind speeds, as shown in the equation (1) or (3). The ranking for water pumped using solar power is as follow: Shavour> MuUs > Tottori, whereas by using wind power: MuUs > Tottori >Shavour. It was considered that pumping hours were practically depend no not only number of windy or sunny days, but also on the stored water in the tank in which the operation of each pump will stop automatically when the tank is full.

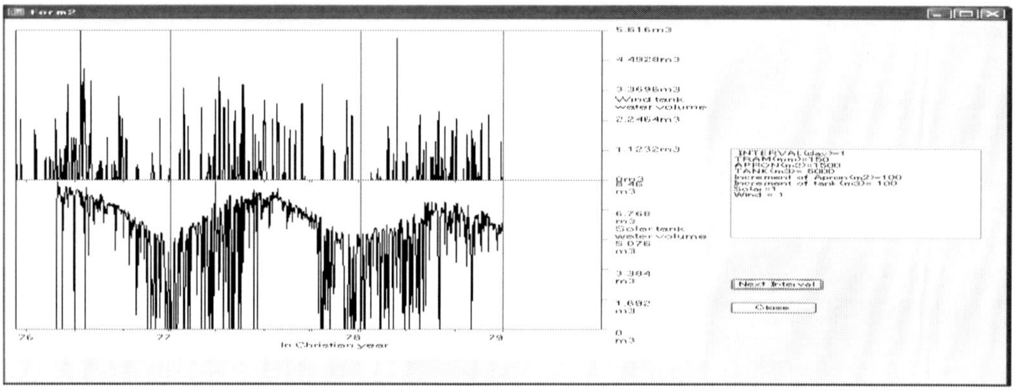

Figure 14. Volume of groundwater pumped by wind and solar energy in Shavour.

1.5.4. Readily Available Soil Moisture Content and Effective Rainfall stored in the Root Zone

Ready available moisture content stored in the TRAM on jth day (RE(j)) is the difference between field capacity (FC) and soil moisture content of irrigation at starting point ($Q_{pF3.0}$). It was calculated by knowing TRAM, ET, Rain and *Ii* values as shown in Figure 5. The TRAM and Max of *Ii* were 32.5 mm and 5 days in Tottori, 35 mm and 5 days in Mu Us, and 150 mm and 9 days in Shavour, respectively. For example, daily variation of RE (j) and PEF (j) are shown in Figure 15 with *Ii* of 1day and Figure 16 for *Ii* of 9 days in Shavour area. Mu Us was considered to decrease PEF (j) with longer *Ii*. However, for Shavour the values of *Ii* for 1-9 days are same, because the farmer had more rainfall during irrigation season, so it contributed to high value of PEF (j). Next, RE (j) varied between 0-6.5mm/day in Mu Us and 0-15mm/day in Shavour when *Ii* was 1day. Furthermore, the variationof RE (j) increased to 35mm/day in Mu Us and to 150mm/day in Shavour when *Ii* was Max. It was concluded that variation in RE (j) values increased with longer *Ii* or TRAM.

1.5.5. Accuracy of Simulation Model

The accuracy of the model is shown in Figure 17 for Tottori and Figure 18 for Mu Us. For both cases, the VOLac (y) was more than 98%. However, the REac (y) was relatively low when the irrigation is seasonly and not for the whole year; such as 94 % for Tottori, 90 % for Mu Us and 98 % for Shavour in the case of 1 day *Ii* using solar and wind pumps, respectively. The later csae could case some errors when the irrigation day or rainy day appeares in the last irrigation cycle. The simulation period was for 3 years in Shavour, 18 years in Mu Us and 31 years in Tottori.

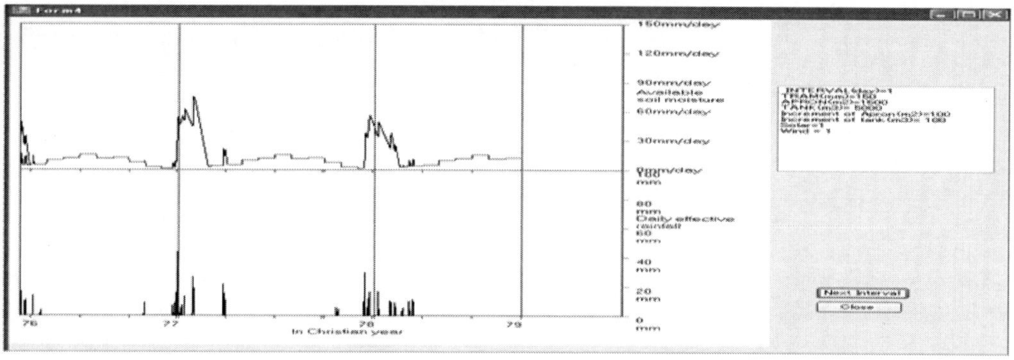

Figure 15. Daily variation of RE (j) and PEF (j) with *Ii* of 1 day in Shavour.

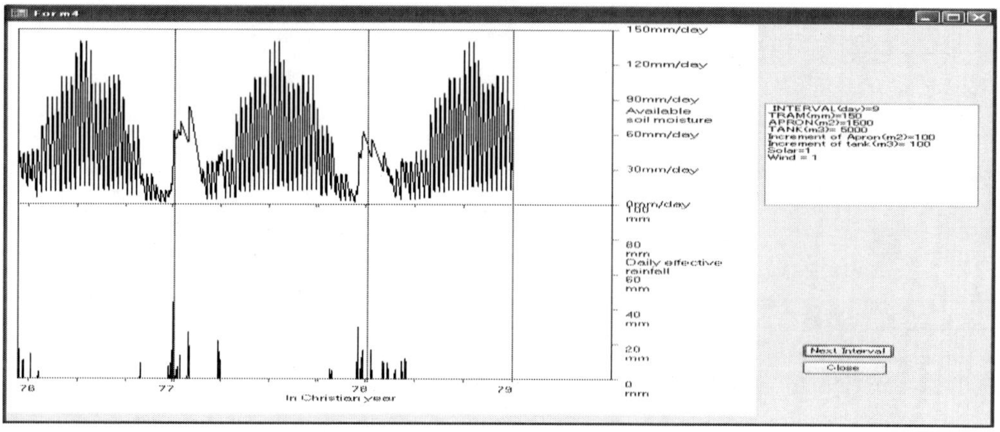

Figure 16. Daily variation of RE (j) and PEF (j) with *Ii* of 9 days in Shavour.

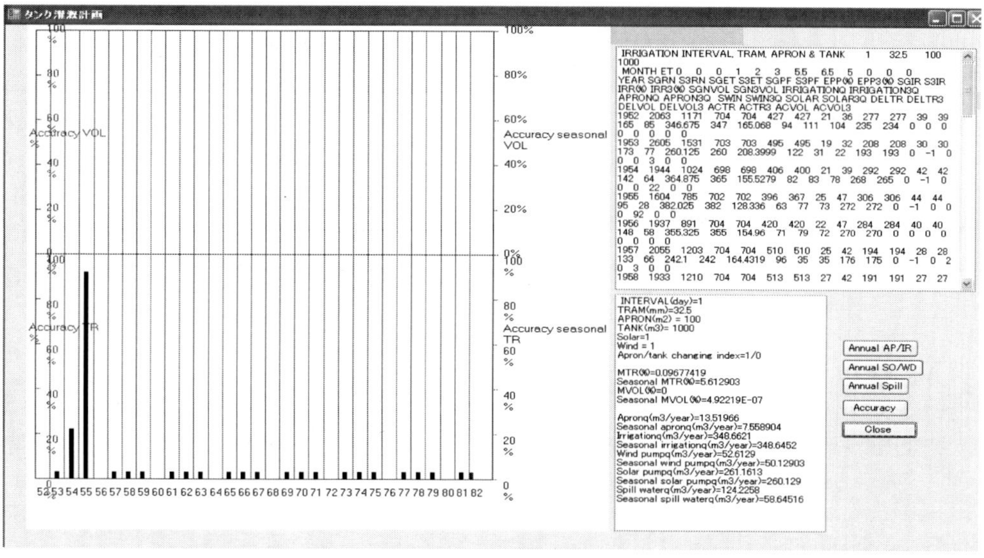

Figure 17. Annual variation error (100-accuracy) of TRac and VOLac during 1952-1982 in Tottori.

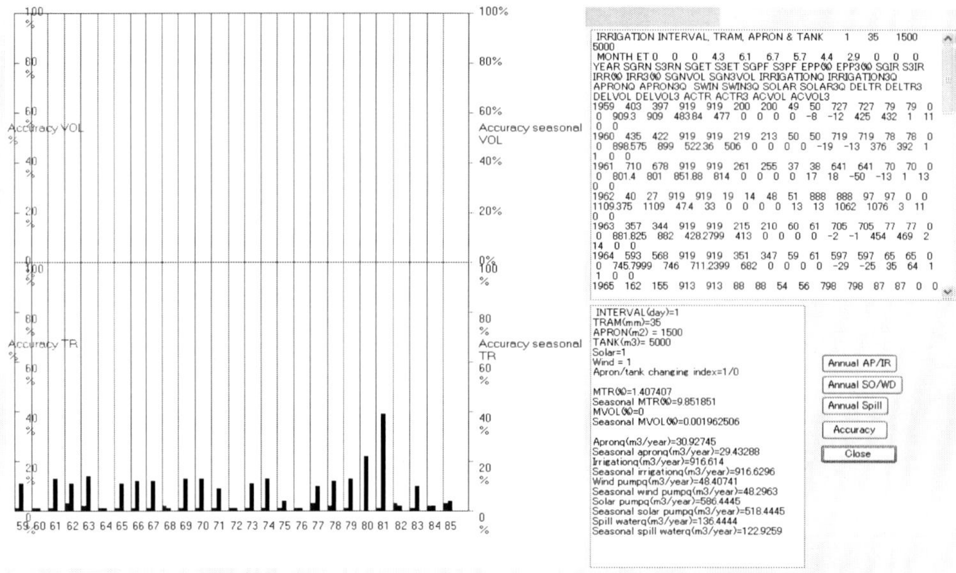

Figure 18. Annual variation error (100-accuracy) of TRac and VOLac during 1952-1982 in Mu Us.

2. MICRO-IRRIGATION SCHEDULING UNDER HARVESTED RAINWATER AND SALINE GROUNDWATER USING CLEAN ENERGY

This part aimed to develop a series of program packages of simulation models on scheduling saline micro-irrigation for small scale irrigation facilities supported by harvested rainwater and pumped saline groundwater using clean energy. The simulation is implemented based on the data from saline and non-saline fields in the following three areas: Mu Us Shamo desert in the Inner Mongolia Autonomous, China; Shavour Soil Institute Station in the Khuzestan Province, Iran; and Arid Land Research Center of Tottori University, Japan.

Initial conditions of simulation models are the main factors of saline irrigation schedules, such as salinity of groundwater (SCGW), reduction of ET (RET), leaching requirement (LR) and salt concentration ratio soil solution (ISCS), salt content ratio soil solution (SCW), crystal salt concentration ratio soil solution (ICSCS), water evaporation coefficient (WEV), and some meteorological data measured by previous fields works, including the factors described in the previous part.

2.1. Theory of the Simulation Model

The initial conditions of the simulation model are shown in the first class of the program, in which main parameters have been completed by filling specific table. The calculation is implemented, and then a suitable size of the catchment with fixed storage tank is estimated under the environment presented in Figure 1. The irrigation parameters were as follows:

- Irrigation interval was on daily basis for the micro-irrigation while on every 2-9 days for the sprinkler irrigation.
- The area of irrigated fields was 1,000 m^2. The catchment was built on an inclined land and the tank and irrigated fields on a flat land, so rainfall was efficiently harvested and irrigation water was applied to crops uniformly.

In addition, groundwater was supplemented by using wind and solar pumps where relatively low salt concentration water was pumped during sunny and windy days.

The following assumptions were made: the efficiency of harvested rain water and water application for irrigation was 80%, the storage tank with capacity of 1,000 m^3 was full at the beginning of the simulation, and ILR ranged between 0.1 and 0.9.

Since evaporation loss from the tank tends to be larger under strong windy or arid lands condition, a mechanism is required to minimize evaporation loss, such as using plastic cover or windbreak trees are planted to create a shady environment over tank surface. Assuming evaporation loss from water stored in the tank, a covering rate of water surface is given in the initial condition as an index number between 0 and 1.

2.2. Soil Salinization due to Irrigation

Daily soil moisture balance equation in the plant root zone is given as follows:

$$RE(j) = RE(j) + Rain(j) - D(j) - ET(j) + RE(j-1) \quad (12)$$

$$\Delta RE(j) = RE(j) - RE(j-1) = IR(j) + Rain(j) - D(j) - ET(j) \quad (13)$$

RE (j-1), RE (j): Ready available soil moisture content reserved on the (j-1) th or jth day (mm).

ΔRE (j): Difference in RE(j) between RE (j-1) and RE (j) (mm).

ET (j): Evapotranspiration on the jth day (mm).

$$D(j) = Dp(j) + Di(j) \quad (14)$$

D (j): Amount of drainage water on the jth day (mm).

Dp (j), Di (j): Amount of drainage water due to rain or irrigation on the jth day (mm).

Iw (j): Amount of irrigation water on the jth day (mm).

Rain (j): Amount of rainfall on the jth day (mm).

When SCS(0) = 0 and ΣSTSc (0) = 0, soil salt balance equation is shown in the plant root zone as follows:

$$\Delta STS(j) = STS(j) - STS(j-1) = Iw(j) * Ic(j) - D(j) * Dc(j) \quad (15)$$

$$STS(j) = SCS(j) * RE(j) \quad (16)$$

$$STS\ (j\text{-}1) = SCS(j\text{-}1) * RE(j\text{-}1) \tag{17}$$

$$SCS\ (j) = (Iw\ (j) * Ic(j) * D(j) * Dc\ (j)) / RE\ (j) + SCS\ (j\text{-}1)/(RE(j) * RE(j\text{-}1)) \tag{18}$$

Crystal salt is assumed to appear in the soil when SCS (j) is more than salt saturation concentration. As for NaCl, the SCS of NaCl is 266 when the SC of groundwater is 1,000 (mg/L). The crystal salt content ratio STSc (j) to the SC of groundwater is defined as follows:

$$STSc(j) = RE(j) * (SCS(j) - SCSst) \tag{19}$$

$$\Sigma STSc(j) = STSc(j) + \Sigma STSc(j\text{-}1) \tag{20}$$

ΔSTS (j): Difference in salt content between the (j-1) th and jth day (g/m^2).
SCS (j): Salt concentration (SC) ratio in soil solution by groundwater.
Dc (j): Salt concentration (SC) ratio in drainage water by groundwater.
Ic (j): Salt concentration (SC) ratio in irrigation water by groundwater.
STS (j): Salt content (ST) ratio in soil solution by groundwater.
SCSst: Saturated SC in soil solution by groundwater.
STSc (j): Crystal salt content ratio mixed with soil by groundwater at the jth day.
ΣSTSc (j): Total salt content ratio crystallized until the jth day by groundwater.
SCSFIRST and MSCSFIRST are the initial value of SCS (j) or STSc (j)

Since frequent irrigation intervals due to micro-irrigation keep soil moisture content high and for longer time, therefore the RE(j) tends to increase toward the TRAM, and the Dc(j) in equation(14) is assumed to be as in the following equation:

$$Dc\ (j) = SCS\ (j\text{-}1) \tag{21}$$

As a result, the next equation is reduced as:

$$Dc\ (j) = SCS\ (j\text{-}1) * TR\ (j\text{-}1) / D\ (j) \tag{22}$$

Also, since Dc is depending on rain or irrigation amount, Dpc (j) and Dic (j) are respectively assumed to be as follows:

$$Dpc\ (j) = STS\ (j\text{-}1) * Dp\ (j) / TR\ (j\text{-}1) \tag{23}$$

$$Dic\ (j) = STS\ (j\text{-}1) * Di\ (j) / TR\ (j\text{-}1) \tag{24}$$

Dpc (j): Salt concentration ratio of drainage water due to rain or groundwater.
Dic (j): Salt concentration ratio of drainage water due to irrigation or groundwater.

When SCS (0) > 0 or ΣSTSc(0) > 0 that is the initial conditions, salt balance in the equation (15) is transformed to the following equation:

$$STS\ (j) = Iw(j) * Ic(j) - D\ (j) * Dc\ (j) + SCS\ (0) * TRAM + \Sigma STSc(0) \tag{25}$$

2.3. Daily Evapotranspiration under Saline Field Conditions

Monthly mean evapotranspiration ETs (j) is shown under the simulated saline fields as follows:

$$ETs (j) = ET (m) * \exp (-SCS (j)) / Etf \quad (26)$$

ETs (j): Monthly mean daily evapotranspiration on the jth day under simulated saline field conditions.

ET (m): Monthly mean evapotranspiration used at the mth month (mm/day).

Etf: Reduction factor for evapotranspiration due to saline condition (ranging from 10 to 15).

TRAMs are assumed as follows:

$$TRAMs = Max.ETs (m) * Dmax \quad (27)$$

ETs (m): ET (m) under saline condition in the mth month (mm/day).

Max.ETs: Maximum of ETs (m) under saline condition in the mth month (mm/day).

Dmax: Maximum of irrigation interval under saline condition (day).

2.4. Scheduling Leaching Water Requirement

Leaching water requirement on the jth day (LR (j)) is shown as follows:

$$LR (j) = D (j) / IR (j) = Ic (j) / Dc (j) \quad (28)$$

$$IR (j) = Iw (j) / (1 - LR (j)) \quad (29)$$

I (j): Daily net irrigation water requirement at the jth day.

IR (j): Daily net irrigation water requirement containing the LR (j).

Dc (j): SCS (j-1) as described in eq. (18).

In order to decrease soil salinity by leaching, the calculation is repeated using LR starting from 0 until 0.9, with an increment of 0.1 until a suitable LR is obtained in which the minimum of SCS (j) is approximately equal to the double of SC of groundwater or irrigation water.

2.5. Water Balance and Salinity of the Irrigation Water in the Storage Tank

The volume ofwater stored in the tank (VOL (j)) on the jth-day could be reduced by using water balance equation. The input is rainwater or/and pumped water, and the output is both irrigation water and evaporation loss.

$$VOL (j) = VVOL + APRON (j) - EVAP(j) - IRVOL (j) \quad (30)$$

Volume of water stored in the tank plus water pumped using solar pump:

$$VOL(j) = VVOL + SOLAR(j) + APRON(j) - EVAP(j) - IRVOL(j) \quad (31)$$

Volume of water stored in the tank plus water pumped using windmill pump:

$$VOL(j) = VVOL + WINQ(j) + APRON(j) - EVAP(j) - IRVOL(j) \quad (32)$$

Volume of water in the tank plus water pumped using both solar and wind pump:

$$VOL(j) = VVOL + WINQ(j) + SOLAR(j) + APRON(j) - EVAP(j) - IRVOL(j) \quad (33)$$

$$APRON(j) = Rc * Rain(j) / 1000 * ANN \quad (34)$$

$$IRVOL(j) = IR(j) * AREA / (1000 * Irrc) \quad (35)$$

$$EVAP(j) = EVc * Ep(j) \quad (36)$$

The SC ratio of stored water (VOLc (j)) in the tank is reduced when the groundwater was added by solar and/or wind pump, assuming Ic equal to VOLc (j-1) as follows:

$$Ic(j) = VOLC(j - 1) \quad (37)$$

$$\left.\begin{array}{l} VOLc(j) = (VVOL * VOLc(j - 1) - Ic(j) * IRVOL) / VOL(j) \\ VOLc(j) = (VVOL * VOLc(j - 1) + SOLAR(j) - Ic(j) * IRVOL) / VOL(j) \\ VOLc(j) = (VVOL * VOLc(j - 1) + WINQ(j) - Ic(j) * IRVOL) / VOL(j) \\ VOLc(j) = (VVOL * VOLc(j - 1) + SOLAR(j) + WINQ(j) - Ic(j) * IRVOL) \\ VOL(j) \end{array}\right\} \quad (38)$$

WINQ (j): Volume of daily water pumped from groundwater by windmill pump.
SOLARQ (j): Volume of daily water pumped from groundwater by solar pump.
AREA: Irrigated area (m^2).
ANN: Apron size for rain harvesting (m^2).
VVOL: Volume of water stored on the (j - 1) th day (m^3).
VOLc (j), VOLc (j-1): Salt concentration (SC) ratio in the water stored in the tank by groundwater on the jth or (j-1) th day.
Rc: Runoff coefficient = 0.8
Irrc: Application efficiency for irrigation = 0.8
EVc: Water evaporation efficiency ranging from 0 to 1.0 when the surface area of the water table is 1000 m^2.
Ep(j): Daily evaporation water from class A pan.

2.6. Accuracy of the Simulation Model under Saline Irrigation

To evaluate the simulation models under saline irrigation, REac(y) (%) and VOLac (y) (%), were estimated and tested the final day of simulation period on the yth year. The accuracy is equal to (1-error). The annual RE(j) was integrated to the TRAM as follows:

$$REac(y) = 100 * (1- (\Delta RE + \sum PEF(j) + \sum Iw(j) - \sum ET(j) - \sum LEQ(j) / TRAM)) \quad (39)$$

When the groundwater was supplied by solar and wind pump, the accuracy of the annual integrated VOL(j) to the tank capacity VOLac (%) is reduced as follows:

$$VOLac(y) = 100 * (1-(\Delta VOL + \sum SOLRQ(j) + \sum WINQ(j) + \sum APRONQ(j) - \sum IwQ(j) - \sum NVOL(j) - \Sigma EVAP(j) / TNN)) \quad (40)$$

REac (y): Accuracy of annual integrated RE (j) to the TRAM that is retained in the plant root zone (%).

VOLac (y): Accuracy of annual integrated VOL (j) to the TNN that is stored in the tank (%).

ΔRE: Difference in RE (j) between the first and final day of irrigation period (mm).

$\sum$PEF (j): Total amount of effective rain during irrigation period in the yth year (mm).

$\sum$Iw (j): Total amount of irrigation water during irrigation period in the yth year (mm).

$\sum$ET (j): Total amount of evapotranspiration during irrigation period in the yth year (mm).

ΔVOL: Difference in VOL (j) between first and final day of the irrigation periods (m^3).

$\sum$SOLRQ (j): Total volume of pumped groundwater by solar pump during irrigation period in yth year (m^3).

$\sum$WINQ(j): Total volume of pumped groundwater by wind pump during irrigation period in yth year (m^3).

$\sum$APRONQ(j):Total volume of runoff water into tank from apron during irrigation period in yth year (m^3).

$\sum$IwQ(j): Total volume of irrigation water going to the field from tank during irrigation period in yth year (m^3).

$\sum$NVOL(j): Total volume of spilled water from tank during irrigation period in yth year (m^3).

ΣEVAP(j): Total volume of evaporation water from the tank in yth year (m^3).

As for salt balance, accuracy of the annual integrated salt content of soil solution to irrigation water RESac(y) (%) is reduced in yth year as in the following equation:

$$RESac(y) = 100 * (1- (\Delta REs + \Sigma(Iw (j) * Ic (j)) - \Sigma(D (j) * Dc (j))/ \Sigma(Iw (j) * Ic (j))) \quad (41)$$

Also, the accuracy of the annual integrated salt content of water stored in the tank VOLSac(y) (%) is reduced in yth year as follows:

$$VOLSac(y) = 100 * (1- (\Delta VOLs + \sum SOLRQ(j) + \sum WINQ(j) - \Sigma(IwQ(j) * Ic(j)) - \Sigma(NVOL(j) * VOLs(j)) / (\Sigma(VOL(j) * VOLs(j)) \quad (42)$$

ΔREs: Difference in annual total salt content ratio containing RE(j) /ground water between first and final day of irrigation period.

ΔVOLs: Difference in annual total salt content ratio containing VOL(j) /ground water between first and final day of irrigation period.

2.7. Model Application

2.7.1. Irrigation Water Requirement and Salt Content Applied by Irrigation Water

Daily variation in irrigation water requirement and salt content applied to the fields are estimated by the button of "rain/irrigation" and "added salt by irrigation". As for Shavour, daily salt content and annual integrated salt contents are shown in Figure 19. The calculations were done following the assumption of using solar and wind pumps, *Ii* of 9 days, TRAMs of 93 mm, max. ETs of 10.3 mm/day, catchment of 250 m^2, recharge tank of 2,000 m^2 and initial LR of 0.1.

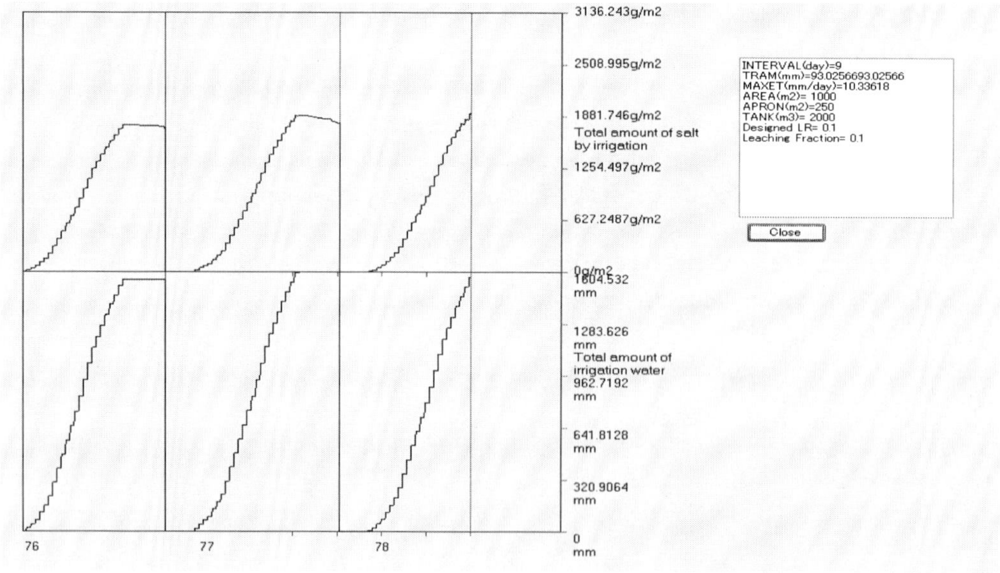

Figure 19. Daily variation in saltsa ccumulated by irrigation water for *Ii* of 9 days in Shavour.

2.7.2. Drainage Water Requirement and Salt Content Leached by Drainage Water

Daily variation in drainage water, salt content due to irrigation water and salt content leached out due to rainfall are estimated by buttons "Volume of leached water", "Leached salt by irrigation" and "Leached salt by rain". As for Shavour, the daily variation of salt content and annual integrated salt content are shown in Figures 20-21 using irrigation water or rainfall, respectively. Both cases used solar and wind pumps with following conditions; *Ii* of 9 days, TRAMs of 93 mm, Max. ETs of 10.3 mm/day, catchment of 250 m^2, recharge tank of 2,000 m^3, and initial LR of 0.1. The annual total salt contents were divided as 1.7-1.8 kg/m^2 for application into the field by irrigation water, and 0.2-0.3 kg/m^2 and 0.7-8 kg/m^2 for leaching out by water and rainfall, respectively. As a result, the rest of salt content was attributed to increase in root zone salinity was approximately 700 g/m^2.

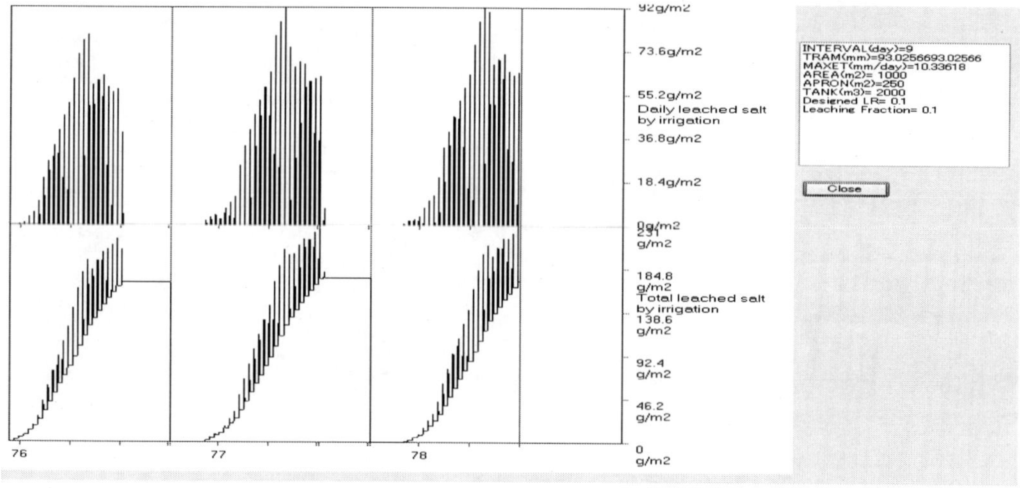

Figure 20. Daily variation in salts leached out by irrigation water for *Ii* of 9 days in Shavour.

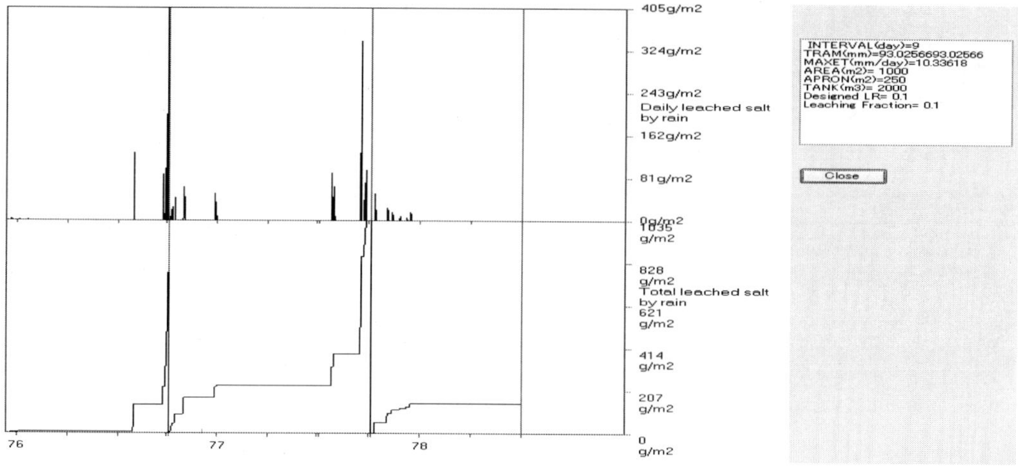

Figure 21. Daily variation in salts leached out by rainfall for *Ii* of 9days in Shavour.

2.7.3. Accuracy of Simulation Model under Saline Irrigation

The REac (y), VOLac (y) and VOLSac(y) showed high values with more than 98% for Shavour, Mu Us and Tottori. The simulation performed relatively good under all three stations as shown in Figures 22 and 23. However, The RESac(y) decreased to 68-98% in Shavour, to 88-89% in Mu Us and to 85-88% in Tottori. This happened because Dc (j) assumed in equation (21) was not enough for the SCS(j-1) and other factors such as rainfall patters. It is suggested that a functional relationship should be found between Dc (j) and SCS (j-1) by conducting leaching studies. However, the RESac was low in Shavour and that happened because of incomplete meteorological data such as from March to December in 1976 (71%) and from January to September in 1978 (25%).

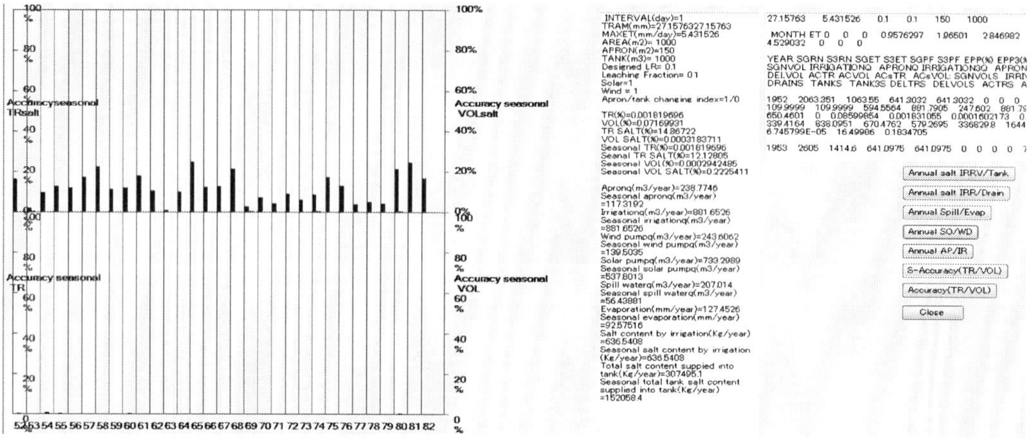

Figure 22. Annual variation error (100-accuracy) of TR and TRs during 1952-1982 in Tottori.

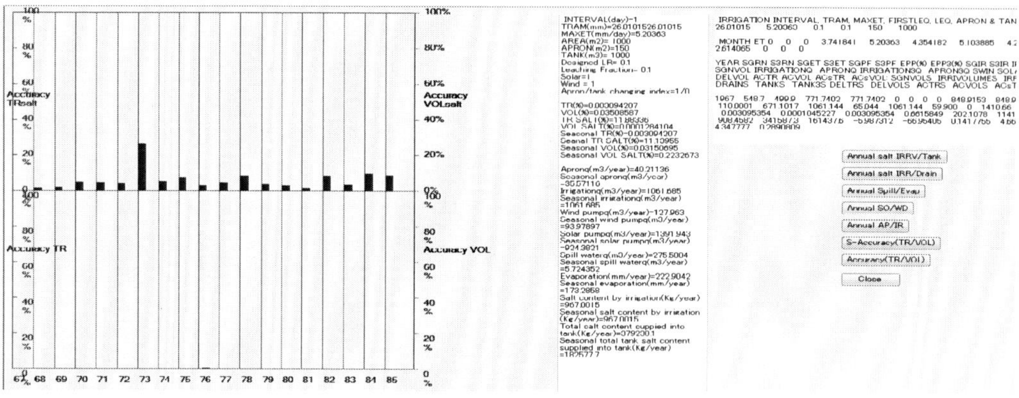

Figure 23. Annual variation error (100-accuracy) of TR and TRs during 1967-1985 in Mu Us Shamo.

2.8. Conclusion

Micro-irrigation has many advantages compared to other irrigation systems. It can provide and keep much water in the root zone for a longer time and reduce salt damage to soil properties and plant production. By using simulation model, the advantage of the micro-irrigation will be recognized and introduced as a guideline to design fresh and saline water irrigation schedules. However, if rain water can be harvested, this could improve irrigation water quality and support salt leaching. Integrating both techniques, harvesting rain water and scheduling micro-irrigation, will improve plant production and save groundwater.

Sustainable irrigation is promoted when clean energy is used for irrigation. This will result in saving water resources and energy, as well as scaling down the irrigation facilities. Since saline irrigation schedule is critical and mis-management of fresh and saline irrigations could lead to salt accumulation and plant death. Therefore, using good models for irrigation scheduling could save water, support plant production and maintain land sustainability.

REFERENCES

Amu-Mensah, F.K., T. Yamamoto and M. Inoue. 2001. Planning for Sustainable Agriculture in Ghana using Tank Irrigation, *Irrigation Engineering and Rural Planning,* No. 40, pp. 79~92.

Japan. JIID. 1990. Engineering Manual for Irrigation and Drainage, Drip Irrigation Planning Guide, *Japanese Institute of Irrigation and Drainage* (JIID), pp.19~36.

USDA, 1954. *Diagnosis and Improvement of Saline and Alakli Soils.* US Government Printing Office, Washington, DC., USA.

Yamamoto, T., F. Amu-Mensah, H. Fujimaki and J. Utsunomiya. 2003. Characteristics of meteorology and sustainable irrigation schedules in the Saherian Region of Africa, Jour. *Japanese Society Irrigation Drainage Reclamation Engineering,* 70(11), pp.1009-1012(in Japanese).

Yamamoto, T., Hossein Dehghani Sanij and Menachem Agassi. 2004. Rehabilitation techniques for degraded, irrigated soil in arid lands, *Journal of International Society for Southeast Asian Agricultural Sciences,* 10(1),pp.41-50.

Yamamoto, T., M. Naruoka, S. Ito, Z. Yang, and J. Zhang.1993. Irrigation Schedules and Conservation Management for a Pilot Farm in the Mu Us Shamo Desert: Control of Desertification and Development of Agriculture in Arid Land Areas in China. *Irrigation Engineering and Rural Planning,* No. 25, pp. 4~15.

In: Irrigation Management, Technologies …
Editor: M. Hossain Ali
ISBN: 978-1-62417-862-7

Chapter 10

Multiple Use of Irrigation Reservoir Water Resource in the Upper East Region of Ghana

*Makarius V. Mdemu**
Department of Regional Development Planning,
School of Urban and Regional Planning, Ardhi University, Tanzania

Abstract

Multiple use of artificial water reservoirs is increasingly becoming an important aspect in planning and development of water resources management in semi-arid areas of sub-Saharan Africa. This paper documents and analyzes the dry season multiple use of medium and small reservoir water resources that are widely distributed throughout the Upper East Region of Ghana. Temporal reservoir water uses and existing reservoir water resources management arrangements are elaborated. In the description and analysis, attention is given both to medium and small reservoirs. Although the later are many in number and spatially distributed, the former are few and site specific. Finally, the constraints and potentials for improving reservoir water resources management are discussed.

Keywords: Multiple water use, reservoir water, water management, Sub-Saharan Africa

1. Introduction

There is a growing interest in closer examining water sources that supply multiple uses as a means of improving water resources planning and management in the semi-arid areas of Sub-Saharan Africa. However, the nature and the extent of multiple water uses is typically

* Corresponding author: Makarius V. Mdemu, Department of Regional Development Planning, School of Urban and Regional Planning, Ardhi University, P.O. BOX 35176, Dar es Salaam, Tanzania, Tel: (mobile) +255 765466058, Email: mmdemu@uni-bonn.de, mak_mdemu@yahoo.co.uk.

neither documented nor understood. Before the advent of piped water supply, people, livestock and the physical environment commonly shared various sources of water, such as natural rivers, lakes and reservoirs or dams, for drinking, cooking, sanitation, livestock watering, and construction purposes.

This long-standing practice of sharing sources of water between uses is still common in many places in the developing world. Population growth that stretches per-capita available water resources in water-scarce countries has necessitated attempts to achieve a balance among water uses, especially those that would have been otherwise neglected because of a perception that certain users or uses are important while others are less so. Multiple water-use systems are considered to be more efficient when the value of water from other uses that were not considered during planning and design are subsequently incorporated in current planning and performance assessments. This is particularly true in the irrigation sector, which usually supplies many other uses (Bakker and Matsuno, 2001; Gowing *et al.*, 2004; Boelee and Laamrani, 2003; Meinzen-Dick; Van der Hoek, 2001). In some irrigation systems, non-crop water use exceeding 50% of the irrigation supply has been reported (Renault and Montginoul, 2003). In Northern Morocco, Boelee and Laamrani (2003) found water being directed from the main irrigation canal into underground house storage tanks via small canals, or by pump to provide water for various purposes such as drinking and cooking, watering livestock, laundry and brick making.

As a result of observed multiple water uses from irrigation systems, the common perception that irrigation systems supply water only to field crops is changing (Gowing *et al.*, 2004; Meinzen-dick and van der Hoek, 2001; Bakker and Matsuno, 2001). Failure to take into account the multiple use aspects of irrigation systems has two major consequences: first, water stored and controlled within irrigation systems and the entire return on investment in irrigation and related infrastructure has been undervalued (Bhattarai *et al.*, 2003; Bakker and Matsuno, 2001) and second, because of such undervaluation, when water scarcity increases in some countries, water is re-allocated from irrigation to more beneficial water sectors such as municipal, industrial or environmental maintenance purposes (Meinzen-Dick and Van der Hoek, 2001).

Undervaluation of irrigation water can be remedied by assessing the full range of benefits from multiple use systems, and by internalizing multiple water use aspects during planning of irrigation systems. However, management complications resulting from positive externalities associated with multiple water use cannot be ignored (Renault and Montiginoul, 2003). The Multiple Use System (MUS) project, a CGIAR-Challenge Program for Water and Food (Penning de Vries *et al.*, 2006) was designed as a response to failures in planning and management of multiple use water resources systems. Apart from the multiple uses of water practiced in large irrigation systems, which are widely documented (Meinzen-Dick and Bakker, 2001, Li *et al.*, 2005, Going *et al.*, 2004, Renwick, 2001), medium and small water reservoirs also provide multiple services. These systems, spatially distributed in the semi-arid regions of Sub-Saharan Africa (SSA), provide water for multiple uses in ways that balance the livelihood interests of competing users, including irrigation, livestock watering, fisheries, domestic uses and other water-related enterprises such as brick making. The medium and small reservoirs, unlike other types of multiple use systems, have a unique feature: a storage facility for capturing runoff during the rainy season for use during the dry season, thus acting as closed surface water systems with little or no surface outflows downstream. This type of multiple water use system was yet to be adequately described in the existing literature. This study, therefore, documents the multiple water uses with reference to medium and small reservoirs in the UER.

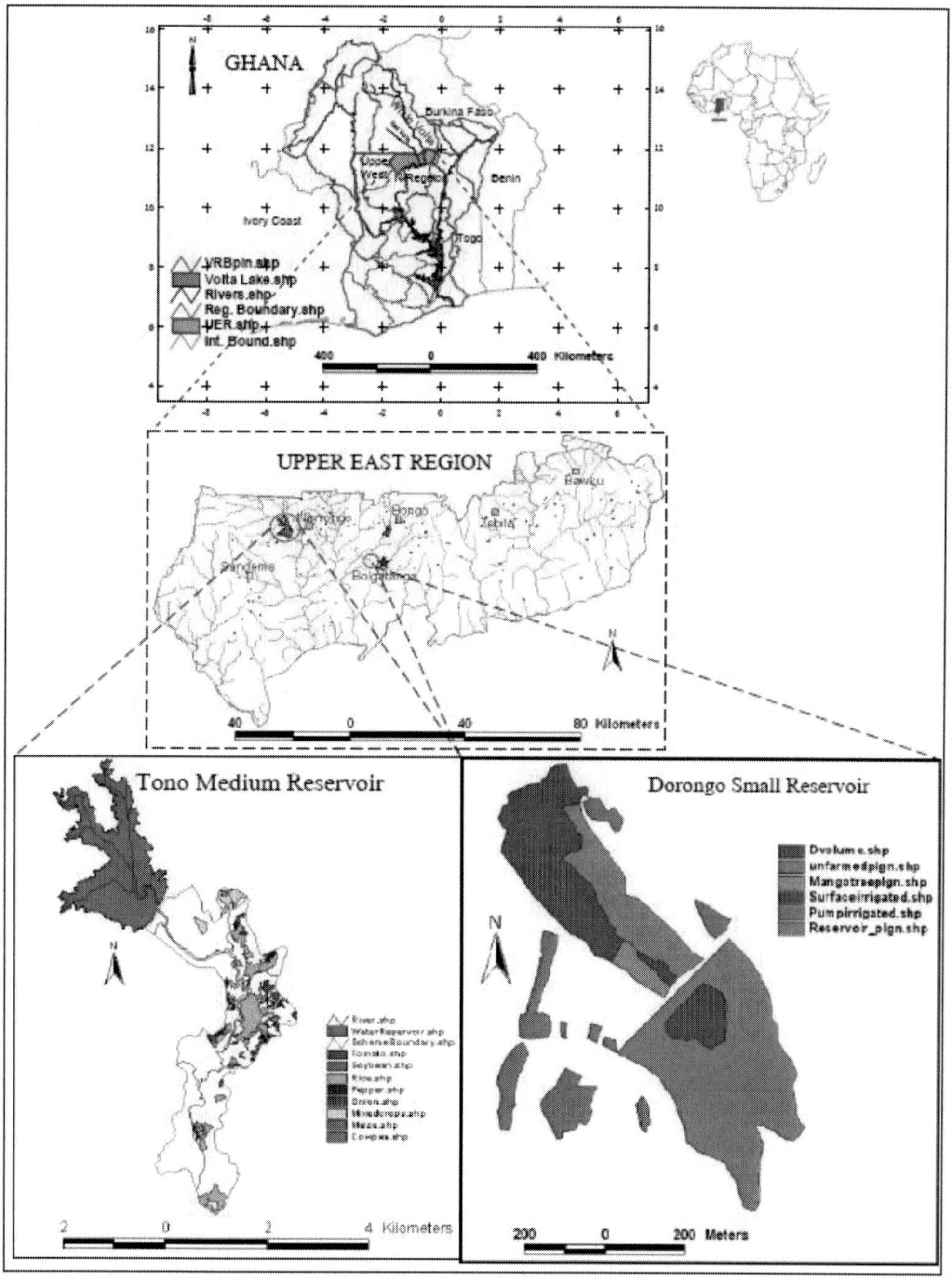

Figure 1. Location of the study area in the Upper East Region, Ghana.

2. Description of the Study Area and Methods

2.1. Geographical Location

This study was conducted in the Upper East Region (UER), which is located in the Northeast corner of Ghana between latitudes 10°30' to 11°15' North and longitudes 0° to 1°30' West (Figure 1). The land surface area of the study region covers 8860 km^2, which is about 4% of the country (238534 km^2).

Administratively, the region is divided into 8 districts with Bolgatanga as its regional capital. Tono[1] and Dorongo, a medium (1860 ha) and small (15 ha) water reservoirs were identified for the analysis of multiple water uses in the study area. Tono reservoir/Irrigation Scheme is located at Navrongo in Kassena-Nankana District. The reservoir and irrigation scheme were constructed between 1975 and 1985. The irrigation system of the scheme consists of a water storage dam with a maximum surface area of 1860 ha and storage volume of 93 million cubic meters. The gross project area covers about 3860 ha, of which 65% (i.e., 2490 ha) has been developed for irrigation. The catchment area, which collects the runoff for the storage dam, is 650 km^2. Canal and road networks with a total length of about 42 km and 120 km, respectively, are important parts of the irrigation scheme infrastructure. The scheme is managed by the Irrigation Company of the Upper Region (ICOUR). ICOUR is a governmental parastatal entity which manages large irrigation schemes in the region, including Vea scheme located at Bolgatanga. The Dorongo reservoir is located about 3 km along the Bolgatanga-Sherigu road, West of the Bolgatanga regional capital.

The small reservoir and irrigation scheme were constructed in 1962 and rehabilitated between June, 2002 and August 2003 under the Land Conservation and Smallholder Rehabilitation Project (LACOSREP) Phase II Project. The dam has an estimated storage capacity of about 0.24 million cubic meters and a maximum surface area of 15 ha with a maximum depth of approximately 6 m at maximum flooding. The catchment area is about 112 ha and the total irrigable area of about 10 ha. The reservoir has a total infrastructural length of about 3.81 km for the dam wall and lined diversion canals. The reservoir is managed by a Water Users' Association (WUA) with about 150 registered members.

2.2. Climate

The climate of the region is influenced by the movement of the harmattan and monsoon winds, which control the climate of the West African sub-region. Temperatures in the region are consistently high, with the hottest months being March or April (40-45°C) and the coolest month being August (26°C). The mean annual temperatures are around 28-29°C, while the absolute minimum temperatures (15-18°C) occur in December. The UER experiences a mono-modal rainfall pattern with the rainy season starting in April-May and lasting in September or the beginning of October. In this region, rainfall is erratic and spatially variable. The average annual rainfall ranges between 700 and 1010 mm per year with the peak rainfall occurring in late August (Figure 2). More than 60% of the annual rain falls between July and

[1] Although Tono and Vea reservoirs are referred to as "large irrigation schemes", in this chapter, they are called "medium schemes" in order to make distinction from common large irrigated schemes elsewhere worldwide.

September, with torrential rains generating high runoff due to inability of the soils to absorb high intensity rains (Kranjac-Berisavljevic *et al.*, 1998). Normally, precipitation in the region exceeds potential evapotranspiration only during the rainy (growing) season, and vice versa is true for the dry season. The annual potential evapotranspiration is estimated to be roughly as much as double the annual precipitation. This translates into a necessity for utilization of water storage reservoirs to ensure availability of water for various uses in the dry season.

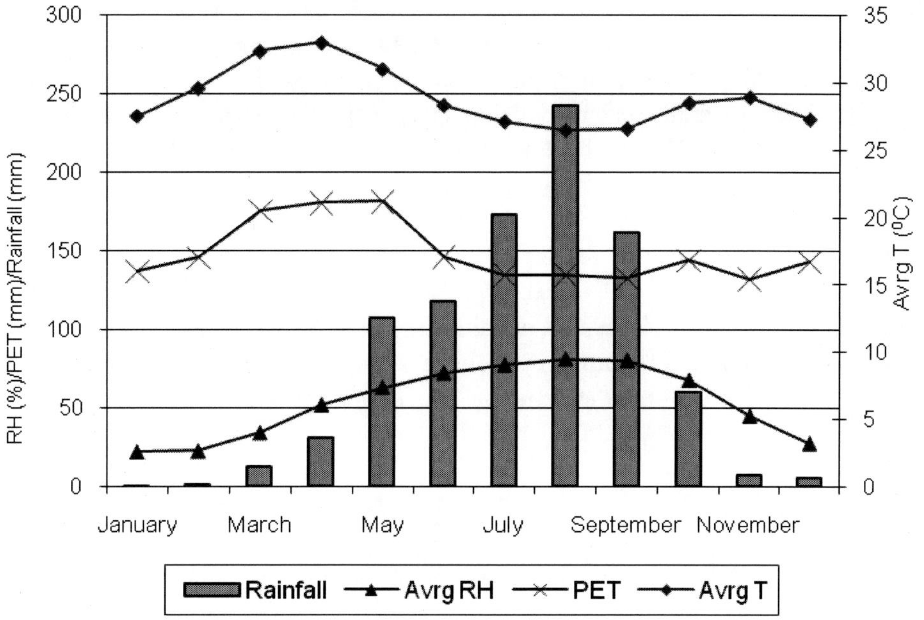

Figure 2. Long-term (1980-2005) monthly average rainfall, relative humidity (RH); potential evapotranspiration (PET) and average temperature (T) for Bolgatanga weather stations (Data source: Ghana Meteorological Service).

2.3. Geology, Soils, Relief and Drainage

Generally, soils developed over granites cover large areas of the UER. The majority of soils in the UER are infertile, except soils occurring in seasonally flooded areas. This is typical of savanna zones, which have low accumulation of organic matter in the surface horizons owing to the high temperatures that cause rapid decomposition rates. The annual burning of the vegetation cover throughout the area also reduces the amount of organic matter in the soils (Boateng and Ayamga, 1992). Large areas of the UER are gently undulating with broad, poorly drained valleys and extensive flood plains adjacent to the Volta River (Kranjac-Berisavljevic *et al.*, 1998). The UER is mainly drained by the White Volta River (WVR), with nearly all small and medium tributaries from various subcatchments in the Northern segment of the region draining Southward into the WVR. Most of the subcatchments in the UER have developed inland valleys of different sizes and shapes. Small and medium-sized reservoirs have been constructed (Liebe, 2002) in these inland valleys to supply water for crop irrigation, livestock and domestic, fishery and other uses during the dry season.

2.4. Socioeconomic Conditions

The UER is among the three poorest regions in Ghana. Agriculture employs above 80% of the population, making this sector key to socioeconomic growth for the region. To improve the difficult natural conditions in the Interior Savanna Zone (ISZ) of Ghana, the Ghanaian government, with assistance from foreign donors, developed ambitious irrigation schemes from the mid-1960s to the 1980s. It was apparently the belief of policy makers that such irrigation developments would replace indigenous soil and water conservation methods which, it was argued, could no longer cope with the pressures from an increasing population, environmental degradation and changes in climatic conditions (Kranjac-Berisavljevic *et al.*, 1998). The majority of these schemes, including Ghana's largest irrigation schemes (Tono and Vea), are located in the UER where population density is high, i.e., about 118 persons per km^2, compared with national population density of 77 persons per km^2 (Bacho and Bonye, 2006), land is scarce and the incidence and depth of poverty is high (Gyasi, 2005). The irrigation schemes are vital for increasing food security and rural income by providing water for dry season farming, livestock and fishery, which are the main sources of cash income for rural households in the northern regions and the UER in particular (Gyasi, 2005). Food crops and high-value crops such as rice, onions, tomatoes, pepper, and traditional leaf vegetables are produced from the irrigation schemes in the area during the dry season.

2.5. The Study Methods

Both secondary and primary data were collected to facilitate documentation of water reservoir multiple uses. Secondary data were collected from the existing literature in the Upper East region, Ministry of Food and Agriculture regional office, and the Glowa Volta databases. Field visits and physical observations, transect walks and discussion with individual farmers and detailed monitoring of Tono and Dorongo reservoirs were used to collect primary data on reservoir water use practices (irrigation, livestock, fishery, other water uses), cropping patterns and farming practices.

3. Reservoir Water Resources on UER

Water reservoirs in the UER are spatially distributed according to topographic controls within the region. More reservoirs are located in the Eastern half of the region (Figure 3) and fewer in the Western half. The Eastern half, which mainly lies above the regional mean elevation, is characterized by drainage boundaries that fold back and forth (Liebe, *et al.* 2005) providing favorable conditions for reservoir construction. Currently, there are about 156 reservoirs in the UER with an estimated total surface area of 3329 ha. Seventy percent (70%) of this area is represented by Tono (1894 ha) and Vea (435 ha). Small reservoirs, managed primarily by WUA and ranging from 1 to 35ha in size, contribute the remaining 30%. At full storage capacities, 293 million cubic meters of water can be captured during the rainy season (Liebe *et al.*, 2005; ICOUR, 1995).

However, the reservoir storage volume decreases gradually due to abstraction, evaporation and other outflow components from maximum storage at the end of the rainy season to minimum or zero storage between February and May, depending on the size of and demands for reservoir water. Reservoirs in the UER were developed mainly to provide water for irrigation in order to increase food crop production, and to supply water for human and livestock needs.

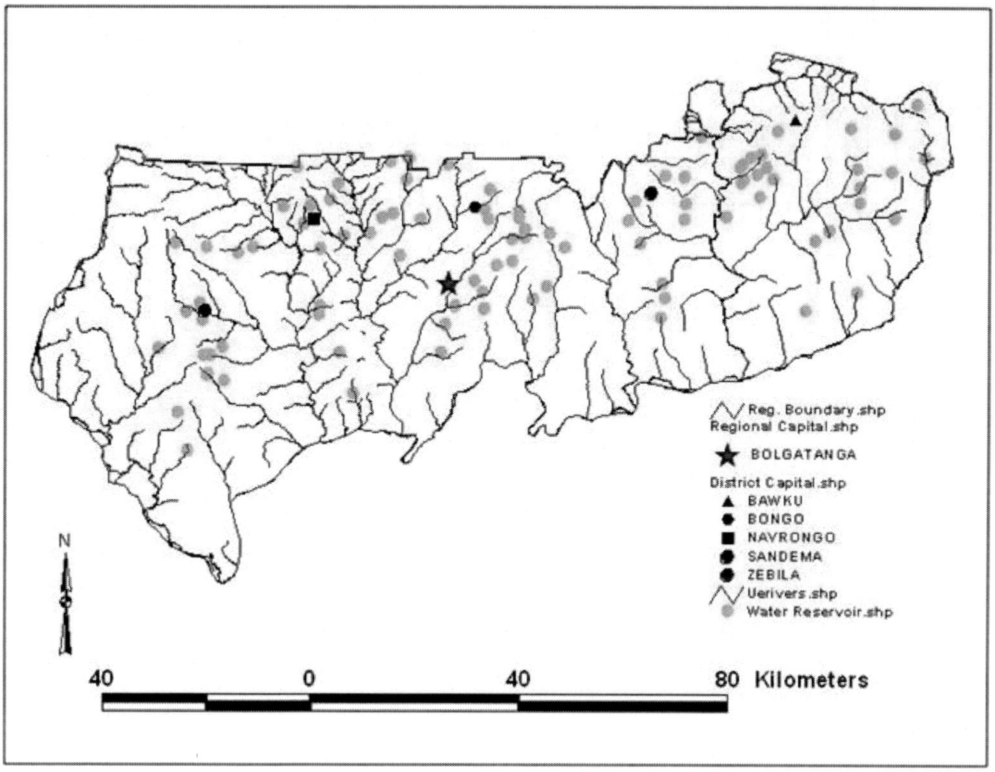

Figure 3. Distribution of small reservoirs with respect to Upper East Region topography (Data source: Glowa Volta Database).

Increased coverage of potable water supply via portable pipe or pumped boreholes in many rural areas of the UER has created the opportunity for fishery water use within the reservoirs, thus designating irrigation, livestock and fishery as the three important uses of reservoir water resources in the region.

3.1. Irrigation Water Use

Water use for agriculture constitutes a large percentage of overall reservoir water use in both medium and small reservoirs. The water in medium reservoirs is used for irrigating rice, soybean, tomatoes, onions, pepper and other leaf vegetables such as cucumber, garden eggs, okra, and kenaf (Figure 4).

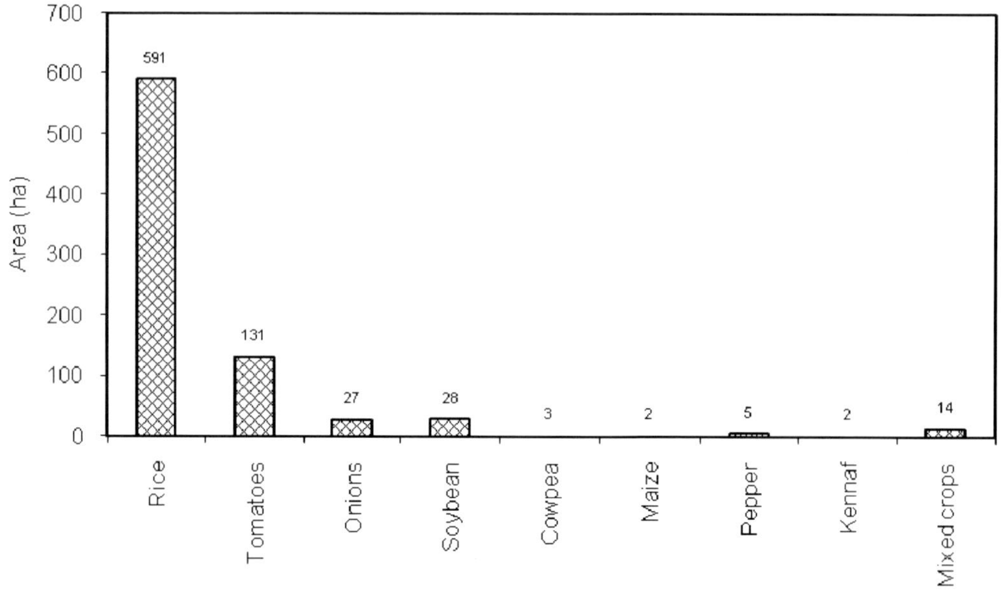

Figure 4. Dry season cropped area at Tono irrigation project (2005/06) (Data source: Field survey).

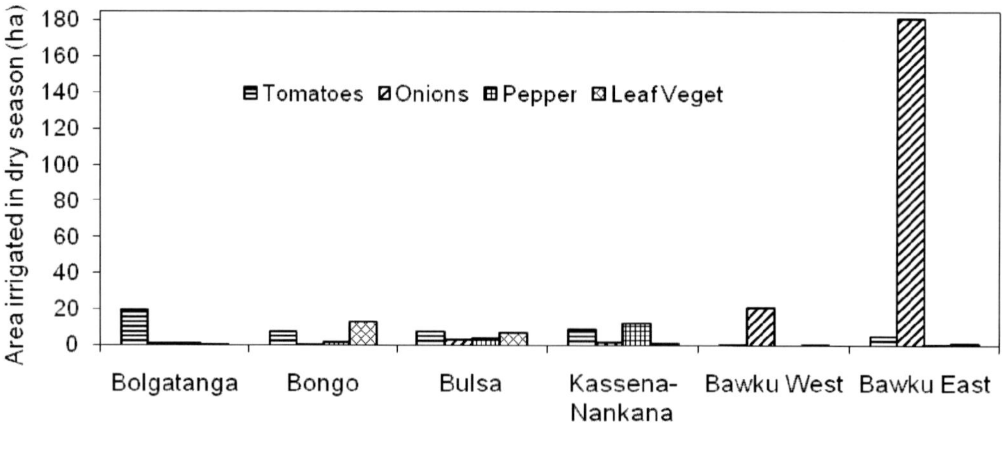

Figure 5. Average dry season crop acreage for small reservoirs in the UER (2000/01-2002/03) (Data source: MOFA Bolgatanga, 2005).

Other crops, particularly in Tono, include perennial crops such as mango trees which are sparsely distributed in marginal areas within the scheme. Tree crops other than fruit production are important for providing shade to farmers during the period of high temperatures from February to April. Non-traditional crops such as cocoyam and sugarcane are also found in marginal areas. In the small reservoirs, four cropping patterns are common (Figure 5): 1) tomato is the major crop and onions, pepper, and leaf vegetables the minor crops at Bolgatanga; 2) at Bongo-Builsa the major crops are tomatoes and leaf vegetables while onions and pepper are minor; 3) tomatoes and pepper are the major crops, while onion and leaf vegetables are minor at Kassena-Nankana; and 4) at Bawku East-West, onion is the major crop, while the remaining crops are minor.

However, due to an increase in diseases affecting tomatoes in Bolgatanga and Kassena-Nankana, and increased emphasis by the Ministry of Food and Agriculture (MOFA) on shifting cropping patterns, a stronger focus on onions was more promising given that environmental conditions in the areas where tomato is grown are similar to those found at Bawku East-West, where onions are extensively produced. The above cropping patterns reflect, among other factors, the different mechanisms used to supply water during irrigating (Table 1).

Table 1. Irrigation water use practices in medium and small reservoirs

Medium reservoirs	Small reservoirs
Plots are supplied with irrigation water from field canals (laterals).	Plots are supplied with irrigation water from laterals or sub-laterals.
Furrows between the ridges for crops like tomatoes, soy bean and cowpea are flooded	Furrows between the ridges for crops like tomatoes flooded.
Small beds are used for onions, water is ponded in the furrows separating the beds and bowls are used to fetch water to sprinkle on top of the beds. Labor intensive but more water efficient than the former, which leads to surface water losses.	Small beds are used for onions, water is ponded in the furrows separating beds and bowls, which are used to fetch water for sprinkling on top of the bed.
Similar practices apply to leaf vegetables such as cucumber, kenaf when they are grown on small beds and are not intercropped.	Similar practices apply to leaf vegetables such as cucumber, kenaf when they are grown on small beds and are not intercropped.
Pump irrigation is practiced along the main drains by pumping drain water to irrigate the adjacent areas.	Pump irrigation is practiced by pumping water directly from the dam to irrigate plots outside the garden area (designed irrigation area).
Continuous or intermittent flooding is applied on lowland rice fields.	No rice is grown during the dry season.

3.2. Livestock Water Use

Livestock watering in medium reservoirs is done from the main reservoir, primary canals and secondary canals. Livestock watering in the secondary canals is practiced only when there are no cropped plots adjacent to the canals, as livestock, particularly cattle, goats, sheep and donkey, are grazed on fallow plots. Whereas the impacts of livestock watering such as physical damage and water pollution due to livestock droppings, tramping and urine are not felt on the main dam because of its large size. Damage can be seen on primary and secondary canals caused by livestock struggling to access water directly from the canals. Livestock are also watered from farm and main drains in medium reservoirs. In small reservoirs, livestock watering is confined to the main reservoir, and few cases of watering from irrigation canals was observed because the water flows in the laterals or canals during irrigation events only. However, few incidences of livestock watering in canals and main drains were also observed in some of the small reservoirs that have relatively longer-duration irrigation events (more than 3 hrs). Thorough livestock watering in irrigation canals, irrigation infrastructure can be destroyed and crops damaged, resulting into negative consequences for the general socioeconomic condition of the community. Table 2 shows the differences in terms of water access points and general perception of water quality between medium and small reservoirs.

Table 2. Livestock watering practices in medium and small reservoirs in UER, Ghana

Medium reservoirs	Small reservoirs
Livestock from the communities around the reservoir get access to water directly in the reservoir.	Livestock access water directly from the reservoir.
Livestock from communities along the main canals and drainage systems get access to water from the canals and drains. Continuous stepping of livestock into the canals when accessing drinking water damages the canals.	On few occasions, livestock can access water from canals. Water is available in canals during irrigation events only (1-5 hrs)
Water quality due to livestock droppings is not a serious problem because of large volumes of water.	As the water volume decreases in the reservoir, the water quality becomes constrained for uses such as bathing and cloth washing because of turbidity due to livestock trampling and droppings.

3.3. Fishery Water Use

Although major fishery activities in medium reservoirs are carried out within the main reservoir (Table 3), considerable activities take place within the primary canals, secondary water storages along primary canals, and in the main river stream that drains the entire irrigation scheme area. While fishing in the main reservoir involves the use of major fishing gear, including canoes and fishing nets, simple fishing gears such as trap nets and fish hooks are commonly employed in the canals and secondary water storages in the project area. During development of the Tono and Vea irrigation reservoirs, a fish aquaculture component was incorporated in the form of fish ponds of 4.8 ha and 3.2 ha, respectively. However, the ponds collapsed because of management problems, although plans were underway to revive fish farming, and some of the fish ponds at Tono had been leased to a private investor.

Table 3. Fishing practices in medium and small reservoirs, UER, Ghana

Medium reservoirs	Small reservoirs
Fishing done in main reservoir; primary canals; secondary recharge storages in canals; drainage streams; Tono River main drain.	Fishing done in main reservoir; diversion boxes after irrigation events.
Fishing canoes and nets are the main fishing gear in the main reservoir.	Set and wait nets and fish hooks are the main fishing gears in the main reservoirs.
Set and wait nets, fish traps and fish hooks are common in the main canals and the main drains.	

Table 4. Other water use in medium and small reservoirs

	Medium reservoirs	Small reservoirs
Domestic use	Communities around the water bodies depend on reservoir and canal water for cooking purposes. Washing of cloth and bathing are carried out both in the main reservoir and irrigation canals. In few occasions, buckets are used to fetch water and wash at short distances from the water body. Quite often, drainage from washing, flow back into the main water body.	Boreholes are mainly used to supply water for cooking purposes Washing of cloth and bathing are done mainly from the main reservoir. Sometime buckets are used to fetch water and wash a short distance from the main reservoir. Quite often, drainage from washing flow back into the main water body.
House construction enterprises	Water is collected from the dam or main canals using pull/push carts or carried on head to construction sites Bricks are sometimes manufactured near the main canals and later carried to the desired destination when dried.	Water is collected from the dam using pull/push carts or carried on head to construction sites. Deposited sediments in the dam are sometimes used to manufacture bricks when water in the dam is almost completely drawn. desilting mechanism serious scouring can increase percolation losses
Environment	The water body in the main reservoir, canals, drainage systems and recharge storages provides favorable environment for water body ecosystems including fish, crocodiles, ducks and water birds.	The water body in the main reservoirs creates favorable habitat for water body ecosystems, including fish, crocodiles, and water birds

In most small reservoirs, fishing activities are confined within the reservoirs themselves, with a small number of fish being trapped in the release-valve boxes, which are normally checked after the valve is closed at the end of each irrigation event. Within the reservoir, fishermen use trap nets and fish hooks as their main fishing gear. Very few fishermen use fishing canoes.

3.4. Other Water Use

In medium reservoirs, domestic water use (cooking, drinking, sanitation, and house construction) is very common for communities living along the secondary canals and surrounding the reservoirs.

People from distant communities, whose farming activities take place within the project, use the canal water for cloth washing and body sanitation. Bathing and cloth washing are normally done in flowing water within the canal or along the edges of canals. Domestic water use in the communities through which the primary canal passes, particularly in Tono, was one of the consideration during the design of the canals, and steps were provided at some locations along the primary canal so that water can be accessed even at low flow levels. For house construction purposes, water containers are commonly used to fetch water from the canal and carried to the construction site, either on the head or using pull-carts with donkeys. In small reservoirs, a similar typology of domestic water use (Table 4) is observed, with the only difference being that water is drawn from the reservoir or bathing and cloth washing are carried out in the water body behind the dam.

4. Land Tenure in the UER

The majority of lands in the UER are under the customary land tenure system. The land is entrusted to the '*tendana*', literally meaning the owner of the land (Kasanga, 1995) who is the custodian. The '*tendana*' is normally the patrilineal descendant of the first family to settle in the area (Gyasi, 2005 and Kasanga, 1995). Tendamba (plural of 'tendana') are spiritual caretakers of land resources and grant usufruct customary rights to community members for free, although ownership rights continue to be vested in the community preventing the granted land from being disposed of by the allocated families (Gyasi, 2005). However, each family's land is secure and controlled by families as long it is used for the purpose it was allocated for. Although a similar land tenure system is applied for irrigated lands, two types of arrangements for land ownership exist: a secure tenure wherein users of irrigation facilities have total control over the command area, and seasonal tenure-ship, common in most small reservoir schemes managed by WUA, whereby land is returned to the original landowning families in the rainy season (Gyasi, 2005). In medium reservoirs, land is owned by the agency managing the schemes, although communities are involved in allocating plots to their farmers from the blocks of farms allocated to the community by the project management.

5. Management of Reservoir Water Resources

The Ghanaian Water Resources Commission (WRC) Act (1996) specifies that ownership and control of all water resources are vested in the President on behalf of the people, and designates the WRC as the body with overall responsibility for water resources management in Ghana (Water Front, 2004). The WRC plans and regulates water resources development in co-operation with local government and planning authorities. Most of the important government agencies involved in water resources management and development, such as Ghana Irrigation Development Authority (GIDA), are represented on the Commission. Other institutions represented on the commission include: Ghana Water Company Limited, which is responsible for portable water, the Hydrological Services Department, Volta River Authority, Water Research Institute, Meteorological Services Department, Environmental Protection Agency, Forestry Commission, and Minerals Commission (Water Front, 2004). The WRC is responsible for the initiation, control and coordination activities connected with development and utilization of water resources and the granting of water rights. Granting of water rights is seen as the main instrument to regulate water resources use (Water Front, 2004).

5.1. Medium Reservoirs

The management of medium irrigation reservoirs in the UER, i.e. Tono and Vea, is the responsibility of ICOUR, a semi-autonomous government agency that provides services to small-holder farmers within the two projects. Participating communities or project villages under ICOUR are organized into irrigators' associations or village committees (VCs) whose functions include land allocation at the village or community level. The village committees elect representatives who act as links between ICOUR management and VCs. The VCs are represented at various ICOUR meetings relating to decisions on operation and management of the schemes, and irrigation levy decisions. An ICOUR consultative meeting is held to reach a consensus on, among other things, the irrigation levy at the start of the dry season for different crops (Figure 6). This levy is charged on a per-hectare basis, and is crop specific. For the VCs to be allocated land, they have to raise and deposit their contributions at the ICOUR project office, and in turn the project manager allocates land consistent with the deposited contributions. The VCs then distribute the land to community members on the basis of each farmer's contribution. The remaining project area (not allocated to VCs) is contracted to anyone from the region (contract farmers) willing to farm in the project upon payment of a water levy consistent with that paid by VCs.

Although it is believed that participation of VCs enables ICOUR to deal collectively with the farmers efficiently, and perform its managerial activities smoothly (Gyasi, 2005), nevertheless there are problems associated with farmers' participation in the project areas, which negatively impacts on reservoir water resources management. For example, allocation of water at tertiary canals was ideally supposed to be performed by VCs. This arrangement proved to be difficult due to violation of water allocation timetables. Lack of project irrigation officers makes proper water allocation on farm laterals nearly impossible, so that during the irrigation season, the gates of farm laterals are often opened even when there are no farmers using the water in some of the laterals.

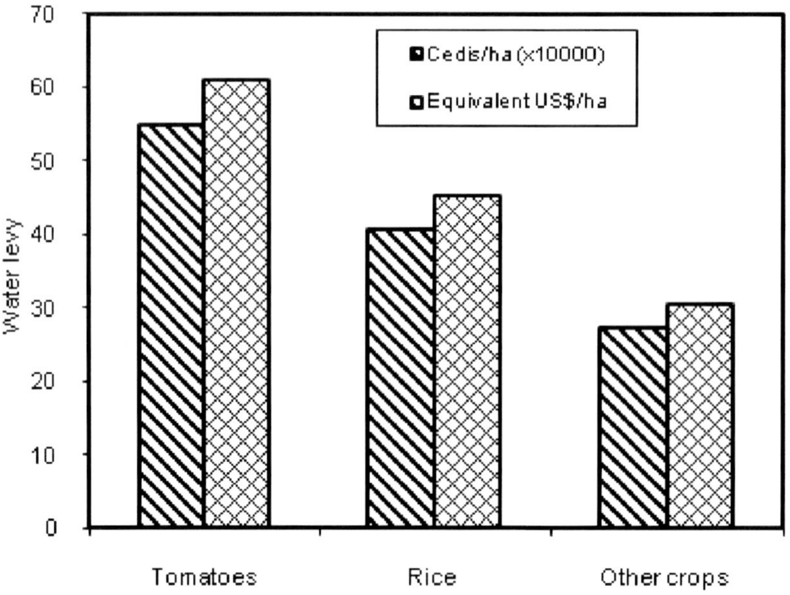

Figure 6. Irrigation levy for 2005/06 dry season at Tono (Source: ICOUR consultative meeting, October 2005, Exchange rate, 1 US$ = 8960 cedis as of Nov 2005, Barclays Bank, Ghana).

As a result, some farmers have been able to irrigate without paying water levies as required by the project management. This not only discourages farmers who abide by the project tenancy agreement, but encourages inefficient use of water resources. Laxity in water supply allocation that does not reflect crop irrigation water requirements have encouraged many farmers to farm on previously uncultivated lands, commonly known as waste lands, which were not fully developed during project development, without paying water levies. These so-called waste lands have expanded in the area in a manner comparable with obvious increases in crop water requirements.

5.2. Small Reservoirs

The Land Conservation and Small Reservoir Rehabilitation Project (LACOSREP), during its first and second phase of operation in the UER, included elements of the Water Users' Association (WUA) as an important strategy for sustainability of rehabilitated small reservoirs upon project completion and the handing over of rehabilitated reservoirs to beneficiary communities.

The formation or prior existence of a functional WUA was, in fact, a prerequisite for a given community to have its dam rehabilitated. During the period of this study, nearly all rehabilitated small reservoirs were being managed by WUAs. A typical WUA consists of a chairman, secretary and a treasurer. The WUA is an umbrella organization that encompasses affiliated groups in the beneficiary community that have stakes in the dam infrastructure and services, and which organize economic interests (Gyasi, 2005). Irrigators, livestock keepers and fishermen are the main economic interest groups, which organize themselves into associations whose representatives constitute the WUA's executive, referred to as Damsite Management Committee (DMC).

WUAs are charged with the overall responsibility for managing the small reservoirs. Their responsibilities include maintenance of dam infrastructure, land allocation, collection of water levies, formulation and enforcement of by-laws, and general administration related to the reservoirs. The DMC, together with the community members, agree on the water levy at the beginning of the season. The levy is a fixed sum, identical for every participating farmer in the community. Livestock keepers and fishermen who are members of the WUA also contribute to the main WUAs. The WUA's contributions are used for repairs and maintenance of irrigation canals, dam walls, valves and spillways. The Ghana Irrigation Development Authority is, in principal, required to provide technical personnel to support WUAs in management of the small reservoirs when the need arises.

6. Constraints and Potentials for Improving Reservoir Water Resources Management

6.1 Medium Reservoirs

The presence of reliable records on the operation of a reservoir can greatly facilitate the planning of water supplies for various demands. Irrigation, the largest user of water reservoirs, certainly plays a key role in guiding the planning of reservoir operations. The presence of deteriorated water conveyance infrastructure, which can no longer properly retain water, thus making control of water a difficult task, is a major constraint to improve reservoir water resources management. Inadequacy of irrigation water management personnel is also a constraint to the management of reservoir water resources. For example, an increased number of irrigation technicians could easily facilitate improved water allocation to irrigated farms depending on the water demand of the cropped fields. Reliance on VCs for water allocation has been shown to be unpractical, because it is difficult for VC members to deny water to fellow farmers who have not paid water levies. Additionally, existence of free-flowing water throughout the system (even in uncultivated plots) due to deteriorated canals makes it difficult for farmers to resist using it. Nevertheless, opportunities exist for improving management. Rehabilitation of irrigation infrastructure could be one of the long-term solutions for efficient reservoir water resources management. Concentrating farmers in a few selected plots could be considered as a short term solution, and could substantially reduce water loss and irrigation water requirement. This is in contrast to the existing practice whereby farmers' plots are scattered throughout the entire project area, limiting the water supply through conveyance losses. The use of project extensionists to provide assistance in water allocation is another short-term solution. An increase in the number of water management personnel at lower levels can help to ease the problems associated with poor water allocation.

6.2. Small Reservoirs

Small catchment sizes combined with unreliable rainfall is a constraint for efficient utilization of small reservoirs. This results in generation of insufficient runoff, which may cause water shortages at the end of the dry season. Water leakage during the rainy season and before the irrigated farming season starts when the hydraulic head in the reservoir is high, hence also a problem. Provision of support by GIDA for planning dry season water-use activities, especially matching command area to available water in the reservoir, could help avoid water deficits before crop harvests. Additional water harvesting wherever possible, by directing runoff from nearby sub-catchments to the dam, could provide a solution to the water shortage caused by small catchment size and unreliable rainfall.

Summary and Conclusion

Water reservoirs are important for socioeconomic development in the UER, without which the livelihood of many communities would be jeopardized. These reservoirs provide water for many uses, ranging from irrigation, livestock watering and fisheries to domestic uses such as cloth washing, bathing, and general construction. The first three are the most important with respect to the volume of reservoir water use. Irrigation is the major reservoir water user. The ICOUR and WUAs are responsible for water management in medium and small reservoirs, respectively. The water levy, charged on the basis of the relative size of plots of land allocated to participating farmers, is used in part to meet the operation and maintenance costs of the schemes. The levy is regarded as an incentive to more responsible use of water, although in certain cases, payments of water levies by farmers have been translated into rights of over- and mis-use of water. This translation is not widely observed in the UER, although there are isolated incidents of water mis-use, particularly in medium reservoirs where water supplies are comparably higher than in small reservoirs. Deteriorated water conveyance infrastructure is a constraint to sustainable reservoir water resources management. This is aggravated by inadequate of water management personnel in medium reservoirs. Rehabilitation of irrigation infrastructure, together with changing management strategies in medium reservoirs, will contribute to improved reservoir water resources management. Insufficient catchment size is a problem at some small reservoirs, resulting in low and uncertain runoff generation. Support by MOFA for planning dry season water-use of small reservoirs, particularly the matching of available water to crop demand, will help farmers to avoid water deficits occurring before crop harvesting. Runoff harvesting from the nearby catchments could also help to substantially increase the volume of reservoir water during the rain season.

Acknowledgement

This study was conducted under the financial support of the German Federal Ministry for Economic Cooperation and Development (BMZ) through DAAD and GTZ. The logistical and technical support of the GLOWA Volta Project in Ghana and at ZEF provided during the conduct of the study is also acknowledged.

REFERENCES

Bacho ZL and Bonye S (2006) From public open access to common property: The prospects and challenges of institutionalizing boundaries for self-governance and management of community irrigation dams in the Upper East Region, Ghana. *Ghana J. Dev. Stud.* 3(1):102-118.

Bakker M and Matsuno Y (2001) A framework for valuing ecological services of irrigation water. A case of an irrigation-wetland system in Sri Lanka. Irrig. *Drainage Syst.* 15: 99-115.

Bhattarai M, Barker R and Narazanamoorthy A (2003) Who benefits from irrigation investments in India?: Implication of irrigation multiplier estimates for cost recovery and irrigation financing. *In proceedings of XI World Water Congress*, Madrid Spain, 5-9 October 2003.

Boateng E and Ayamga T (1992) Soils and Land evaluation studies at Tono-Navirongo, Wiaga, Zuarungu and Manga Agricultural Stations, UER. IFAD/MOFA LACOSREP Report, 24 pp.

Boelee E and Laamrani H (2003) Multiple use of irrigation water in Northern Morocco. In: International Symposium on Water, *Poverty and Productive uses of Water at the Household Level*, Muldersdrift, South Africa, 21-23 January 2003.

Gowing JW, Li G and Gunawardhana T (2004) Multiple-use management in a large irrigation system: Benefits of distributed secondary storage. *Irrig. Drainage Syst.* 18: 57-71.

Gyasi KO (2005) Determinants of Success of Collective Actions on Local Commons: *An Empirical Analysis of Community-Based Irrigation Management in Northern Ghana.* Peter Lang, Vol. 49.

ICOUR (1995) *Information hand book*, Upper East Region, Ghana.

Kasanga R. K., (1995) "Land tenure and regional investment prospects: the case of the tenurial systems of Northern Ghana", *Property Management*, Vol. 13 Iss: 2, pp.21 – 31.

Kranjac-Berisavljevic' G, Blench RM, Bayorbor TB, Turton CN, Abdulai AS, Boyd C, Obeng F and Drake E (1998) Rethinking Natural Resources Degradation in Semi-Arid sub Saharan Africa: the Case of Semi-Arid Ghana. *Faculty of Agriculture, University for Development Studies, Tamale, Ghana, and Overseas Development Institute* (ODI), London, United Kingdom.http://www.odi.org.uk/rpeg/soil_degradation/ghlit.pdf. Cited 15 May 2005.

Liebe J, van de Giesen N and Andrein M (2005) Estimation of small reservoir storage capacities in a semi-arid environment. A case study in the Upper East Region of Ghana. *Phys. Chem. Earth* 30: 448-454.

Liebe, J (2002) Estimation of water storage capacity and evaporation losses of small reservoirs n the Upper East Region of Ghana. Diploma Thesis. Geographische Institute der Rhinischen Friedrich-Wilhelms-Universitat Bonn, 106p.

Meinzen-Dick, RS and Bakker M (2001) Water rights and multiple water uses: Issues and examples from Kirindi Oya, Sri Lanka. *Irrigation and Drainage Systems* 15(2), 129-148.

Meinzen-Dick RS and Van der Hoek W (2001) Multiple use of water in irrigated areas. Irrig. *Drainage Syst.* 15:93-98.

Penning de Vries FWT, Boelee E, Butterworth J, Cousins T, Duran A, Hagmann J, Mintesinot B, Morardet S, Moriarty PB, Restrepo I, Ruaysoongnern S, Scott C, Suryarwanshi S, Smits S, Van Koppen B and Yoder B (2006) Learning Alliances for the broad implementation of an integrated approach to multiple sources, multiple uses and multiple users of water. Manuscript for the International Conference on 'Integrated Assessment of Water Resources and Global Change: A North-South Analysis', February 2005 Bonn Germany. http://www.zef.de/watershed2005. Cited 2 August 2006.

Renault, D and Montginoul M, (2003) Positive externalities and water service management in rice-based irrigation systems of humid tropics. *Agric. Water Manage.* 59: 171-189.

Renwick ME (2001) Valuing water in a multiple-use system: Irrigated agriculture and reservoir fisheries. Irrig. *Drainage Syst.* 15: 149–171.

Water Front (2004) Issue No. 1, Oct-Dec 2004, Ghana.

In: Irrigation Management, Technologies …
Editor: M. Hossain Ali
ISBN: 978-1-62417-862-7

Chapter 11

CAPITAL BUDGETING METHODS FOR IRRIGATION TECHNOLOGY ADOPTION DECISIONS: A REVIEW

Chi Huu Truong[*]
Department of Applied Finance and Actuarial Studies,
Macquarie University, Sydney

ABSTRACT

Policies to encourage the uptake of efficient irrigation technology have played an important role in achieving water conservation goals in arid area and in reducing polluted return flows from irrigation farms. A good policy depends significantly on the method used to determine farmers' irrigation technology adoption decision. This chapter provides a critical review of capital budgeting methods applicable to modeling the irrigation technology adoption decision. Elements affecting the value of investment, including the roles of risk preferences, risk transformation, the value of information, are analyzed and real option frameworks are discussed. The chapter is concluded with future research suggestions.

INTRODUCTION

Policies to encourage the uptake of efficient irrigation technology have played an important role in achieving water conservation goals in arid area and in reducing polluted return flows from irrigation farms (Dinar and Zilberman 1991; Koundouri *et al.* 2006; Brennan 2007). A good policy depends significantly on the method used to determine farmers' irrigation technology adoption decision, i.e. how such decision is modelled. This chapter provides a critical review of capital budgeting methods used in the literature for irrigation technology with a view to suggest directions for future research.

Studies that examine irrigation technology adoption can be classified into deterministic and stochastic. In the deterministic studies, the investment decision maker knows the value

[*] Tel. (02) 98508481, Fax: + 612 9850 8586. E-mail: chi.truong@mq.edu.au.

that each relevant variable will take in each future period. In the stochastic studies, some variables are allowed to vary stochastically according to a well-defined probabilistic rule. The decision maker does not know exactly what value the random variable is going to take in a future period. This chapter is composed of three sections with the first section discussing deterministic studies, the second discussing stochastic studies and the third section focusing on future research.

1. Deterministic Studies

Deterministic studies may be static studies where economic variables are assumed to take the same value every period, or dynamic studies where variables may take different values across time. Deterministic investment studies were initiated when Fisher (1930) proposed the net present value (NPV) criterion: a project should be undertaken if its NPV is positive. Since then, the approach has been further developed to incorporate more sophisticated dynamics. Details of this approach and its application in general investment studies can be found in Arrow and Kurz (1970), Takayama (1985), Seierstad and Sydseter (1987), and Leonard and Long (1992).

In the literature of irrigation technology adoption, the static approach is popular. Examples of studies adopting this approach are Caswell and Zilberman (1986), Lichtenberg (1989), Caswell et al. (1990), Moore and Negri (1992) and Schaible (1997). This approach is popular because it is simple and can be used to provide clear illustrations of theoretical insights. For example, Caswell et al. (1990) perceived that enhanced efficiency of irrigation technology is a substitute for land quality. They use the static approach to illustrate that low landquality farms have higher incentives to adopt modern irrigation technologies compared to high landquality farms. Lichtenberg (1989) illustrated that when a modern irrigation technology is adopted, the cropping pattern may change. These results are likely to hold in more realistic frameworks. Furthermore, static frameworks are often used as the foundation upon which more realistic models are built. For example, the irrigation efficiency modelling method where a more efficient irrigation technology is modelled as an increase in the proportion of applied irrigation water absorbed by the plants, as first proposed by Caswell and Zilberman (1986), has been adopted in many stochastic studies (Carey and Zilberman 2002; Koundouri *et al.* 2006; Baerenklau and Knapp 2007).

Static studies are often valued for the qualitative results they provide. However, to provide persuasive policy recommendations, quantitative results are often needed. When some of the considered variables vary across periods, i.e. they are dynamic, as is often the case with output prices, the quantitative results provided by the static approach may become irrelevant in later periods. To avoid this problem, static studies often provide sensitivity analysis for several levels of a dynamic variable. For example, Caswell et al. (1990) illustrated the impacts of policy measures on the uptake of more efficient irrigation technologies for two levels of output prices. However, in doing so, it is assumed that the investor has static expectations about the evolution of the dynamic variable. Thus, if the variable is currently low, it is expected to stay low forever in the future and vice versa. In reality, variables such as output prices often behave in the way that if the current prices are low, they might reasonably be expected to increase to higher levels in future periods. Ignoring the dynamics of variables may yield incorrect results.

Explicit incorporation of the dynamics of variables increases the complexity of the problem considerably. Dynamic problems must be solved by one of three approaches, namely: the optimal control approach, the dynamic programming approach, and the calculus of variation approach (Arrow and Kurz 1970). Random variables are often assumed to take their expected values with certainty in future periods. For example, Wu et al. (1994) assumed that output prices took the expected values provided by a price forecast model when they evaluated the modern irrigation technology adoption decision.

As demonstrated by Dixit and Pindyck (1994 p138), when exogenous variables are deterministic but vary over time, investing in a project based on the NPV computed in a similar fashion as in the static case may not be optimal. The static approach to investment analysis only compares the decision to invest in the project in the current period with the decision to ignore the project forever. However, because exogenous variables change over time, a project that is not worthy in the current period may become valuable in future periods. For investment in the current period to be optimal, apart from being positive, the NPV of the project invested in the current period must be higher than the NPV calculated for starting the project in any other period. The optimal investment behaviour may involve delaying the investment in the current period and waiting for the optimal time to invest. Delaying the start (and so timing of cashflow) of a project with a positive NPV would decrease the NPV if the cash flows remain the same other than being translated in time. However, with dynamically changing variables, it is possible that a delayed project start could have a higher NPV. Essentially, a decision maker does not have just a single "do or not do" investment project, but confronts a set of mutually exclusive projects defined by their starting dates; and the best of these projects is the one with the highest positive NPV.

2. Stochastic Studies

In reality, the NPV of a new irrigation technology (that to be adopted) is often stochastic, changing randomly when water price and/or irrigation output prices change. The NPV rule derived assuming a deterministic environment has been found to perform poorly in describing the decisions of farmers in adopting modern irrigation technology (Carey and Zilberman 2002). It is therefore necessary to consider the investment decision in stochastic environment.

Stochasticity gives rise to three aspects that may affect the investment decision. These include the decision maker's attitude to income risk, the translation of risk from prices and physical factors to income, and the value of new information learned over time.

Role of Risk Attitudes

The impacts of risk attitudes on resource allocation decisions were examined by Sandmo (1971). Sandmo argues that most people are risk averse and shows that a risk averse producer maximising his expected utility will produce less when the output price is risky compared to when output price is fixed at the expected level. Later studies extend Sandmo's framework to allow for more complex production functions and different utility functions. Detail discussion of this literature can be found in Just and Pope (2003) and Just (2003).

The approach of Sandmo (1971) is criticised by McDonald and Siegel (1985) as inappropriate for the valuation of claims on future profits generated by investment projects. Valuation of these claims is of interest, for instance, when a farmer wants to choose investment projects that maximise the value of his farm, or when a traded firm considers an investment project, or when a government auctions the right to exploit a coal mine. In these cases, decisions are aimed at maximising the market value of the firm. In Sandmo's model, it is asserted that any increase in the variation of a firm's income will reduce the value of the firm. However, according to the Capital Asset Pricing Model (CAPM) (see for example, Sharpe (1964)), the value of the firm only decreases when the increased risk co-varies with the economy. Independent risks posed by individual firms (or projects) will always be diversified away in an optimal portfolio.

McDonald and Siegel (1985) also showed that when decisions are aimed at maximising the market value of the firm, the financial economic framework used to value investment projects does not change when investors' risk attitudes change. Risk attitudes only change the discount rate. With an appropriate discount rate, investment projects can be valued as if investors are risk neutral even though they are risk averse. The valuation framework is called 'risk neutral valuation'.

In dealing with risk, McDonald and Siegel (1985) assumed that the claim on the cash-flow of the considered project is a traded asset so that the CAPM can be applied to value its price. Smith and McCardle (1999), however, argued that claims on cashflows of investment projects are usually non-traded assets even when the firm is listed in the stock market. In this case, risk neutral valuation can still be used if a set of traded assets that spans the risk of a project's cashflow is available, i.e. the market is complete. Smith and McCardle illustrated the use of a futures contract, a call option and a risk-free bond to span the three possible movements of oil price over one time period.

Ewald and Yang (2008) argued that each investment project has its own unsystematic risks independent of others and that, often no traded assets or portfolios of traded assets that have returns perfectly correlate with the project's cashflow, and therefore usable in hedging the project, exist. They argued that for this reason, a risk averse investor should be risk averse towards the unsystematic risks and the project investment decision should depend on the investor's risk attitude.

On the other hand, Yin and Turvey (2003) argued that in the farming sector, only the systematic risk component of an investment project is relevant to investment decisions even when the cashflow of the project is not spanned by traded assets. In their framework, the annual cashflow of a farm is a non-traded asset. Because the farm itself is traded, to avoid risk-free arbitrage, the market will value the annual cashflow of the farm and a traded asset with the same systematic risk equally. In evaluating the market value of the project, traded assets with systematic risks similar to those of the project's cashflow can be used as the spanning asset. Therefore, the risk neutral valuation framework can be used for investment projects in the farming sector.

Despite the controversy regarding the risk neutral valuation framework, the approach initiated by McDonald and Siegel (1985) has enhanced the understanding of the value of flexibility in a risky environment, regardless of the discount rate.

Transformation of Risk from Output Price to Income

Risk attitudes only affect resource allocation via their influence on how the decision maker values a risky income. However, income risk is often sourced from risky output prices, risky input prices or risky input supplies. The mechanism that translates the risk from these original sources to income may mean resource allocation is affected by risk even when the decision maker is risk neutral.

The transformation of risky variables to profit is often non-linear. For example, short run profit obtained when capital inputs are fixed and variable inputs are optimally chosen in response to observed output prices is convex in output price when the production function is concave (Hartman 1976) or when the production function is linear but the production can be temporarily suspended (McDonald and Siegel 1985). The expected short term profit is larger than the short term profit at the certainty equivalent output price, according to Jensen's inequality. The firm is benefited from risk in the short run, regardless of its risk attitude.

To have accurate valuation of investment projects, the valuation framework needs to incorporate the risk transformation mechanism.

Value of Information

In deterministic settings, variables may change over time, but in a way that is known at the commencement of the decision problem. The decision maker can make the optimal decisions for all time periods at the very first time period. There is no need (or opportunity), by definition, to obtain additional information.

When the benefits of investment are uncertain, investment decisions may depend on whether additional information can be obtained. The value of information (and/or value of flexibility) has traditionally been recognised and captured in sequential decision models of investment. For example, Arrow and Fisher (1974) illustrated the significance of option value involved with irreversible development. Conrad (1980) showed that the value of the option lost when an irreversible investment occured was equal to the value of information. Using a sequential decision model, Dias (2004) illustrated that in petroleum exploration, two oil prospects that were not worth drilling when considered separately might become worthy if they were drilled sequentially. In this case, sequential drilling allows the decision maker to obtain further information regarding the chance of success regarding the second drilling by drilling the first prospect.

In the same fashion, Laughton (1998) demonstrated that an investor who can invest in an oil project in the current period or the next two periods may defer the investment decision to the following period to obtain more information regarding the investment benefits. If the decision maker defers the investment decision and preserves the flexibility of starting the project, he can avoid the downside risk of the low oil price and can invest to benefit the upside risk when the oil price goes up.

Information can only be valuable to a decision maker if he has the capacity to respond to it (Hartman 1976). When all the variables have been selected before the information arrives, information does not change resource allocation, nor the profit, and therefore is worthless.

With the passage of time, opportunities to learn information usually arise, and as well, opportunities to react to emerging information. That is, in risky time-related projects, the investor potentially has a sequential decision problem in which initial period decisions need to be taken with due recognition for what might happen and how one can react in the future. Retaining options or having flexibility may become important. The ability to react in the best possible way to a stochastic variable, as it is revealed in the future, leads to a non-linear relationship between any stochastic variable and income (or NPV). Stochasticity in a variable with the decision maker having an ability to respond will mean that the expected NPV is different from that of the deterministic case. Even a risk neutral decision maker needs to account for this risk, if he is to make optimal decisions.

Standard Real Option Model

A separate body of literature, financial options theory, also deals with sequential decision problems, but quite specialised ones, namely problems of how long to repeatedly delay exercising an option on a financial stock to obtain the highest expected value. A number of authors, most notably Dixit and Pindyck (1994), recognised that the timing of investment question could be posed as a "real option" problem. One of the differences between the financial and real options literature is that the financial options literature is preoccupied with valuing options, not with when to exercise them, whereas the real options literature is mainly concerned with whether to exercise now or not and less concerned with the value of the option per se (Hirschauer and Musshoff 2008).

The decision to invest in a project is similar to the decision to exercise a financial call option. An investor who holds a call option has to consider whether he should exercise the option now or delay until the next period when he can consider the question again. The sunk cost of making an investment is equivalent to the exercise price of the option and the value of the invested project acquired by making the investment is equivalent to the stock price gained by exercising the option. To create a perfect analogy with the call option on a financial stock, the real options theory adopts several assumptions even though these are not necessarily realistic. It is assumed that commodity prices are random walks and that holding a unit of the commodity provides a constant, positive convenience yield per unit of time. The convenience yield then resembles the dividend rate paid by the financial stock. If the convenience yield is not positive, it is always optimal to defer the investment decision and the project will never be invested (Dixit and Pindyck 1994 p138). It is also assumed that the investment decision is totally irreversible.

Assumptions made in the standard real option model are for the convenience of implementing the financial economic models in investment analysis. Some of the assumptions have been criticised as unrealistic. First, the positive net convenience yield (net of storage cost) implies that only backwardation in the commodity price is possible. In reality, commodity prices do exhibit backwardation in some periods and contango (negative net convenience yield) in some other periods due to seasonal effects (Samuelson 1965; Laughton 2000). Data on futures prices also indicate that one commodity can experience backwardation in some years and contango in other years (Schwartz 1997). Second, the random walk process allows the spot price and futures price to drift to extreme values. This contradicts the economic law of supply and demand. In competitive markets, high prices will attract high cost producers to enter the market, increasing supply and driving price down. Similarly, as price decreases to low levels, high cost producers exit the market, putting upward pressure on prices (Samuelson 1965). Third, the term-structure of the commodity price implied by a random walk process is different from the term-structure observed in reality. A random walk price process implies that the riskiness of holding a futures contract increases as the contract's term increases. However, as observed by Samuelson (1965), prices of longer term futures contracts often have lower variances.

Mean Reverting Process Models

Given the limitations of the standard real option model, recent real option studies have extended the theory by considering commodity prices that follow mean reverting processes (Laughton and Jacoby 1993; Smith and McCardle 1999; Schwartz and Moon 2000; Metcalf and Hassett 1995). Metcalf and Hassett (1995) have shown that the values of real options and their exercise thresholds are very different when a random walk process rather than a mean reverting process is used. The behavior of commodity price under a mean reverting process is consistent with the law of commodity supply and demand. When the current price is above the steady state price, the process tends to pull the next price down towards the steady state price. On the other hand, if the current price is below the steady state price, the process pushes the price up towards the steady state price.

The above argument for a mean reverting commodity price process also applies to water. Although the capacity to produce more or less water in the short run is limited, stochastic variation in water supply due to random rainfall and river flows is often buffered by water stocks in storage facilities. High water prices due to low rainfall and/or low river flows can be reduced by water release from storage. On the other hand, high river flows can be stored, which will increase water prices. It is, therefore, more appropriate to use a mean reverting process rather than a random walk process to model water price. Details on the mean reverting behavior of water prices under the impacts of storage can be found in Truong (2011) for surface water and in Tsur and Graham-Tomasi (1991) and Knapp and Olson (1995) for groundwater. A mean reverting process will result in a lower real option value and an earlier investment time. Details about an irrigation technology investment modeling framework in which a mean reverting process is used to model water price can be found in Baerenklau and Knapp (2007).

Multiple Process Models

Accurate modelling of commodity prices plays an important role in the determination of optimal investment decisions. Recent real options studies have investigated the suitability of single stochastic processes versus multiple stochastic processes in modelling commodity prices.

Theoretical arguments for using a multiple stochastic process to model oil prices have been put forwards by Pindyck (1999). Conditional on a fixed level of long run marginal production cost, oil prices are demonstrated to follow a mean reverting process. With stochastic events such as technological progress and discoveries of new oil reserves, the long run marginal cost of oil production is stochastic and a multiple stochastic process is more appropriate in that it provides one stochastic process to model long run shocks and another process to model temporary shocks. With a single stochastic process model, permanent shocks that affect the long run marginal cost cannot be distinguished from temporary shocks. Analysis with the model may suggest increases or decreases in the longterm production capacity when the oil price is changed only by temporary shocks. Consequently, investment decisions advised by one factor models may not be optimal.

Schwartz (1997) examined the performance of multiple stochastic process models with a single stochastic process model in representing the behaviour of oil, copper and gold prices. He found that multiple stochastic process models fit data significantly better than the one factor mean reverting model. The data show that the volatility of futures returns decreases sharply as the term structure increases up to some level after which it remains relatively constant. This pattern cannot be attained using the one factor model.

Based on the above theoretical arguments and empirical evidence, multiple process models have been proposed in more recent studies to evaluate commodity derivatives and the option to invest in commodity production capacity (Schwartz and Smith 2000; Lucia and Schwartz 2002; Manoliu and Tompaidis 2002; Burger et al. 2004; Cartea and Villaplana 2008). Despite the large number of studies that apply multiple stochastic process models for other commodities, multiple process models have not been used to model optimal irrigation technology adoption decision in previous studies.

3. Future Research

The range of the capital budgeting methods is wide and no model is best *a priori*. Model selection should be based on the empirical settings. In regions where irrigation output prices are stable and water allocation is stable (e.g. regions supplied by aquifers) or has low correlation between years (e.g. regions in which water storage facility is not available), the value of adopting an irrigation technology is rather certain and the NPV rule will suffice. In regions where irrigation output prices are stable, but the stochastic water supply is buffered by water storage, a mean reverting real option model is more appropriate.

On the other hand, when irrigation output prices are stochastic and water supply is buffered by water storage, a multiple stochastic process model is more suitable. Variation in irrigation output prices may change the long term capitals such as irrigation land areas and change the long term water demand while variation in rainfalls, water inflows to water storage facilities will change the current water demand and water supply. There are, therefore, short run shocks that result in temporary changes in water prices and long run shocks that result in water price changes that last many periods. These different shocks have different impacts on the NPV of adopting a new technology and should be distinguished. In that sense of information preservation, two stochastic processes, with one to model short run shocks and another to model long run shocks, are more appropriate than a single stochastic process in modelling water prices. In addition to the stochasticity of irrigation output prices, climate change that may change the distribution of water inflow to the region and result in changes in longterm capitals is another reason for using a multiple process model.

Multiple stochastic processes open up new research questions such as how such processes can be estimated and what difference in investment decisions and investment values will be resulted in when those processes are used. Research in this direction is both exciting and significant in terms of policy implications.

ACKNOWLEDGMENT

The author is grateful to Associate Professor Ross Drynan for his valuable comments and editing efforts on earlier versions of this work.

REFERENCES

Arrow, K.J. and Kurz, M. (1970). *Public Investment, the Rate of Return, and Optimal Fiscal Policy*. Published for Resources for the Future by the Johns Hopkins Press, Baltimore.

Arrow, K.J. and Fisher, A.C. (1974). Environmental Preservation, Uncertainty, and Irreversibility, *Quarterly Journal of Economics* 88, 312-319.

Baerenklau, K.A. and Knapp, K.C. (2007). Dynamics of Agricultural Technology Adoption: Age Structure, Reversibility, and Uncertainty, *American Journal of Agricultural Economics* 89, 190-201.

Brennan, D. (2007). Policy Interventions to Promote the Adoption of Water Saving Sprinkler Systems: The Case of Lettuce on the Gnangara Mound, *Australian Journal of Agricultural and Resource Economics* 51, 323-341.

Burger, M., Klar, B., Müller, A. and Schindlmayr, G. (2004). A Spot Market Model for Pricing Derivatives in Electricity Markets, *Quantitative Finance* 4, 109-122.

Carey, J.M. and Zilberman, D. (2002). A Model of Investment under Uncertainty: Modern Irrigation Technology and Emerging Markets in Water, *American Journal of Agricultural Economics* 84, 171-183.

Cartea, Á. and Villaplana, P. (2008). Spot Price Modeling and the Valuation of Electricity Forward Contracts: The Role of Demand and Capacity, *Journal of Banking & Finance* 32, 2502-2519.

Caswell, M., Lichtenberg, E. and Zilberman, D. (1990). The Effects of Pricing Policies on Water Conservation and Drainage, *American Journal of Agricultural Economics* 72, 883-890.

Caswell, M.F. and Zilberman, D. (1986). The Effects of Well Depth and Land Quality on the Choice of Irrigation Technology, *American Journal of Agricultural Economics* 68, 798-811.

Conrad, J.M. (1980). Quasi-Option Value and the Expected Value of Information, *Quarterly Journal of Economics* 94, 813-820.

Dias, M.A.G. (2004). Valuation of Exploration and Production Assets: An Overview of Real Options Models, *Journal of Petroleum Science and Engineering* 44, 93-114.

Dinar, A. and Zilberman, D. (1991). The Economics of Resource-Conservation, Pollution-Reduction Technology Selection: The Case of Irrigation Water, *Resources and Energy* 13, 323-348.

Dixit, A.K. and Pindyck, R.S. (1994). *Investment under Uncertainty*. Princeton University Press, Princeton, New Jersey.

Ewald, C.O. and Yang, Z.J. (2008). Utility Based Pricing and Exercising of Real Options under Geometric Mean Reversion and Risk Aversion toward Idiosyncratic Risk, *Mathematical Methods of Operations Research* 68, 97-123.

Fisher, I. (1930). *The Theory of Interest: As Determined by Impatience to Spend Income and Opportunity to Invest It*. Macmillan, New York.

Hartman, R. (1976). Factor Demand with Output Price Uncertainty, *The American Economic Review* 66, 675-681.

Hirschauer, N. and Musshoff, O. (2008). A Note: Are (Positive) Real Options Values a Contradiction in Terms?, *Briefing Notes in Economics* 77, 1-8.

Just, R.E. (2003). Risk Research in Agricultural Economics: Opportunities and Challenges for the Next Twenty-Five Years, *Agricultural Systems* 75, 123-159.

Just, R.E. and Pope, R.D. (2003). Agricultural Risk Analysis: Adequacy of Models, Data, and Issues, *American Journal of Agricultural Economics* 85, 1249-1256.

Knapp, K.C. and Olson, L.J. (1995). The Economics of Conjunctive Groundwater-Management with Stochastic Surface Supplies, *Journal of Environmental Economics and Management* 28, 340-356.

Koundouri, P., Nauges, C. and Tzouvelekas, V. (2006). Technology Adoption under Production Uncertainty: Theory and Application to Irrigation Technology, *American Journal of Agricultural Economics* 88, 657-670.

Laughton, D. (1998). The Management of Flexibility in the Upstream Petroleum Industry, *Energy Journal* 19, 83-114.

Laughton, D. (2000). *Modern Asset Pricing and Project Evaluation in the Energy Industry*. Western Centre for Economic Research. University of Alberta, Western Centre for Economic Research Bulletin.

Laughton, D.G. and Jacoby, H.D. (1993). Reversion, Timing Options, and Long-Term Decision-Making, *Financial Management* 22, 225-240.

Leonard, D. and Long, N.v. (1992). *Optimal Control Theory and Static Optimization in Economics*. Cambridge University Press, Cambridge; New York.

Lichtenberg, E. (1989). Land Quality, Irrigation Development, and Cropping Patterns in the Northern High-Plains, *American Journal of Agricultural Economics* 71, 187-194.

Lucia, J.J. and Schwartz, E.S. (2002). Electricity Prices and Power Derivatives: Evidence from the Nordic Power Exchange, *Review of Derivatives Research* 5, 5-50.

Manoliu, M. and Tompaidis, S. (2002). Energy Futures Prices: Term Structure Models with Kalman Filter Estimation, *Applied Mathematical Finance* 9, 21-43.

McDonald, R.L. and Siegel, D.R. (1985). Investment and the Valuation of Firms When There Is an Option to Shut Down, *International Economic Review* 26, 331-349.

Metcalf, G.E. and Hassett, K.A. (1995). Investment under Alternative Return Assumptions: Comparing Random Walks and Mean Reversion, *Journal of Economic Dynamics and Control* 19, 1471-1488.

Moore, M.R. and Negri, D.H. (1992). A Multicrop Production-Model of Irrigated Agriculture, Applied to Water Allocation Policy of the Bureau-of-Reclamation, *Journal of Agricultural and Resource Economics* 17, 29-43.

Samuelson, P.A. (1965). Proof That Properly Anticipated Prices Fluctuate Randomly, *Imr-Industrial Management Review* 6, 41-49.

Sandmo, A. (1971). Theory of Competitive Firm under Price Uncertainty, *American Economic Review* 61, 65-73.

Schaible, G.D. (1997). Water Conservation Policy Analysis: An Interregional, Multi-Output, Primal-Dual Optimization Approach, *American Journal of Agricultural Economics* 79, 163-177.

Schwartz, E.S. (1997). The Stochastic Behavior of Commodity Prices: Implications for Valuation and Hedging, *Journal of Finance* 52, 923-973.

Schwartz, E.S. and Moon, M. (2000). Rational Pricing of Internet Companies Revisited, *Financial Analysts Journal* 56, 62.

Schwartz, E. and Smith, J.E. (2000). Short-Term Variations and Long-Term Dynamics in Commodity Prices, *Management Science* 46, 893-911.

Seierstad, A. and Sydseter, K. (1987). *Optimal Control Theory with Economic Applications.* Advanced Textbooks in Economics ; V 24, North-Holland, Amsterdam.

Sharpe, W.F. (1964). Capital-Asset Prices - a Theory of Market Equilibrium under Conditions of Risk, *Journal of Finance* 19, 425-442.

Smith, J.E. and McCardle, K.F. (1999). Options in the Real World: Lessons Learned in Evaluating Oil and Gas Investments, *Operations Research* 47, 1-15.

Takayama, A. (1985). *Mathematical Economics.* 2nd edn, Cambridge University Press, Cambridge [Cambridgeshire] ; New York.

Truong, C. (2011). An Analysis of Storage Capacity Reallocation Impacts on the Irrigation Sector, *Environmental and Resource Economics* 51, 141-159.

Tsur, Y. and Graham-Tomasi, T. (1991). The Buffer Value of Groundwater with Stochastic Surface-Water Supplies, *Journal of Environmental Economics and Management* 21, 201-224.

Wu, J., Mapp, H.P. and Bernardo, D.J. (1994). A Dynamic Analysis of the Impact of Water Quality Policies on Irrigation Investment and Crop Choice Decisions, *Journal of Agricultural and Applied Economics* 26, 506-525.

Yin, S.H. and Turvey, C.G. (2003). The Pricing of Revenue Assurance: Comment, *American Journal of Agricultural Economics* 85, 1062-1065.

In: Irrigation Management, Technologies ...
Editor: M. Hossain Ali

ISBN: 978-1-62417-862-7

Chapter 12

IRRIGATION WITH WASTEWATER: IMPACTS, SCOPE AND CHALLENGES

M. G. Mostofa Amin**, *Mohammed Mizanur Rahman and Moudud Hasan

Department of Irrigation and Water Management
Bangladesh Agricultural University, Mymensingh, Bangladesh

ABSTRACT

To combat the growing water demand in agriculture, wastewater from different sources is considered as a potential supplement to fresh irrigation water. Use of wastewater in irrigation in a sustainable manner can increase water security in the agricultural sector. Irrigation by wastewater can increase the crop yield because of its nutrient and organic matter content, which will reduce the agricultural production cost. On the other hand, it can spread harmful contaminants to soil and water environment if not managed properly.

Wastewater application can also affect the soil hydraulic properties. Protecting sources of freshwater is critically important to the future of agriculture because freshwater available for agricultural use is becoming scarce due to population growth and increasing pollution in many parts of the world. Judicious use of wastewater in irrigation and treatment before field application, when necessary, can help to improve water security for agriculture. However, more studies should be conducted to assess the overall impacts of wastewater irrigation on the quality and quantity of agricultural produce as well as the long-term consequences on soil and water environment. Impacts of wastewater irrigation on the soil and water environments, strategic measures for using wastewater in irrigation, and challenges of wastewater use in irrigation are discussed in this chapter.

* E-mail: aminmgmbau@yahoo.com.

1. INTRODUCTION

The world's water resources have become increasingly threatened as numerous water consuming sectors compete for the limited usable water of the earth. Agriculture is the largest sector, using about 70% of total extracted water (surface and subsurface). Wisser et al. (2008) estimated that about 2200 to 3800 km^3 of water is used in a year globally for irrigation purpose. It is projected that there will be an additional 3.5 billion people by 2050 with most of the growth in developing countries that already suffer water stress (UN, 2005), which will cause increased water scarcity unless there are corresponding augmentations in water conservation and recycling of this vital resource. It takes around 3000 liters of water, converted from liquid to vapor, to produce enough food to satisfy one person's daily dietary need. To meet the food demands of the growing population, the irrigated cropping intensity will need to be increased considerably (Faurés et al., 2012). This growth in irrigated cropping intensity is more crucial in developing countries where the irrigated area is expected to jump from 202 million ha in 1997/99 to 242 million ha by 2030 (Faurés et al., 2012). Excessive withdrawal of groundwater, which often provides a freshwater source for irrigation in areas where surface water is scarce, is reported to cause groundwater mining and land subsidence. Scientists and water resources managers have been working to determine how this inevitable demand for irrigation water will be met in a sustainable manner.

To combat the growing water demand in agriculture, wastewater from different sources (animal farms, household, industrial, irrigation return flow, etc.) is considered as a potential replacement for fresh irrigation water. In addition to offsetting the increasing demand for freshwater, wastewater may contain organic matter beneficial to the soil as well as many major and minor nutrients required by plants. Nonetheless, the use of untreated wastewater is subject to many limitations because it can result in the contamination of the soil environment and nearby water bodies by undesirable and harmful elements that can also cause human health hazards through direct contact with the wastewater and through consumption of the fresh agricultural produces (Yuan, 1993). There is no global recommended wastewater standard for safe irrigation. The goal of this chapter is to synthesize the knowledge gained from the use of wastewater irrigation worldwide to estimate its pros and cons for sustainable water management in agriculture.

2. IMPACT OF IRRIGATION

To sustain food security for the ever increasing population, measures should be taken to increase total food production. The potential for horizontal expansion of arable land is limited. The remaining option is to increase production per unit area by providing better input, such as improved irrigation, fertilizer and seeds. Irrigation is necessary to maintain required water content in soil for better growth and development of crops. Adequate water content in soil is necessary to facilitate the germination of seed and the uptake of nutrients from soil. Nutrients from soil reach the root surface by mass flow and diffusion process that are also related to the water content of the soil. Furthermore, the movement of nutrients through the plant body by physiological activities is also associated with the uptake of soil water. Thus, the availability of soil water for crop influences different yield attributes including dry matter production, plant height, leaf area, duration of grain filling and protein content of grain (Thomson and Chase, 1992).

3. Necessity and Scope of Wastewater for Irrigation

Although many parts of the world have sufficient surface water and groundwater resources, in other parts of the world irrigation water for basic food production is a limited resource during the major cropping seasons. Either non-uniform distribution of rainfall or scarce rain creates a long dry season in many countries. For example, the main crop season in Bangladesh does not receive enough rainfall to produce rain-fed crops. Moreover, during the dry winter season, a significant amount of the surface water dries. Because of the lack of a reliable supply of surface water, the irrigation in Bangladesh is primarily dependent on groundwater. However, groundwater aquifers in some areas are being depleted beyond the pumping capacity of suction pumps. For all these reasons, a chronic shortage of water is felt in Bangladesh and the surrounding region during the dry season irrigation period. Use of additional sources of water for irrigation may be an important solution to this problem. Wastewater has been used successfully for irrigation of various crops in different parts of the world (Feigin et al., 1990). Land application of wastewater has been practiced in India since prehistoric period (Chakrabarti, 1995). The potential benefits of wastewater use in agriculture includes increased crop production, reduction of uncontrolled pollution of environment, cost reduction in wastewater treatment, availability of nutrients in the effluent for further use, and saving input for fertilizers. We can also achieve some environmental and socio-economic benefits by using wastewater for irrigation because it usually reduces effluent disposal in receiving water bodies and improves crop production during dry season (Coppola et al., 2004). On the other hand, disposal of wastewater on agricultural land may create some hazards for soil and water environments. The hazardous effects of wastewater use depend on the physical, chemical and biological properties of the wastewater. The proper management and judicious use of wastewater is important as it is usually rich in nutrients such as nitrogen, potassium and phosphorus in addition to containing large numbers of chemical contaminants and microorganisms. It is important to consider what is necessary for protecting the soil and water as well as how to reuse the wastewater safely and economically. To attain both objectives, knowledge about the possible effects of wastewater uses in agriculture is necessary. Wastewater without harmful contaminants can be a preferred unconventional water source since both the supply and the demand of wastewater are increasing due to population growth.

4. Origin and Occurrence of Wastewater

Wastewater is produced by various sources, including municipalities, industries, agricultural fields, and livestock farms. In Europe and North America, animal slurry– a mixture of urine, feces and water from concentrated livestock is collected in a tank or lagoon and applied to agricultural land as a nutrient and organic matter source. The hydraulic load of this slurry application can be as high as one irrigation application. In developing countries, agricultural farms also produce a large quantity of wastewater, which is usually not managed properly and often drained to earthen pond. The treatment of effluents from municipalities and industries prior to disposal to a surface water body is necessary for protecting human health and the local environment.

However, due to the lack of implementation of laws and limited public awareness in many countries, the effluents are not treated properly before disposal. Although the utilization of this wastewater, treated or untreated, for irrigation could minimize water scarcity significantly, the possible environmental consequences of untreated wastewater use in irrigation should be assessed beforehand. The physical, biological and chemical characteristics of wastewater vary widely depending on the source of wastewater. The wastewater from animal farms may contain large quantity of organic matter, nutrients, pathogenic microorganisms, tetracycline resistant bacteria, endocrine disrupting hormones, chloride and other minerals (Amin, 2011). A list of some common constituents and contaminants is given in Table 1. The contaminants in animal wastewater also vary considerably with the types of feed, region of farms, species, sex and life stage of animals. Wastewater from municipalities and industries may contain wide variety of harmful contaminants, including heavy metals, active chemicals and pathogens in addition to nutrients and organic matter. Quantitative and qualitative data of irrigation-suitable wastewater are not available in many developing countries.

Table 1. List of some common constituents/ contaminants in wastewater of different sources

Type	Constituents/ Contaminants
Chemicals	Na, Ca, Mg, K, Cl, SO_4, CO_3, HCO_3, NH_4, NO_3, P, Mn, Cu, Zn, etc.
Antimicrobials	Chlortetracycline, Doxycycline, Methacycline, Minocycline, Oxytetracycline, Tetracycline, Penicillin, cloxacillin, Sulfachloropyridazine, Sulfadiazine, Sulfadimethoxine, Sulfaguanidine, Sulfamerazine, Sulfameter, Sulfamethazine, Sulfamethoxazole, Trimethoprim, Erythromycin, Roxithromycin, Tylosin, Ciprofloxacin, Norfloxacin, Ofloxacin, and Salinomycin.
Steroid Hormones	Estrone, 17α-estradiol, 17β-estradiol, Androstenedione, Dehydroepiandrosterone, Dihydrotestosterone, Testosterone, Pregnenolone, Progesterone, Ethynyl-estradiol, etc.
Viruses	*Enteroviruses*
Bacteria	Pathogenic *E. coli, Salmonella* spp., *Shigella* spp., *Vibrio cholerae*
Protozoa	*Entamoebahistolytica*
Helminths	*AscarisLumbricoides,* Hookworms, *Schistosomamansoni, Taeniasaginata, Trichuristrichiura*

Sources: Amin (2011), Lee at el. (2007), Blumenthal and Peasey (2002), and Blumenthal et al. (2000).

For planning and proper management of wastewater irrigation system, it is vital to know the information of the dependable amount of wastewater to be available for irrigation. Wastewater is generally produced throughout the year by various sources, but irrigation is required for specific crop seasons. Storage facilities or disposal system for seasonal surplus wastewater are needed for better management. If the availability of wastewater is not enough, the conventional sources of irrigation that the wastewater would replace also require management during the dry season to be ready for peak irrigation season. Additionally the transport of wastewater from sources to agricultural field needs to be considered during feasibility analysis.

5. IMPACTS OF WASTEWATER IRRIGATION

5.1. Impacts on the Crop Production

The use of wastewater in irrigation can increase yield and yield attributes of different agricultural crops variably depending on the physical, chemical and biological properties of wastewater. Wastewaters from all sources and origins contain a considerable quantity of nutrients and organic matter. Information regarding the feasibility and potential benefits of wastewater irrigation for crop production is available in literature (e.g., Chakrabarti and Chakrabarti, 1988). Wastewater irrigation, particularly with municipal and livestock farms, increases crop yield because of the additional supply of organic matter and nutrients (Hasan, 2011). However, wastewater containing more nitrogen than the dose recommended for optimal crop yield may increase physiological growth of crop rather than yield (Hussain et al., 2012; Misra and Behera, 1991). It is still a perplexing phenomenon to scientists that wastewater may increase crop yield even though it contains only the required nutrients at recommended dose. Day and Tucker (1977) noted that wastewater might have several non-fertilizer elements responsible for higher crop production. The chemical composition of wastewater is significantly different from most of the natural waters used in irrigation. It is, therefore, critically important to assess the potential disadvantages of wastewater reuse for irrigation. The pathogenic contamination of ready-to-eat type crops can cause disease outbreak. Bioaccumulation of contaminants in food grain also needs to be investigated.

5.2. Impacts on the Soil Environment

Irrigation with wastewater may have both short-term and long-term effects on the soil properties. The long-term and short-term effect on the soil properties can be considerably different (Tiarks et al., 1974). A number of research studies were conducted to assess the long-term effect of manure application on soil properties (Edmeades, 2003; Gilley and Risse, 2000; Mosaddeghi et al., 2009; Ndayegamiye and Cote, 1989; Schjonning et al., 1994). The long-term effect is generally moderate and homogeneous due to the temporal redistribution and degradation of organic matter. Wastewater application can change the physicochemical properties of soil, so the initial convective redistribution of wastewater constituents can be considerably different compared to what expected in un-amended soil (Bradford et al., 2008;

Unc and Goss, 2004). Olesen et al. (1997) conducted an experiment to investigate the short-term effect and reported that the water holding capacity and solute diffusivity can be affected in the presence of animal wastewater in soil. There is a lack of information in literature concerning the mechanistic short-term effect on soil properties. The short-term effect during the first month following wastewater application is expected to be more pronounced and important because it is the time when major physical, chemical and biological processes take place in the applied wastewater constituents. Organic matter in wastewater can increase the water holding capacity of amended soil. Solids in wastewater may seal the soil interstices and consequently reduce hydraulic conductivity of the soil. Persistently increased soil water content in the wastewater-injected slit was observed even after seven weeks of wastewater application in a study (Amin, 2011). The soil water dynamics and solute transport in the wastewater-treated soil may not be described accurately by the intrinsic soil hydraulic properties (Amin et al., 2012). In addition, a temporal and spatial variation of soil hydraulic properties may exist in wastewater-applied soil (Amin, 2011). Soil scientists use different pedotransfer functions to estimate soil hydraulic properties using soil textural data (e.g., Schaap et al., 2001). The pedotransfer functions could be improved by integrating the effect of wastewater organic matter.

Agroecologists have investigated the nitrogen dynamics in wastewater-treated soils (e.g., Petersen and Andersen, 1996; Petersen et al., 2003; Thomsen et al., 2010), and some such studies show that soil-wastewater incorporation can influence nitrification-denitrification processes (Markfoged et al., 2011; Olesen et al., 1996). Wastewater may also cause salinity problems in the soil. Excess soluble salts in water may result in low crop yields. Baat (1989) reported that soils with high cation exchange capacity and organic matter bind ions and make them less bioavailable. The presence of wastewater can influence the attenuation process of different contaminants in soil. The initial spreading of contaminants in soil is primarily controlled by the soil hydraulic properties, but these properties of soil can be changed significantly following the wastewater application. The increased water holding capacity can prolong the residence time of the contaminants in the wastewater-applied soil. The sorption/attachment or entrapment of organic contaminants, nutrients, and microorganisms to soil particles usually increased with the residence time (Casey et al., 2004). The increased water holding capacity decreases gradually, but some biochemical activities in relation to the fate of the wastewater constituents vary considerably with the initial soil water content. Zitnick et al. (2011) suggested that the presence of wastewater can change the inactivation, degradation and mineralization of some organic contaminants. The solute transport parameters in soil can also be influenced by the wastewater irrigation. The changed physicochemical properties of soil can alter the infiltration and runoff of rainwater and other environmental processes, such as green house gas emission (Gilley and Risse, 2000). The wastewater organic matter remaining in soil can increase the hydrophobicity of the soil, and as a result infiltrating water can tend to bypass this hydrophobic portion. It can also enhance the hydrophobic interaction between soil and contaminants. The changed pore structure in the presence of very fine and colloidal wastewater organic matter can affect diffusivity of the soil (Olesen et al., 1997), which in turn will affect uptake of nutrients by plant, sorption-desorption processes, and degradation and leaching of wastewater constituents.

The effects of wastewater irrigation on soil hydraulic properties are expected to depend on soil textures, wastewater types and the application rate of the wastewater. Spatial and temporal variations of soil hydraulic properties are predicted in the wastewater-applied land.

For better on-farm wastewater management and for risks assessment of water quality issues, it would be useful to investigate the influence of soil-wastewater incorporation on the soil properties. It was suggested (e.g., Amin, 2011; Laegdsmand et al. 2010; Schjonning et al., 1994) that the presence of wastewater would:

1) increase water holding capacity due to organic matter content,
2) increase sorption/ attachment ability because of added sorption sites and increased hydrophobic interaction, and
3) reduce diffusivity due to higher tortuosity.

However, the overall effect of wastewater application on soil would decrease gradually due to degradation and movement of wastewater constituents.

5.3. Impacts on the Water Environment

Wastewater is used as nutrient and organic-matter source to the agricultural land worldwide, but it is also releasing many emerging contaminants, such as heavy metals, pathogens and organic contaminants, to the environment (Lee et al., 2007; Unc and Goss, 2004). There is a potential risk of contamination of both shallow groundwater and surface water via the land-applied wastewater (Laegdsmand et al., 2009; Kjaer et al., 2007). The contaminants once released from agricultural soils, take different pathways to surface water and groundwater bodies via surface runoff, subsurface flow, and subsurface drainage (Kjaer et al., 2007; Lee et al., 2007). Figure 1 illustrates possible pathways of contaminant spreading from land-applied wastewater. Wastewater is usually applied to agricultural field by a surface application or by injection at around 6–10 cm soil depth (Hadrich et al., 2010). Wastewater remaining on top of the soil can contaminate nearby surface water bodies quickly during subsequent precipitation events if overland flow dominates the hydrology of the land compared to the condition where precipitation joins surface water mainly through subsurface flow or groundwater flow (Lee et al., 2007).

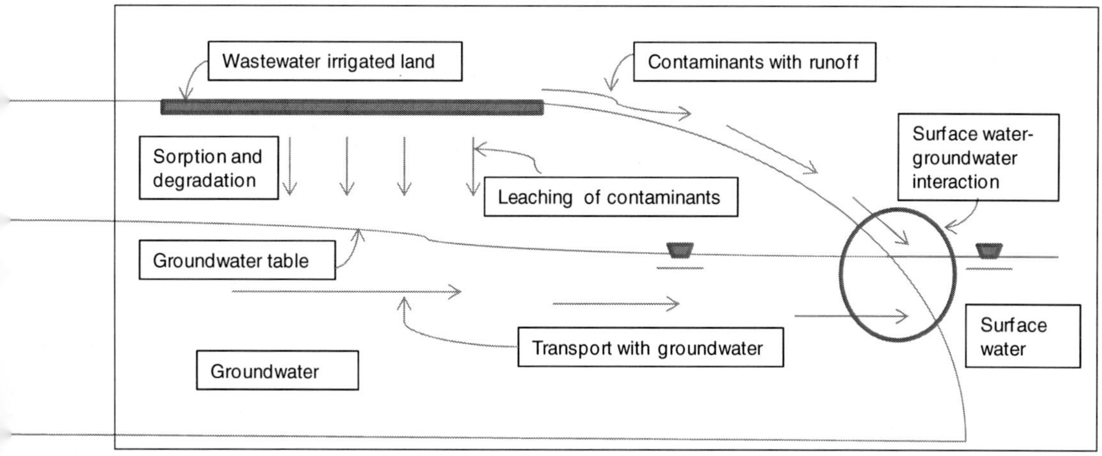

Figure 1. Schematic pathway diagram of constituents/ contaminants from land applied wastewater.

Soil-injected wastewater, on the other hand, could increase the persistence and leaching of wastewater-borne contaminants to groundwater because the injection method reduces the length of leaching path as well as the opportunity for incorporation into the bulk soil (Cameron et al., 1996; Hutchison et al., 2004; Semenov et al., 2009). Contamination of surface water via leaching is much more prominent than that by surface runoff where farmed-lands are drained systematically by sub-surface drainage system (Kjaer et al., 2007). While the fate and transport of various contaminants in soil and water has received substantial attention (Unc and Goss, 2004; Hanselman et al., 2003), their movement and persistence in wastewater-treated soil still need more investigations (Lee et al., 2007). The attenuation of pollution in soil can be affected by the presence of wastewater. Wastewater constituents may remain concentrated due to poor distribution in soil, which can increase the persistence of strongly sorbing organic contaminants (Zitnick et al., 2011). It has been observed that microorganisms can survive in the wastewater-applied soil more than two months (Amin, 2011). The persistence of the wastewater-borne contaminants in soil before commencing subsequent precipitation determines the potential leaching risk and groundwater contamination. Similarly, nitrification and denitrification processes can influence the risk of nitrate leaching.

Table 2. Possible impacts of wastewater irrigation on human health [Reviewed by Carr et al. (2004)]

	Nematode infection	Bacterial/Viruses	Protozoa
Consumers	Significant risks of Ascaris infection for both adults and children with untreated wastewater	Cholera, typhoid and shigellosis outbreaks reported from use of untreated wastewater; sero-positive responses for Helicobactor pylori (untreated); increase in non-specific diarrhea when water quality exceeds 10^4 FC/100 ml	Evidence of parasitic protozoa found on wastewater-irrigated vegetables surfaces but no direct evidence of disease transmission
Farm workers and their families	Significant risks of Ascaris infection for both adults and children with contact with untreated wastewater	Increased risk of diarrheal disease in young children with wastewater contact if water quality exceeds 10^4 FC/100 ml; elevated risk of Salmonella infection in children exposed to untreated wastewater; elevated seroresponse to Norovirus in adults exposed to partially treated wastewater	Risk of Giardia intestinalis infection was insignificant for contact with both untreated and treated wastewater, increased risk of ameobiosis observed from contact with untreated wastewater

Sources (direct citation): Armon et al. (2002), Blumenthal and Peasey (2002), Blumenthal et al. (2000).

Soil and water environments are generally not suitable for the survival of wastewater-borne microorganisms. However, some of those microorganisms can still survive 2–3 months in soil environment (Amin, 2011) and they can contaminate fresh agricultural produce, bathing facilities or even drinking water if not managed properly. The possible health hazards created by the wastewater-borne microorganisms are described in Table 2 [Reviewed by Carr et al. (2004)]. Wastewater from animal farms is a major source of different organic contaminants. Considerable concentration of steroid hormones has been detected in natural aquatic environment (Barel-Cohen et al., 2006; Raman et al., 2004). Very low concentration of these hormones can harm the reproduction system of aquatic animal. A detectable amount of hormones can be transported considerable length of stream (Barel-Cohen et al., 2006). Snow et al. (2008) stated that the hormones released into a stream can migrate to river bed sediments, into soil, and probably into groundwater. The excessive input of nutrients especially nitrogen and phosphorous to water bodies are responsible for eutrophication of aquatic ecosystems. Irrigated agriculture which accounts for over two-third of the world's current freshwater use degrades water environment through the excessive use of fertilizer and agricultural chemicals as well as the loss of top soil (Bratt, 2002). The process of nutrients over-enrichment of natural water- 'eutrophication' is a threat to sustainable water resources management all over the world. Certain amount of nutrients enrichment in water is useful for the growth of aquatic plants and animals but excessive nutrient enrichment can enhance rapid plant growth and lead to troublesome algal bloom. All these problems will be more acute in developing countries because of the simultaneous effect of population growth and rising living standard.

6. Management of Wastewater Irrigation

Since wastewater use in irrigation has both beneficial and harmful effects, proper management is crucially important for its sustainable use. Treatment of wastewater before disposal or any field application is mandatory in many countries. Wastewater from some specific sources like livestock farms may be applied to agricultural land without treatment in accordance with the existing regulations in some countries. Farmers and other wastewater users are not always persuaded to treat or manage wastewater properly due to lack of enforcement of laws and unawareness of environmental consequences. The necessity for treatment and management system significantly varies with the properties of wastewater as well as the desired uses of the wastewater. Wastewater from rural household and livestock farms may contain less persistent contaminants in comparison to industrial and municipal effluents, which can contain many hazardous contaminants, including heavy metals and antimicrobials. It is thus important to develop site-specific monitoring and management systems. The following strategic measures can be adopted for the safe use of wastewater:

6.1. Dilution

Wastewater diluted with conventional irrigation water can be a viable option for the application to agricultural land if the wastewater contains excess nutrients. In this way concentration of harmful contaminants can be lowered below the lowest effect level.

Comparatively low solid content in the diluted wastewater can also facilitate the infiltration and redistribution of irrigation water in the root zone area.

6.2. Blending

Water from conventional sources can also be lower quality, e.g., saline groundwater. A blend of saline groundwater and non-saline treated wastewater could produce a blended water of acceptable level of salinity and contaminant concentration, for instance. A suitable ratio of wastewater and conventional water can be identified based on the properties of the waters (Ali, 2010).

6.3. Alternative Use of Fresh-Water and Wastewater

To reduce the problem of over-enrichment of nutrients, we can alternately use wastewater and water from conventional source. Major portion of wastewater-borne organic contaminants could be degraded and microorganisms inactivated before subsequent wastewater application, so this management technique would reduce the accumulation of contaminants in soil.

6.4. Use of Wastewater at Less Sensitive Stage(s) of Crop

Certain growth stages of crop may be more sensitive to contaminated water while some other may not. One or more less sensitive growth stages should be identified, and then the wastewater could be used at that stage(s) in order to have a less detrimental effect to crop. Wastewater should be applied very carefully to the ready-to-eat types of agricultural produce. Harvesting period is also very sensitive to low quality water.

6.5. Reducing Nutrient Load before Field Application

A portion of nutrient-rich dry matter can be removed from slurry before field application to reduce the nutrient load to the agricultural land and protect water environment from excess nutrients. Slurry management techniques including solid-liquid separation are being used for improved nutrient management, easier handling of slurry and reduction of obnoxious odour emission after field application.

A major portion of the separated solid fraction can be used in biogas plants. To synchronize the increasing slurry production of a growing animal industry with the nutrient and water demand of crop production, slurry separation technology has become a solution for slurry management through recycling on farms. The separation technology is becoming increasingly popular in Europe, e.g., from 4 plants in 2002 to 50 plants in 2010 in Denmark. This solid-liquid separation also reduces the concentration of strongly sorbing contaminants in the separated liquid fraction.

6.6. Filtering Wastewater before Application

To reduce specific hazardous element, wastewater can be filtered using low-cost filter. In addition, different appropriate coagulant can be applied to reduce the load of particular contaminants.

6.7. Proper Scheduling of Wastewater Irrigation

Scheduling of wastewater irrigation is also very important for increasing water use efficiency and reducing the spreading of harmful contaminants. Irrigation return flow from wastewater-applied land can carry nutrients and contaminants to nearby streams. Soil environment is not favorable for most of the wastewater-borne pathogens, and some organic contaminants degrade rather quickly after field application, so the risks of leaching and transport of these contaminants decrease rapidly with the time between the field application and the subsequent precipitation (Amin, 2011). Scheduling of irrigation should be made in such a way that wastewater irrigation will be applied in the beginning of a dry spell so as to provide time for degradation of contaminants and inactivation of pathogens in surface soil. It will thus reduce the opportunity of spreading of pollutants in water environment.

7. Challenges in Using Wastewater

Because wastewater is not a conventional source of irrigation water, there are some challenges that must be addressed for sustainable use of wastewater for irrigation. There is still debate about the social acceptance of wastewater irrigation. Social acceptance is a dynamic process. It will increase if a population is severely affected by an acute shortage in their water supply. Moreover, if the frequency of drought and water scarcity condition increases with global warming, it will dramatically augment the necessity of water reuse. Society as a whole needs to be prepared well in advance to cope with this condition. Management of wastewater in terms of quality and quantity is another important dispute. It is still somewhat unknown how much wastewater will be available for irrigation considering the requirements of quality, conveyance, and social acceptance. It is a great challenge to develop a scenario and strategy for wastewater reuse in any specific region of a developing country. Slurry from animal farms is being treated before field application in many developed countries. One of the current challenges in animal wastewater management is to increase the slurry separation efficiency. Phosphorous removal is the main indicator used in determining the efficiency of slurry separation technology. In Europe, more than 65% of animal manure is handled and stored in liquid form, as a mixture of feces, urine, and water. This slurry usually contains 3–12% dry matter, from which on average 40–60% dry matter is removed by mechanical separation. Slurry particles primarily range from dissolved/ colloidal material to several cm. Suspended and dissolved particles can increase the adsorption/ attachment of sorbing nutrients in the liquid phase, which will reduce the removal efficiency of nutrients with the solid fraction. It is important to increase the removal of heavy metals as well as excess nutrients from wastewater. To do so accurately and precisely, the partitioning of nutrients and heavy metals among different particle size grades of slurry need to be investigated. It was observed by Amin et al. (2011) that the partitioning of organic contaminants among different particle size grades of manure was significantly different. The results obtained from similar experiments would provide information to decide how many particles from raw wastewater should be removed to achieve desired contaminant-removal efficiency. It will also be necessary to study the effect of pre-treatment, such as ozonation, acidification, and coagulation, before or after mechanical separation, on the partitioning of nutrients and other contaminants.

8. General Discussion

As discussed above, irrigation by wastewater, because of its nutrient and organic matter content, can increase crop yield, but it can also spread harmful contaminants to soil and water environment. Protection of water environment is critically important as water resources are becoming scarce because of population growth and increasing pollution in many parts of the world (O'Geen et al., 2010; Lee et al., 2007). Judicious use of wastewater and treatment of wastewater before field application if necessary can help to improve water security for agriculture. However, more studies should be conducted to assess the overall impacts of wastewater irrigation on the quality and quantity of agricultural produce as well as the long–term consequences on soil and water environment.

References

Ali, M.H. 2010. Water– An element of irrigation. *In*: Fundamentals of Irrigation and On-farm Water Management. *Springer Science Business* Media, NY.

Amin, M.G.M. 2011. Environmental fate of pathogens and hormones applied to agricultural land in treated slurry. PhD Thesis, *Aarhus Faculty of Science and Technology*, Aarhus University. ISBN: 978-87-91949-94-4. August 2011.

Amin, M.G.M., S.O. Petersen, and M. Lægdsmand. 2012. Sorption of 17β-estradiol to pig slurry separates and soil in soil-slurry environment. *J. Environ. Qual.* 41:179–187.

Armon, R., D. Gold, M. Brodsky, and G. Oron. 2002. Surface and subsurface irrigation with effluents of different qualities and presence of Cryptosporidium oocysts in soil and on crops. *Wat. Sci. Technol.* 46:115–122.

Baat, E. 1989. Effects of heavy metals on soil microbial processes and populations (a Review). *Wat. Air Soil Pollut.*, 47:353–379.

Barel-Cohen, K., L.S. Shore, M. Shemesh, A. Wenzel, J. Mueller, and N. Kronfeld-Schor. 2006. Monitoring of natural and synthetic hormones in a polluted river. *J. Environ. Manage.* 78:16–23.

Blumenthal, U., and A. Peasey. 2002. *Critical Review of Epidemiological Evidence of the Health Effects of Wastewater and Excreta Use in Agriculture.* World Health Organization (WHO), Geneva, Switzerland. (Unpublished document available on request).

Blumenthal, U., D.D. Mara, A. Peasey, G. Ruiz-Palacios, and R. Stott. 2000. Guidelines for the microbiological quality of treated wastewater used in agriculture: recommendations for revising WHO guidelines. Bullet. *World Health Organiz.* 78(9):1104–1116.

Bradford, S.A., E. Segal, W. Zheng, Q. Wang, and S.R. Hutchins. 2008. Reuse of Concentrated Animal Feeding Operation Wastewater on Agricultural Lands. *J. Environ. Qual.* 37:S97–S115.

Bratt, A. 2002. Farmers' Choices: Management Practices to Reduce Nutrient Leakage within a Swedish Catchment. *J. Environ. Plann. Manage.* 45(5):673–689.

Cameron, K.C., A.W. Rate, M.J. Noonan, S. Moore, N.P. Smith, and L.E. Kerr. 1996. Lysimeter study of the fate of nutrients following subsurface injection and surface application of dairy pond sludge to pasture. *Agric. Ecosys. Environ.* 58:187–197.

Carr R.M., U.J. Bluementhal, and D.D. Mara. 2004. Health guidelines for the use of wastewater in agriculture: Developing realistic guidelines. Waste water use in irrigated agriculture, *CAB International Publishing*. 41–58.

Casey, F.X.M., H. Hakk, J. Simunek, and G.L. Larsen. 2004. Fate and transport of testosterone in agricultural soils. *Environ. Sci. Technol.* 38:790–798.

Chakrabarti, C. and T. Chakrabarti. 1988. Effects of sewage and sludge application on the growth, yield, enzyme activity and micronutrient uptake of cotton crop plants. *Int. J. Environ. Stud.*, 34:169–178.

Coppola, A., A. Santini, P. Botti, S. Vacca, V. Comegna, and G. Severino. 2004. Methodological approach for evaluating the response of soil hydrological behavior to irrigation with treated municipal wastewater. *J. Hydrol.*, 292:114–134.

Day, A. D., and T. C. Tucker. 1977. Effects of treated wastewater on growth, fibre, protein and amino acid content of sorghum grains. *J. Environ. Qual.*, 6:325–327.

Edmeades, D.C. 2003. The long-term effects of manures and fertilisers on soil productivity and quality: a review. *Nutri. Cycl. Agroecosys*., 66:165–180.

Feigin, A., I. Ravina, and J. Shalhevet. 1990. Irrigation with treated sewage effluent. *Eological Series*, Springer Verlag, New York.

Faurés, J.M., J. Hoogeveen, and J. Bruinsma. Access in 2012. ftp://ftp.fao.org/agl/aglw/ docs/ fauresetalagadir.pdf.

Gilley, J.E., and L.M. Risse. 2000. Runoff and soil loss as affected by the application of manure. *Trans*. ASAE 43:1583–1588.

Hadrich, J.C., T.M. Harrigan, and C.A. Wolf. 2010. Economic Comparison of Liquid Manure Transport and Land Application. *Appl. Engineer. Agric*., 26:743–758.

Hanselman, T.A., D.A. Graetz, and A.C. Wilkie. 2003. Manure-borne estrogens as potential environmental contaminants: A review. *Environ. Sci. Technol.* 37:5471–5478.

Hasan, M. M. 2011. Growth and yield of wheat under irrigation by dairy wastewater. Master of Science Thesis, Department of Irrigation and Water Management, *Bangladesh Agricultural University*, Mymensingh-2202, Bangladesh. December 2011.

Hussain, I., L. Raschid, M. A. Hanjra, F. Marikar, and W. Van der Hoek. 2012. Wastewater use in agriculture: Review of impacts and methodological issues in valuing impacts. *International Water Management Institute*, Working paper 37.

Hutchison, M.L., L.D. Walters, A. Moore, K.M. Crookes, and S.M. Avery. 2004. Effect of length of time before incorporation on survival of pathogenic bacteria present in livestock wastes applied to agricultural soil. *Appl. Environ. Microbiol.* 70:5111–5118.

Kjaer, J., P. Olsen, K. Bach, H.C. Barlebo, F. Ingerslev, M. Hansen, and B. Halling-Sorensen. 2007. Leaching of estrogenic hormones from manure-treated structured soils. *Environ. Sci. Technol.* 41:3911–3917.

Laegdsmand, M., H. Andersen, O.H. Jacobsen, and B. Halling-Sørensen.2009. Transport and fate of steroid estrogens in slurry-amended soil monoliths. *J. Environ. Qual.* 38:955–964.

Laegdsmand M., P. Moldrup, and P. Schjonning. 2010. Multitracer and filter-separated half-cell method for measuring solute diffusion in undisturbed soil. *Soil Sci. Soc. Ame. J.* 74:1084–1091.

Lee, L.S., N. Carmosini, S.A. Sassman, H.M. Dion, and M.S. Sepulveda. 2007. Agricultural contributions of antimicrobials and hormones on soil and water quality. *Advanc. Agron.*, 93:1–68.

Markfoged, R., L.P. Nielsen, T. Nyord, L.D.M. Ottosen, and N.P. Revsbech. 2011. Transient N2O accumulation and emission caused by O_2 depletion in soil after liquid manure injection. *Euro. J. Soil Sci.* 62:541–550.

Misra, R.N., and P.K. Behera. 1991. The effect of paper industry effluent on growth, pigment, carbohydrate and protein of rice seedlings. *Environ. Pollut.*, 72:159–167.

Mosaddeghi, M.R., A.A. Mahboubi, and A. Safadoust. 2009. Short-term effects of tillage and manure on some soil physical properties and maize root growth in a sandy loam soil in western Iran. *Soil Til. Res.* 104:173–179.

Ndayegamiye, A., and D. Cote. 1989. Effect of long-term pig slurry and solid cattle manure application on soil chemical and biological properties. *Can. J. Soil Sci.* 69:39–47.

O'Geen, A.T., R. Budd, J. Gan, J.J. Maynard, S.J. Parikh, and R.A. Dahlgren. 2010. Mitigating Nonpoint Source Pollution in Agriculture with Constructed and Restored Wetlands. *Advan. Agron.*108:1–76.

Olesen, T., B.S. Griffiths, K. Henriksen, P. Moldrup, and R. Wheatley. 1996. Modeling diffusion and reaction in soils: V. Nitrogen transformations in organic manure amended soil. *Soil Sci.* 162:157–168.

Olesen, T., P. Moldrup, and K. Henriksen. 1997. Modeling diffusion and reaction in soils .6. Ion diffusion and water characteristics in organic manure-amended soil. *Soil Sci.* 162:399–409.

Petersen, S.O., and M.N. Andersen. 1996. Influence of soil water potential and slurry type on denitrification activity. *Soil Biol. Biochem.* 28:977–980.

Petersen, S.O., H.H. Nissen, I. Lund, and P. Ambus. 2003. Redistribution of slurry components as influenced by injection method, soil, and slurry properties. *J. Environ. Qual.* 32:2399–2409.

Raman, D.R., E.L. Williams, A.C. Layton, R.T. Burns, J.P. Easter, A.S. Daugherty, M.D. Mullen, and G.S. Sayler. 2004. Estrogen content of dairy and swine wastes. *Environ. Sci. Technol.*, 38:3567–3573.

Schaap, M.G., F.J. Leij, and M.T. van Genuchten. 2001. ROSETTA: a computer program for estimating soil hydraulic parameters with hierarchical pedotransfer functions. *J. Hydrol.* 251:163–176.

Schjonning, P., B.T. Christensen, and B. Carstensen. 1994. Physical and chemical properties of a sandy loam receiving animal manure, mineral fertilizer or no fertilizer for 90 years. *Euro. J. Soil Sci.* 45:257–268.

Semenov, A.V., L. van Overbeek, and A.H.C. van Bruggen. 2009. Percolation and Survival of Escherichia coli O157:H7 and Salmonella enterica Serovar Typhimurium in Soil

Amended with Contaminated Dairy Manure or Slurry. *Appl. Environ. Microb.,* 75:3206–3215.

Snow, D.D., S.L. Bartelt-Hunt, S.E. Saunders, S.L. Devivo, and D.A. Cassada. 2008. Detection, Occurrence and Fate of Emerging Contaminants in Agricultural Environments. *Wat. Environ. Res.* 80:868–897.

Thomson, J.A., and D.L. Chase. 1992. Effect of limited irrigation on growth and yield of some dwarf wheat in southern New South Wales. Aust. *J. Expt. Agric.,* 32:725–730.

Thomsen, I.K., A.R. Pedersen, T. Nyord, and S.O. Petersen. 2010. Effects of slurry pre-treatment and application technique on short-term N_2O emissions as determined by a new non-linear approach. *Agric. Ecosys. Environ.* 136:227–235.

Tiarks, A.E., A.P. Mazurak, and L. Chesnin. 1974. Physical and Chemical Properties of Soil Associated with Heavy Applications of Manure from Cattle Feedlots. *Soil Sci. Soc. Ame. J.* 38:826–830.

Unc, A., and M.J. Goss. 2004. Transport of bacteria from manure and protection of water resources. *Appl. Soil Ecol.* 25:1–18.

UN (United Nations). 2005. World population to reach 9.1 billion in 2050, UN projects. *Retrieved* 2009-03-12.

Wisser, D., S. Frolking, E.M. Douglas, B.M. Fekete, C.J. Vorosmarty, and A.H. Schumann. 2008. Global irrigation water demand: Variability and uncertainties arising from agricultural and climate data sets. *Geophys. Res. Lett.* 35, L24408, doi:10.1029/2008GL035296.

Yuan, Y. 1993. Etiological study of high stomach cancer incidence among residents in wastewater-irrigated areas, Environmental Protection Science 19(1), 70-73, (in Chinese). In: World Resources Institute, A Guide to the Global Environment: *Environmental Change and Human Health*, Oxford University Press, New York, USA, pp. 122.

Zitnick, K.K., N.W. Shappell, H. Hakk, T.M. DeSutter, E. Khan, and F.X.M. Casey. 2011. Effects of liquid swine manure on dissipation of 17beta-estradiol in soil. *J. Hazard. Mater.* 186:1111–1117.

INDEX

#

21st century, 35

A

ABA, 71, 74, 75, 76, 82
abstraction, 178, 261
access, ix, 33, 47, 181, 182, 183, 184, 185, 186, 188, 190, 264, 271
accessions, 224, 226
accounting, 32, 46, 47, 119, 201
acetic acid, 207
acid, 71, 74, 82, 194, 205, 207, 208, 213, 218, 219, 221, 222, 226, 297
acidic, 15, 166
adaptability, 135
adaptation, 66, 125, 140
adjustment, 143
adsorption, ix, 159, 160, 163, 176, 295
adults, 292
adverse effects, 18, 19, 88, 89, 155
aerosols, 16
Africa, 1, 2, 3, 42, 137, 179, 182, 183, 191, 192, 253, 255, 256, 271
age, ix, 23, 49, 181, 184, 185, 186, 190
agencies, 267
agricultural sector, x, 160, 285
air temperature, 235
alfalfa, 223
alimentation, 66, 68
almonds, 84
alters, 82
amino, 297
ammonium, 46, 96, 124, 207, 208
anatomy, 161
anchorage, 146
anisotropy, 161
anther, 14
antioxidant, 225
APC, 35
apple(s), viii, 44, 67, 71, 72, 73, 74, 75, 76, 77, 78, 79, 80, 81, 82, 83, 84, 85, 86
application techniques, 31
aquaculture, 265
aquifers, 163, 183, 280, 287
Arabidopsis thaliana, 226
arbitrage, 276
Argentina, 137, 222
ARS, 127, 193
ascorbic acid, 207
Asia, 1, 2, 3, 15, 20, 34, 35, 37, 38, 130, 137, 141, 155, 156, 192
Asian countries, 129
aspiration, 50
assessment, ix, 118, 124, 137, 156, 159, 160, 162, 163, 170, 174, 176, 177, 178, 179, 224, 291
assets, 276
assimilation, 73, 86, 199, 220
atmosphere, 16, 63, 65, 140, 199, 201
attachment, 290, 291, 295
attitudes, 275, 276, 277
authorities, 267
aversion, 32
awareness, 28, 88, 132, 155, 288

B

bacteria, 288, 297, 299
Bangladesh, 1, 4, 17, 20, 24, 27, 36, 37, 141, 178, 192, 285, 287, 297
barriers, 31, 34, 182
base, 11, 12, 23, 139, 141, 171, 201
batteries, 48

beneficial effect, 194
beneficiaries, 152
benefits, 24, 28, 31, 32, 33, 82, 135, 137, 140, 152, 155, 182, 184, 195, 230, 256, 271, 277, 287, 289
bicarbonate, 160
biochemistry, 223
biogas, 294
biological processes, 290
biomass, 43, 72, 116
bioregulators, 82
biosynthesis, 199, 220
biotic, 218, 219, 220
birds, 266
blends, 89
blood, 225
bone, 225
boreholes, 183, 185, 261
Brazil, 2, 4, 137, 193
breakdown, 164
breeding, 128
Bulgaria, 225
Burkina Faso, 192
Burma, 2
buttons, 236, 250

C

Ca^{2+}, 167
Cairo, 35
calcium, 160, 164, 165, 169, 173
calculus, 275
calibration, viii, 44, 49, 50, 61, 62, 64, 65, 67, 207
cambium, 50
Cambrian, 183
canals, 30, 152, 256, 257, 263, 264, 265, 266, 268, 269
cancer, 225, 299
capacity building, 152
capital budgeting methods, vii, x, 273, 280
capital expenditure, 32
capital input, 277
capital intensive, 182
capital markets, 188
carbohydrate(s), 194, 220, 298
carbon, 2, 73, 84, 86, 156
Caribbean, 4
case study(ies), 38, 123, 157, 160, 165, 177, 178, 179, 271
cash, 90, 260, 275, 276
cash flow, 275
catchments, 28, 270
categorization, 171
cation, 164, 290
cattle, 264, 298
cell membranes, 199
Census, 38, 39
challenges, vii, x, 35, 182, 183, 271, 285, 295
chemical(s), ix, 30, 50, 82, 132, 136, 159, 163, 165, 167, 169, 174, 176, 177, 221, 287, 288, 289, 290, 293, 298
chemical characteristics, 177, 288
chemical properties, 298
chicken, 194
children, 292
China, vii, x, 2, 3, 4, 6, 24, 34, 36, 37, 39, 83, 125, 128, 191, 229, 230, 235, 244, 253
chlorine, 167
city(ies), 127, 152
civilization, 230
classes, 129, 172, 174, 175, 176
classification, 6, 7, 165, 171, 172, 174, 176, 192, 236
clean energy, 229, 230, 232, 244, 252
climate, 32, 34, 35, 45, 69, 72, 83, 84, 88, 111, 131, 132, 165, 192, 236, 240, 258, 280, 299
climate change, 88, 131, 280
climates, 7, 72, 160, 230, 235
climatic factors, 67
closure, 21, 55, 58, 71, 74
CO2, 86
coal, 276
collaboration, 139
Colombia, 4
color, 80, 81, 207, 208
combined effect, 106, 114, 219
commercial, 90, 101, 111, 119, 134, 193, 199
commodity, 28, 134, 278, 279, 280
common symptoms, 17
community(ies), 107, 135, 141, 149, 185, 264, 265, 266, 267, 269, 270, 271
compaction, 30
compensation, 44, 67, 124
competition, viii, 19, 72, 74, 129, 147
competitive markets, 278
complexity, 62, 275
complications, 256
composition, ix, 83, 85, 126, 163, 167, 176, 177, 193, 194, 199, 206, 219, 220, 221, 222, 224, 225, 226, 227, 289

compounds, 160, 194, 199, 221
computation, 187, 188
computer, x, 48, 201, 203, 229, 230, 298
computing, 35, 47, 66, 122, 221
concordance, 54, 58, 59
conductance, 47, 59, 67, 76
conduction, 50, 64
conductivity, 62, 64, 89, 159, 163, 166, 174, 290
conference, 144
conflict, 130
Congo, 2
Congress, 35, 122, 123, 126, 271
consensus, 267
conservation, vii, viii, x, 1, 32, 33, 35, 37, 38, 129, 130, 134, 137, 139, 155, 156, 157, 260, 273, 286
conserving, 33, 34, 149, 157, 191
constituents, 20, 22, 163, 194, 222, 288, 289, 290, 291, 292
construction, 28, 32, 256, 260, 265, 266, 270
consumption, vii, 1, 7, 41, 42, 43, 44, 60, 65, 88, 90, 152, 286
containers, 11, 265
contaminant, 291, 294, 296
contaminated water, 294
contamination, 6, 286, 289, 291, 292
cooking, 256, 265, 266
cooling, 41, 48, 61
coordination, 267
copper, 280
correlation, 52, 53, 56, 62, 65, 161, 167, 219, 280
correlation coefficient, 52
cost, 25, 28, 31, 32, 33, 131, 132, 134, 138, 140, 142, 181, 183, 189, 190, 202, 218, 223, 271, 278, 279, 285, 287, 295
cost saving, 33
cotton, viii, 86, 122, 123, 124, 126, 130, 132, 134, 137, 138, 139, 140, 143, 148, 149, 151, 152, 297
covering, 141, 245
creep, 26
critical period, 19, 112
crop production, viii, ix, 19, 82, 87, 88, 89, 118, 126, 130, 139, 149, 229, 230, 236, 261, 287, 289, 294
crop residue, 134, 140
cross-validation, 162, 169
crown, 7, 197
crust, 165, 166
crystalline, 183
cultivar(s), viii, 6, 9, 10, 36, 39, 46, 47, 61, 71, 72, 74, 80, 84, 85, 88, 90, 96, 193, 194, 198, 199, 205, 218, 221, 222, 224, 225, 227
cultivation, vii, ix, 1, 2, 6, 7, 15, 19, 34, 35, 36, 38, 42, 68, 72, 90, 96, 134, 139, 140, 181, 184, 189, 190, 229, 231
cultural conditions, 12
cultural practices, viii, 3, 87, 90, 141
culture, 11, 34, 38, 39, 66
cumulative distribution function, 184
current prices, 274
cycles, 48, 61
Cyprus, 68
cytokinins, 84

D

damages, 264
danger, 88, 122
data analysis, 206
data collection, 184, 202
data set, 299
decay, ix, 193, 195, 198, 224, 225
decomposition, 259
defense mechanisms, 219
deficiency(ies), 7, 88, 235
deficit, vii, viii, 17, 19, 27, 42, 62, 68, 71, 72, 73, 74, 78, 80, 81, 82, 83, 84, 85, 87, 88, 89, 94, 95, 96, 99, 101, 102, 106, 107, 108, 110, 111, 112, 114, 115, 116, 117, 118, 122, 123, 124, 125, 127, 128, 152, 201
deficit irrigation practices, viii, 87
degradation, vii, ix, 6, 88, 89, 122, 159, 219, 230, 260, 271, 289, 290, 291, 295
dehydration, 72, 77
Delta, 47, 48, 49, 205, 209, 210, 211, 212, 214, 215, 216, 217
denitrification, 290, 292, 298
Denmark, 294
Department of Agriculture, 38, 157, 179
Department of Commerce, 38, 39
deposits, 168
depression, 25, 55
depth, 7, 10, 19, 20, 21, 23, 24, 25, 28, 34, 45, 75, 90, 101, 116, 119, 124, 126, 143, 145, 148, 201, 203, 205, 234, 236, 257, 260, 291
derivatives, 280
desorption, 290
detectable, 293
developed countries, 295

developing countries, 156, 157, 286, 287, 288, 293
deviation, 58, 62, 162, 171
diarrhea, 194, 292
diffusion, 2, 286, 298
diffusion process, 286
diffusivity, 290, 291
digestion, 207
diploid, 6
disaster, 149
discharges, 165
discontinuity, 161
discs, 143
diseases, 15, 24, 138, 194, 199, 220, 224, 225, 262
dispersion, 162, 164
distribution, 7, 12, 32, 43, 64, 67, 68, 72, 73, 80, 83, 90, 108, 119, 126, 130, 160, 162, 168, 177, 184, 192, 202, 280, 287, 292
distribution function, 184
divergence, 55
diversification, 36, 131, 138
diversity, 2, 6
DOI, 84, 177, 179
domestication, 3
donors, 182, 260
dough, 5, 12, 14
drainage, vii, 1, 20, 21, 30, 31, 32, 34, 67, 88, 116, 119, 123, 124, 126, 127, 131, 139, 146, 163, 165, 200, 222, 236, 245, 246, 250, 260, 264, 265, 266, 291, 292
drawing, 162
drinking water, 264, 293
drought, ix, 6, 11, 17, 19, 25, 26, 27, 28, 36, 37, 62, 63, 73, 76, 78, 84, 123, 124, 131, 154, 165, 182, 193, 194, 195, 200, 219, 220, 221, 222, 226, 236, 295
dry matter, 72, 84, 112, 116, 117, 118, 120, 207, 223, 287, 294, 295
drying, viii, 11, 23, 34, 36, 62, 71, 74, 75, 76, 77, 78, 79, 81, 82, 83, 84, 85, 86, 124, 131, 208

E

Easter, 298
economic development, 34, 130
economic efficiency, 37
economic evaluation, 81
economic growth, 130, 182
economic incentives, 129, 134
economics, 191
ecosystem, 25, 41, 65
Ecuador, 4
editors, 36, 37
education, ix, 155, 181, 182, 184, 185, 186, 190
efficiency level, 143
effluent, 124, 126, 287, 297, 298
effluents, 288, 293, 296
Egypt, 2, 35, 38
elaboration, vii
electrical conductivity, ix, 89, 159
elongation, 5, 13, 14, 17, 18
emission, 39, 90, 290, 294, 298
emitters, 47, 90, 101
empowerment, 129
endocrine, 288
endosperm, 7, 12, 13, 39
energy, x, 1, 16, 30, 32, 33, 82, 137, 142, 152, 157, 183, 200, 218, 229, 230, 232, 241, 242, 244, 252
energy expenditure, 33
enforcement, 269, 293
engineering, 177
entrapment, 290
environment, vii, x, 4, 5, 8, 14, 34, 35, 38, 74, 89, 123, 124, 163, 178, 192, 194, 216, 217, 218, 224, 244, 245, 256, 266, 271, 275, 276, 285, 286, 287, 288, 291, 293, 294, 295, 296
environmental conditions, 12, 58, 73, 76, 229, 262
environmental control, 126
environmental degradation, 260
environmental factors, 219
environmental impact, 30
environmental issues, 155
Environmental Protection Agency, 267
environmental stress, 195
environmental sustainability, 72, 130
environmental variables, 44
environments, viii, 5, 38, 65, 80, 85, 123, 214, 215, 222, 226, 231, 285, 287, 293
enzyme(s), 35, 219, 297
equilibrium, 61, 166
equipment, 30, 32, 33, 141, 204
equity, 129, 192
erosion, 31, 38, 137, 203, 235
EU, 42
Europe, 2, 287, 294, 295
evacuation, 116

evaporation, 20, 21, 24, 30, 42, 43, 55, 66, 71, 81, 100, 101, 102, 116, 119, 160, 169, 235, 240, 244, 245, 247, 248, 249, 261, 272
evapotranspiration, viii, 20, 22, 35, 42, 43, 45, 46, 60, 61, 64, 66, 69, 72, 73, 77, 87, 89, 90, 119, 122, 128, 165, 201, 221, 225, 229, 231, 234, 236, 237, 238, 247, 249, 259
evidence, viii, 2, 130, 280, 292
evolution, 3, 36, 167, 274
excision, 63
exercise, 278, 279
expenditures, 32, 33, 136
expertise, 32, 33
exploitation, 82, 88, 131, 154
exposure, 64, 219
externalities, 256, 272
extraction, 238

F

facies, 168
families, 135, 266, 292
farm size, ix, 131, 181, 184, 185, 186, 189, 190
farmer-based organization, ix, 181, 185, 188, 190
farms, x, 2, 32, 33, 90, 107, 122, 131, 132, 142, 182, 235, 267, 269, 273, 274, 286, 287, 288, 289, 293, 294, 295
fat, 6
fatty acids, ix, 193, 194, 199, 206, 207, 209, 211, 214, 216, 221
feces, 287, 295
fertility, 123, 140, 146
fertilization, 5, 12, 14, 86, 126
fertilizers, viii, 30, 74, 87, 96, 194, 287
fiber, vii, 12
field crops, 21, 72, 138, 256
field trials, 145
financial, 4, 33, 183, 190, 191, 271, 276, 278
financial performance, 183, 190
financial resources, 33
financial support, 4, 271
fish, 264, 265, 266
fisheries, 256, 270, 272
fishing, 264, 265
fixation, 226
fixed costs, 190
fixed rate, 33
flatulence, 194
flavo(u)r, 6, 219
flesh firmness, 80, 81
flexibility, 276, 277, 278
flooding, 20, 24, 30, 35, 146, 226, 257, 263
floods, 18, 131, 132
flour, 199, 219
flow value, 62
flowers, 7, 12
food, vii, viii, ix, 34, 36, 88, 107, 129, 130, 135, 139, 154, 155, 156, 181, 182, 199, 230, 260, 261, 286, 287, 289
food production, viii, 88, 129, 230, 286, 287
food security, viii, 34, 129, 130, 135, 139, 155, 182, 260, 286
force, 49, 232
forecasting, 27
formation, 20, 39, 62, 117, 183, 224, 269
formula, 69, 119, 165
France, 56, 57
freezing, 235
freshwater, 19, 34, 285, 286, 293
fruit growth, viii, 71, 72, 73, 74, 80, 83, 84
fruit quality, viii, 71, 72, 76, 80, 81, 83, 84, 85, 86
fruit size, 42, 74, 80
fruit weight, 80
fruits, 73, 74
funding, 155
fungi, 199
fungus, ix, 193, 195, 199, 220
furrow irrigation/bed planting, viii, 129

G

genes, 18, 194, 226
genetic background, 194
genetics, 128
genome, 6
genotype, ix, 74, 194, 206, 218, 219, 224
genus, 6
geometry, 143, 144, 177
Georgia, 207, 223
Germany, 272
germination, 4, 7, 13, 17, 198, 224, 226, 286
GIS, 177
global climate change, 88
global scale, 88
global warming, 295
governance, 156, 192, 271
grades, 296
grading, 136
grain size, 5
granules, 12

grass, 6
gravity, 29, 30, 131
grazing, 235
Greece, 2, 122
green revolution, 156
Green Revolution, 3, 130
greenhouse, 125, 156
guidance, 201
guidelines, 35, 221, 297
Gulf Coast, 2

H

habitat, 266
hairless, 12
harmful effects, 293
harmony, 38
harvesting, ix, 20, 25, 36, 143, 178, 191, 198, 205, 206, 219, 221, 229, 231, 248, 252, 270
hazards, 160, 286, 287, 293
health, 132, 156, 178, 194, 200, 224, 286, 288, 292, 293
heavy metals, 288, 291, 293, 295, 296
hedging, 276
height, 4, 5, 12, 17, 18, 32, 47, 68, 287
herbicide, 138
heterogeneity, 41, 60, 64, 65
heterosis, 3
highlands, 123
high-value crops, 260
histogram, 162
historical data, 149
history, vii, 1, 4, 5, 6, 149
HM, 214, 215, 216, 217
Holocene, 166
hormones, 82, 288, 293, 296, 298
horticultural crops, 83
House, 266
HPV, 44
human, 1, 194, 221, 261, 286, 288, 292
human health, 194, 286, 288, 292
humidity, 46, 55, 102, 165, 195, 198, 259
husband, 66
hybrid, 3, 4, 12
hydrophobicity, 290
hypothesis, 185, 188

I

ideal, 140
imagery, 85
impact assessment, 137, 178
improvements, viii, 6, 32, 33, 43, 80, 81, 87, 90, 155, 203
in vitro, 225
in vivo, 225
incidence, 18, 138, 260, 299
income, 135, 260, 275, 276, 277, 278
India, 1, 2, 4, 7, 17, 20, 24, 123, 138, 141, 144, 155, 156, 157, 178, 179, 181, 191, 192, 225, 271, 287
indirect measure, 166
Indonesia, 2, 4, 7
induction, 73
industrialization, 34
industries, 287, 288
industry, vii, 134, 144, 156, 194, 294, 298
inefficiency, 143
inequality, 277
inequity, 132
INF, 194, 205, 208, 213, 218, 219
infection, ix, 193, 195, 197, 198, 199, 206, 218, 219, 220, 221, 222, 226, 227, 292
infrastructure, 182, 256, 257, 264, 269, 270
inhibition, 199, 220
initiation, 4, 5, 7, 11, 13, 14, 94, 106, 112, 114, 124, 267
initiation time, 11
inoculum, 205
insects, 15, 16, 19
insecurity, ix, 181, 182
institutions, 139, 267
insulation, 61, 63
integration, viii, 88, 122
integrity, 199, 219, 220, 221
interest groups, 269
internalizing, 256
internode, 5, 13, 14
investment, x, 32, 34, 190, 256, 271, 273, 274, 275, 276, 277, 278, 279, 280
investment model, 279
investments, 32, 129, 134, 271
investors, 276
ions, 160, 163, 164, 166, 167, 168, 169, 176, 290
Iowa, 223
Iran, vii, x, 177, 229, 230, 235, 244, 298
iron, 143, 183
irrigation application, vii, 1, 24, 26, 111, 132, 135, 136, 137, 141, 142, 144, 145, 151, 287
Islam, 26, 29, 36, 38

Islamabad, 156
isoflavone, 208, 210, 212, 213, 215, 217, 218, 220, 223, 224, 225, 226
isoflavonoid, 223
issues, 2, 141, 155, 189, 291, 297
Italy, 35, 67, 71, 122, 123, 124, 126, 178

J

Japan, vii, x, 2, 5, 6, 20, 37, 126, 229, 230, 236, 237, 244, 253

K

Kenya, 137
kill, 195
knots, 64
Korea, 2, 6

L

lakes, 235, 256
land tenure, 184, 266
Latin America, 1, 4, 37
laws, 192, 269, 288, 293
leaching, ix, 87, 88, 89, 93, 94, 96, 100, 101, 102, 108, 109, 112, 116, 120, 122, 146, 159, 174, 176, 236, 244, 247, 250, 251, 252, 290, 292, 295
lead, 73, 162, 199, 200, 252, 293
leaf area, 4, 17, 44, 69, 73, 78, 112, 113, 287
Leaf conductance, 47, 59
leakage, 270
learning, 4
Lebanon, 177
legislation, 149
lens, 5
lesions, 198
life cycle, 16
light, 16, 21, 22, 23, 73, 80, 235
lignin, 194, 199, 220, 222
linoleic acid, 194, 208, 218
lipids, 219, 225
liquid phase, 295
livestock, 256, 260, 261, 264, 269, 270, 287, 289, 293, 297
local conditions, 140
local government, 267
logging, 131, 138
Louisiana, 2, 3
love, 66
low temperatures, 16
LSD, 100, 110, 118, 120, 209, 210, 211, 212, 214, 215, 216, 217
luminescence, 2
Luo, 37, 83
lying, 17, 30

M

machinery, 32, 134, 139, 140, 141
magnesium, ix, 159, 160, 165, 173, 176
majority, 88, 135, 259, 260, 266
Malaysia, 4, 20
Malus domestica, 84
man, 18
Manju, 221
manufacturing, 134, 141
manure, 39, 96, 189, 289, 295, 297, 298, 299
mapping, 163, 169, 177
Marani, 123
marginal product, 279
marketing, 12
marsh, 3
mass, 49, 78, 80, 155, 165, 286
materials, 199
matrix, 12, 45, 201, 224
matter, 12, 72, 84, 112, 116, 117, 118, 120, 140, 154, 207, 223, 259, 285, 286, 287, 288, 289, 290, 291, 294, 295, 296
measurement(s), 41, 43, 44, 46, 47, 49, 51, 54, 56, 58, 59, 61, 62, 63, 64, 67, 73, 76, 83, 84, 112, 149, 161, 166, 200, 201, 203, 222
Mediterranean, 2, 42, 45, 69, 72, 85, 124, 159
membership, ix, 181, 184, 190
membrane permeability, 199, 220
membranes, 199
metabolism, 199, 220, 222, 223
metals, 288, 291, 293, 296
meter, 4, 90, 100, 135, 201
methodology, 94, 114, 125, 201
Mexico, 156, 157
microclimate, 73, 200
microorganisms, 287, 288, 290, 292, 293, 294
microscope, 5
Microsoft, x, 229, 230
Middle East, 177
milligrams, 11
mineralization, 166, 167, 176, 183, 290
minors, 135, 152

Missouri, 3, 195
modelling, 176, 177, 274, 279, 280
models, ix, 169, 170, 185, 186, 190, 201, 223, 225, 229, 230, 231, 244, 249, 252, 274, 277, 278, 280
modernization, 130, 156
modifications, 32, 62
modules, 48
moisture, 21, 23, 29, 39, 43, 53, 54, 55, 58, 65, 71, 73, 75, 81, 82, 101, 104, 106, 112, 125, 142, 145, 160, 199, 200, 206, 223, 229, 231, 234, 238, 242, 245, 246
molecular biology, 223
Mongolia, x, 229, 230, 244
Montana, 126
Moon, 279, 283
Morocco, 256, 271
morphology, vii, 1, 6, 36
mother cell, 16
Mozambique, 2
MR, 194, 199, 205, 206, 208, 209, 210, 211, 212, 213, 214, 215, 216, 217, 218, 219, 220
multiplier, 271
multi-stage sampling, 185
Myanmar, 4

N

Na^+, 167
NaCl, 169, 246
National Academy of Sciences, 127
National Research Council, 37
Native Americans, 6
natural resources, 134, 230
Natural Resources Conservation Service (NRCS), 203, 225
negative consequences, 264
negative effects, ix, 181, 190, 220
Nepal, 141, 192
Netherlands, 36
neutral, 15, 276, 277, 278
New South Wales, 3, 5, 38, 137, 299
NGOs, 190
Nigeria, 178
NIR, 193, 205, 207, 220
nitrification, 290, 292
nitrogen, viii, 5, 39, 87, 96, 97, 98, 99, 123, 124, 125, 126, 127, 223, 226, 287, 289, 290, 293
nodes, 12, 13, 199
nodules, 226
North Africa, 42
North America, 6, 35, 287
nucellus, 12
null, 185
null hypothesis, 185
nutrient(s), 11, 30, 31, 68, 80, 83, 111, 126, 140, 147, 194, 219, 220, 221, 285, 286, 287, 288, 289, 290, 291, 293, 294, 295, 296, 297
nutrition, 39, 194, 221, 224

O

oil, ix, 15, 21, 38, 42, 46, 67, 101, 115, 119, 124, 193, 194, 199, 206, 207, 208, 209, 211, 213, 214, 216, 218, 219, 220, 223, 225, 235, 276, 277, 279, 280
oleic acid, 194, 208, 213, 218, 219, 220
olive oil, 42
operations, 31, 36, 136, 139, 269
opportunities, viii, 72, 84, 125, 130, 132, 182, 191, 192, 269, 277
opportunity costs, 33
optimism, 156
optimization, 73
organ, 73
organic matter, 20, 140, 154, 259, 285, 286, 287, 288, 289, 290, 291, 296
organic soils, 174
organize, 269
organs, 4, 7, 73, 78
ownership, 266, 267
oxidation, 219
ozonation, 296

P

Pacific, 31, 38, 86
Pakistan, v, vii, viii, 1, 129, 130, 131, 132, 135, 136, 137, 138, 139, 140, 141, 142, 143, 149, 154, 155, 156, 157, 178, 192
parallel, 3, 165
parents, 66
Partial root-zone drying (PRD), viii, 71, 74
pasture, 297
pathogens, ix, 193, 219, 288, 291, 295, 296
pathways, 291
PCA, 178
peace, 129
pedicel, 12
peel color, 80

percolation, 20, 21, 22, 24, 27, 30, 119, 266
peri-urban, 152
permeability, ix, 10, 159, 164, 165, 172, 176, 199, 220
permit, 17, 29, 186
Perth, 129
PET, 259
petroleum, 277, 281, 282
pH, ix, 74, 159, 160, 166, 167
phenol, 194, 199, 222, 225, 226
phenolic compounds, 194
Philippines, 2, 4, 24, 36, 38, 39
phosphate, 11
phosphorus, 208, 287, 293
photosynthesis, 73, 86
physical environment, 256
physical properties, 156, 298
physical structure, 164
physicochemical properties, 289, 290
physiological mechanisms, 75
physiology, 39, 82, 85, 86
pigmentation, 12
pistil, 12
plant growth, 16, 21, 85, 124, 160, 164, 293
plants, vii, 1, 2, 5, 6, 7, 11, 17, 20, 22, 24, 44, 65, 74, 75, 76, 78, 82, 85, 86, 96, 100, 112, 117, 140, 142, 146, 160, 164, 170, 194, 199, 206, 219, 220, 224, 229, 274, 286, 293, 294, 297
platform, 130
Pliocene, 165, 166, 168
PLS, 207
policy, x, 32, 37, 38, 156, 182, 183, 186, 190, 260, 273, 274, 281
policy makers, 182, 183, 260
policy options, 38
policymakers, 182
political aspects, 130
politics, 156
pollen, 12, 16
pollination, 14
pollutants, 295
pollution, 154, 160, 264, 285, 287, 292, 296
ponds, 30, 31, 265
population, vii, 1, 3, 4, 34, 44, 123, 130, 131, 132, 138, 139, 142, 144, 154, 230, 260, 285, 286, 287, 293, 295, 296, 299
population density, 260
population growth, 34, 285, 287, 293, 296
portfolio, 276
Portugal, 123
positive correlation, 52, 62, 219
positive externalities, 256
positive relationship, 63
potassium, 287
potato, viii, 30, 87, 89, 90, 91, 94, 95, 96, 97, 98, 99, 100, 101, 104, 105, 106, 107, 123, 125, 127, 138
potential benefits, 24, 182, 287, 289
poverty, ix, 141, 156, 181, 182, 191, 260
precipitation, 30, 32, 119, 201, 235, 259, 291, 292, 295
prediction models, 169
preparation, 18, 19, 20, 21, 22, 23, 25, 28, 136, 138, 143, 189
present value, 274
preservation, vii, 280
president, 267
price changes, 280
primary canal, 264, 265
primary data, 185, 260
principles, 163
private sector, 140
probability, ix, 25, 42, 162, 181, 184, 185, 186, 187, 188, 189, 190
probability distribution, 162
probe, 43, 44, 47, 48, 49, 51, 56, 61, 62, 64, 65, 68
Probe calibration, 50
probit models, 185, 186, 190
producers, 33, 278
production costs, 24
production function, 276, 277
profit, 19, 28, 72, 199, 277
profitability, 33, 124, 156
programming, x, 230, 275
project, 19, 20, 36, 135, 139, 140, 141, 144, 155, 235, 256, 257, 262, 264, 265, 267, 268, 269, 270, 274, 275, 276, 277, 278
protection, 178, 299
proteins, 219
pruning, 46, 64, 74, 80, 82
public awareness, 288
public policy, 156
publishing, 139
pumps, 181, 182, 189, 190, 191, 230, 236, 242, 245, 250, 287
PVC, 25, 27

Q

quality improvement, 81
quantification, 73
Queensland, 34

R

race, 1
radiation, 15, 16, 46, 48, 57, 58, 63, 69, 90, 199
radicle, 12, 13
radius, 11
rain forest, 67
random walk, 278, 279
rate of change, 49
reactions, 164
reading, 25, 61
reality, 156, 274, 275, 278
recall, 20
recognition, 278
recommendations, 25, 123, 274, 297
recovery, 13, 17, 271
rectification, 85
recycling, 286, 294
redistribution, 220, 289, 294
reduction division, 16, 17
reform(s), 132, 141
regions of the world, viii, 88
regression analysis, 56, 65
regulations, 293
rehabilitation, 132
reinforcement, 122
rejection, 185
relative size, 270
relevance, 188
reliability, 32
remediation, 37
remote sensing, 85
repair, 26, 58, 134
reproduction, 293
reproductive organs, 4
requirements, vii, 1, 16, 20, 22, 31, 32, 35, 42, 60, 61, 66, 68, 88, 96, 112, 118, 121, 122, 123, 135, 136, 137, 149, 152, 154, 199, 204, 221, 268, 295
researchers, 20, 24, 96, 201, 218, 219
reserves, 201, 279
residues, 154
resilience, 157
resistance, ix, 11, 37, 39, 48, 62, 194, 199, 206, 209, 210, 211, 212, 214, 215, 216, 217, 218, 219, 220, 221, 222, 225
resolution, 43, 44, 65
resource allocation, 275, 277
resource management, 139
resources, viii, x, 33, 35, 42, 72, 87, 88, 89, 118, 122, 130, 131, 134, 138, 140, 149, 183, 190, 192, 200, 229, 230, 252, 255, 256, 261, 266, 267, 268, 269, 270, 286, 287, 293, 296, 299
response, viii, 16, 18, 53, 55, 56, 62, 63, 64, 71, 72, 76, 77, 78, 80, 84, 85, 95, 96, 97, 98, 99, 106, 114, 123, 125, 126, 127, 182, 199, 218, 220, 256, 277, 297
restoration, 62
restrictions, 32
revenue, 190
rice field, 18, 21, 23, 24, 34, 39, 263
rights, 34, 266, 267, 270, 272
risk(s), x, 19, 32, 33, 74, 89, 102, 122, 182, 225, 273, 275, 276, 277, 278, 291, 292, 295
risk aversion, 32
river flows, 279
root growth, 7, 11, 73, 298
root hair, 11
root system, 7, 11, 36, 58, 74, 75, 76, 82
root zones, 124
rooting behavior, vii, 1, 8
roots, 7, 10, 11, 13, 24, 44, 55, 64, 71, 74, 75, 76, 82, 84, 85, 101, 140, 146, 197, 200
routes, 3
rowing, 78, 117
Royal Society, 36
rubber, 189
rules, 24
runoff, 19, 30, 31, 202, 204, 234, 235, 249, 256, 257, 259, 270, 290, 291, 292
rural areas, 28, 261
rural development, 36

S

safety, 232
saline water, vii, viii, 87, 88, 89, 90, 94, 96, 97, 98, 101, 102, 106, 107, 109, 111, 114, 115, 119, 121, 122, 123, 125, 126, 127, 131, 142, 149, 174, 176, 235, 252
salinity levels, 102, 108
salmonella, 288, 292, 298
salt accumulation, 90, 94, 101, 112, 115, 252

salt concentration, 117, 142, 235, 244, 245
salt tolerance, 89, 114, 124, 127
salts, 30, 94, 96, 101, 102, 108, 109, 116, 120, 122, 145, 160, 163, 164, 170, 171, 236, 251, 290
saltwater, 168, 175, 176
Sap flow, 43, 47, 67, 68, 69, 73
SAS, 205, 206, 225
saturated fat, 194
saturated fatty acids, 194
saturation, 21, 246
savings, 19, 24, 31, 33, 34, 35, 73, 74, 81, 99, 138, 145, 151, 152, 154, 188
scaling, viii, 44, 66, 129, 252
scarcity, 72, 88, 122, 130, 132, 256, 286, 288, 295
scattering, 16
schooling, 186
science, 39, 177, 191
scientific method, 200
scope, vii, 18, 130, 141
security, viii, x, 34, 129, 130, 135, 139, 155, 156, 182, 260, 285, 286, 296
sediment(s), 31, 37, 266, 293
seeding, 14, 18, 19, 134, 205
seedlings, 18, 20, 23, 298
seminars, 139
senescence, 5
sensing, 85
sensitivity, 72, 73, 74, 76, 114, 209, 210, 211, 212, 214, 215, 216, 217, 274
sensors, 41, 46, 47, 49, 53, 54, 55, 56, 57, 59, 62, 63, 65, 201, 202, 223
sequencing, 6
service provider, 152
services, 140, 141, 256, 267, 269, 271
sewage, 297
sex, 288
shade, 46, 262
shape, 32, 46, 47, 143
sheep, 264
shoot(s), 7, 10, 11, 14, 52, 64, 73, 74, 78, 82, 83, 84, 85
shortage, 25, 57, 58, 59, 63, 83, 94, 106, 114, 117, 131, 132, 141, 143, 150, 151, 152, 154, 155, 270, 287, 295
showing, 51, 62, 76, 77, 124, 143, 170, 171, 172, 176
signalling, 82
signals, 48, 49, 53, 62, 71, 74, 75, 80, 81
signs, 185, 186
simulation, vii, ix, 108, 229, 230, 231, 232, 233, 234, 236, 239, 242, 244, 245, 249, 251, 252
siphon, 29
skimming, 131
sludge, 297
smoothing, 136
SO42-, 167
social acceptance, 295
social class, 129
sodium, ix, 159, 160, 163, 164, 165, 167, 171, 174, 176
software, 48, 169, 207
soil erosion, 31, 137, 203, 235
soil particles, 290
soil type, 18, 23, 32, 34, 140, 194
solar energy, x, 16, 229, 230, 241, 242
solubility, 166
soluble solids, 80, 81
solution, 11, 50, 152, 207, 208, 244, 246, 249, 269, 270, 287, 294
sorption, 290, 291
South Africa, 137, 179, 271
South America, 2
South Asia, 15, 129, 130, 137, 141, 156, 192
Southeast Asia, 2, 3, 253
sowing, 137, 138, 140, 141, 142, 143, 144, 145, 146, 147, 148, 149
soybean(s), 199, 222, 223, 226, 263
Spain, 6, 44, 69, 123, 177, 178, 271
species, 2, 6, 43, 62, 65, 72, 74, 88, 288
spectroscopy, 226
spiritual care, 266
spore, 206
Spring, 93, 147, 207
Sri Lanka, 2, 4, 20, 37, 38, 191, 192, 271, 272
SSA, 256
stability, 15, 194
stakeholders, 139, 182, 190
stamens, 12
standard deviation, 162, 186
standard of living, 129
starch, 12
starch granules, 12
starvation, 74
state(s), x, 3, 166, 178, 201, 279
statistics, 166, 167, 170, 171, 185, 186
sterile, 4, 12
stigma, 12
stochastic processes, 279, 280
stock, 276, 278

stock price, 278
stomach, 299
stomata, 55, 58
stomatal conductance, 76
storage, ix, 21, 28, 32, 33, 62, 80, 83, 136, 145, 160, 224, 229, 230, 231, 232, 236, 244, 245, 256, 257, 259, 261, 271, 272, 278, 279, 280
stress, 17, 19, 24, 25, 36, 38, 61, 67, 72, 73, 74, 84, 85, 88, 112, 115, 117, 120, 124, 138, 142, 145, 194, 195, 200, 218, 220, 221, 223, 226, 286
stress factors, 194
stress response, 221
structure, 12, 64, 164, 165, 176, 279, 280, 290
sub-Saharan Africa, x, 182, 183, 255
subsidy, 132, 139
substitution, 121
subsurface flow, 291
sucrose, 194, 208, 213, 219, 220
sugar beet, 223, 235
sugarcane, viii, 130, 134, 136, 138, 143, 146, 147, 151, 152, 154, 262
sulfate, 167, 169
Sun, 38
surface area, 75, 248, 257, 261
surplus, 28, 289
survival, 293, 297
susceptibility, 205, 214, 215, 216, 217
sustainability, 72, 82, 130, 143, 156, 157, 252, 268
swelling, 14, 73, 164
Switzerland, 296
symptoms, 17, 198
synchronize, 294
synthesis, vii, viii, 82, 129, 220
Syria, 128

T

Taiwan, 20
tanks, ix, 136, 229, 230, 231, 256
Tanzania, 137, 255
target, 90, 115, 140
Task Force, 4
taxonomy, 6
technical support, 271
techniques, vii, ix, 2, 11, 22, 27, 31, 34, 43, 44, 85, 88, 99, 138, 155, 162, 169, 176, 178, 194, 199, 200, 252, 253, 294
technological progress, 279
technologies, ix, 19, 31, 32, 33, 36, 131, 132, 134, 135, 138, 139, 141, 155, 156, 157, 181, 182, 183, 184, 185, 186, 187, 188, 189, 190, 274
temperature, 5, 13, 15, 16, 17, 32, 46, 48, 49, 53, 58, 61, 67, 80, 90, 165, 194, 198, 199, 202, 208, 222, 226, 259
temporal variation, 290
tension, 226
tenure, 184, 266, 271
test statistic, 185
testing, 150
testosterone, 297
texture, 5, 12, 94
Thailand, 1, 2, 7, 17, 20, 37
thermal properties, 62
thinning, 80, 82
Third World, 156
time periods, 277
time resolution, 44, 65
tissue, 11, 199, 219, 224
tones, 134
total product, 134
total revenue, 190
traditional practices, 33
traditions, 32
training, ix, 134, 141, 152, 155, 181, 182, 190
traits, 123, 194
transformation(s), x, 129, 273, 277, 298
translation, 270, 275
transmission, 292
transpiration, viii, 11, 20, 21, 41, 42, 43, 44, 56, 58, 60, 61, 65, 66, 67, 74
transplant, 26
transplantation, 25, 28
transport, 84, 189, 289, 290, 292, 295, 297
transportation, 182
treatment, ix, 31, 49, 61, 75, 77, 80, 89, 90, 91, 94, 95, 100, 101, 106, 107, 108, 109, 110, 111, 112, 114, 115, 116, 117, 118, 120, 121, 152, 194, 206, 213, 285, 287, 288, 293, 296, 299
trial, 31, 46, 150
trunk growth, 78
turgor, 73
Turkey, 127, 222
typhoid, 292

U

U.S. Department of Agriculture, 38
U.S. Department of Commerce, 38, 39

uniform, 11, 29, 44, 72, 136, 202, 287
United Kingdom (UK), 84, 271
United Nations (UN), 178, 191, 221, 223, 286, 299
United States, 1, 2, 3, 7, 39
urban, 34, 130, 151, 152
urbanization, 160
urine, 264, 287, 295
USDA, 31, 46, 127, 178, 193, 236, 253

V

validation, 62, 66, 162, 169
valuation, 221, 276, 277, 281, 282, 283
valve, 90, 100, 265
vapor, 62, 65, 286
variable costs, 190
variables, 44, 45, 46, 160, 161, 178, 185, 186, 274, 275, 277
variations, 13, 43, 53, 54, 55, 57, 58, 59, 62, 63, 64, 84, 160, 167, 177, 194, 290
varieties, 4, 5, 6, 7, 11, 13, 14, 15, 16, 17, 18, 35, 39, 72, 80
vascular system, 63, 65
vector, 161, 184
vegetable oil, 221
vegetables, 88, 122, 123, 151, 152, 235, 260, 262, 263, 292
vegetation, viii, 44, 66, 259
vegetative growth, viii, 5, 13, 71, 73, 74, 78, 80, 83, 84, 96, 115
vegetative stages, 5, 68
velocity, 31, 32, 44, 55, 58, 64, 67, 232, 235
vessels, 48, 50, 63, 64
Vietnam, 1, 2, 4, 7
virus infection, 227
vision, 36
vitamins, 6
volatility, 280

W

wage rate, 18
Wales, 3, 5, 38, 137, 299
Washington, 34, 178, 179, 192, 253
waste, vii, 268
wastewater, vii, x, 131, 151, 152, 285, 286, 287, 288, 289, 290, 291, 292, 293, 294, 295, 296, 297, 299
wastewater irrigation, x, 285, 286, 289, 290, 292, 295, 296
water absorption, 11
water chemistry, 177
water conductive stem section, viii, 50, 65
water evaporation, 244
water policy, 32, 37, 38
water quality, 36, 37, 88, 89, 96, 97, 98, 99, 122, 125, 127, 132, 146, 160, 162, 164, 171, 172, 174, 175, 176, 177, 179, 236, 252, 264, 291, 292, 298
water requirement, 23, 68, 151, 199
water resources, vii, viii, ix, x, 35, 42, 72, 87, 88, 89, 118, 122, 130, 131, 134, 138, 140, 192, 200, 229, 230, 252, 255, 256, 261, 267, 268, 269, 270, 286, 293, 296, 299
water rights, 34, 267
water shortages, 88, 270
water supplies, 33, 34, 58, 89, 102, 132, 134, 143, 154, 269, 270
watershed, 178
wavelengths, 16
weather patterns, 182
web, 201
wells, ix, 32, 88, 107, 159, 166, 170, 171, 172, 174, 175, 183, 185, 189
West Africa, 2, 6, 192, 258
Western Australia, 137
wetting, vii, 20, 23, 34, 35, 62, 67, 74, 90, 145
wind power, 236, 242
wind speed, 62, 242
wireless technology, 223
withdrawal, 15, 286
wood, 44, 48, 50, 53, 62, 64
workers, 292
working hours, 232
World Health Organization (WHO), 296, 297
worldwide, 130, 224, 257, 286, 291

X

xylem, 50, 63, 65, 67, 74, 82, 84

Z

Zimbabwe, 125, 137